Discoverin
Mathemat...

A Course for Secondary Schools
Third (Metric) Edition

H A Shaw

Head of Mathematics Department,
Gloucestershire College of Education

F E Wright MSc

Department of Mathematics
Bishop Lonsdale College of Education, Derby

Illustrated by AG Standley

1

Edward Arnold

First published 1960
by Edward Arnold (Publishers) Ltd.,
41 Bedford Square, London WC1B 3DQ

Reprinted 1960, 1961, 1962, 1963, 1964, 1965, 1967

Second (Decimal) Edition 1970
Reprinted 1971, 1973

Third (Metric) Edition 1975

Reprinted 1976, 1978, 1979, 1982

ISBN 0 7131 1917 9

Printed and bound in Great Britain
at The Pitman Press, Bath

PREFACE

How many times have you heard "I hated Maths", "I was never any good at Maths"? Why should this be so? We would suggest that one reason is that in the past there has been far too much mechanical teaching, the children have not always been encouraged to discover the rules for themselves. So often the mathematics in the Secondary Schools has consisted of pounding away at material experienced in the Junior School.

Mastery of the Basic Principles is very important, but there are some children who will find it very difficult, and to continue to pound away at this elementary work with such children may quickly make them hate Mathematics. We feel that it is far more important that a liking for Mathematics should be induced, and the Basic Principle practised in some of the more interesting and wider fields of this subject.

It is apparent that centuries ago much research was made by man in order that he might progress: he counted his possessions, measured his land, recorded time, learned to navigate in order that he might travel and trade in other lands. Mathematics is not a static thing, it is progressive and plays a vital part in our lives and we should try to think of it in these terms when we are teaching. It is important that Mathematics should be presented as a series of tools and should show clear statements of measurements, quantities and relationships.

In this course each book will contain Geometry, Arithmetic and Algebra in separate sections. It is not, however, our idea that Mathematics should be kept in separate compartments, but set out in this fashion it does allow the teacher to present it as a unified subject linking the three sections as he so desires.

Geometry is put first in this book as it might well be a new subject to the children, thus providing an incentive and help in creating enthusiasm. Revision of the four rules applied to number has been given in solution of crosswords, and Fractions and Decimals have been included as separate chapters for the benefit of children who have not at this stage experienced them.

The teaching of mathematics should be carried out in all streams and not confined to the A's. Algebra and Geometry can be made just as exciting to the D's and F's as to the A's. Many of the lower streams in a six stream entry school have worked through most of the exercises in this book. Do not condemn the D's and the F's to the laborious task of learning the four rules of money, weights and measures for a further four years of their lives, give them a wider experience of mathematics in as attractive a form as possible.

H.A.S.
F.E.W.

PREFACE TO THE METRIC EDITION

The use of metric units in this country was made legal in 1897, and today over 50% of our exports go to countries using a metric system.

With the development of world communication and trade it is obvious that our International System of units would have many advantages.

The International System of units (S.I. units) recommended by the International Conference in 1960 was accepted by this and other countries throughout the world as a basis for change to a metric system. S.I. units have been introduced in science, technology, industry, economics and our daily lives. *Discovering Mathematics* has been completely revised bearing in mind the recommendations of the Metrication Board. Book 1 is now available complete with answers.

Contents

8. **Number** 74
The counting numbers, odd, even, composite, four rules applied to number, H.C.F. and L.C.M.

9. **Revision** 79
Puzzles devised to test number operations. Patterns in number.

10. **Fractions** 84
Four rules applied, Changing to decimals.

11. **Decimals** 99
Four rules, Averages. Metric units.

12. **Area** 115
Measurement in squares, Rules for finding area of rectangle, triangle and trapezium.

13. **The Circle** 135
Circumference, Area,
Summary of what has been discovered.

Algebra

14. **Use of Letters** 151
Introduction to use of letters, Collection of terms, Positive and Negative numbers.

15. **Indices** 158
Multiplication and division of terms.

16. **Substitution** 163
Solution of codes, Finding numerical values for literal statements.

17. **Simple Equations** 167
Making equations from everyday problems, Finding the unknown.

18. **Transformation of Formulae** 180
Simple changing of the subject, Substitution in equations.

19. **Graphs** 183
Illustration of statistics, Columns, Travel, Conversion, Straight-lines, Reading of results from straight-line graphs, Solution of equations, Areas of figures formed by straight lines.

20. **Revision** 208
Extra exercises covering the work done through the book, Tests.

Answers 225

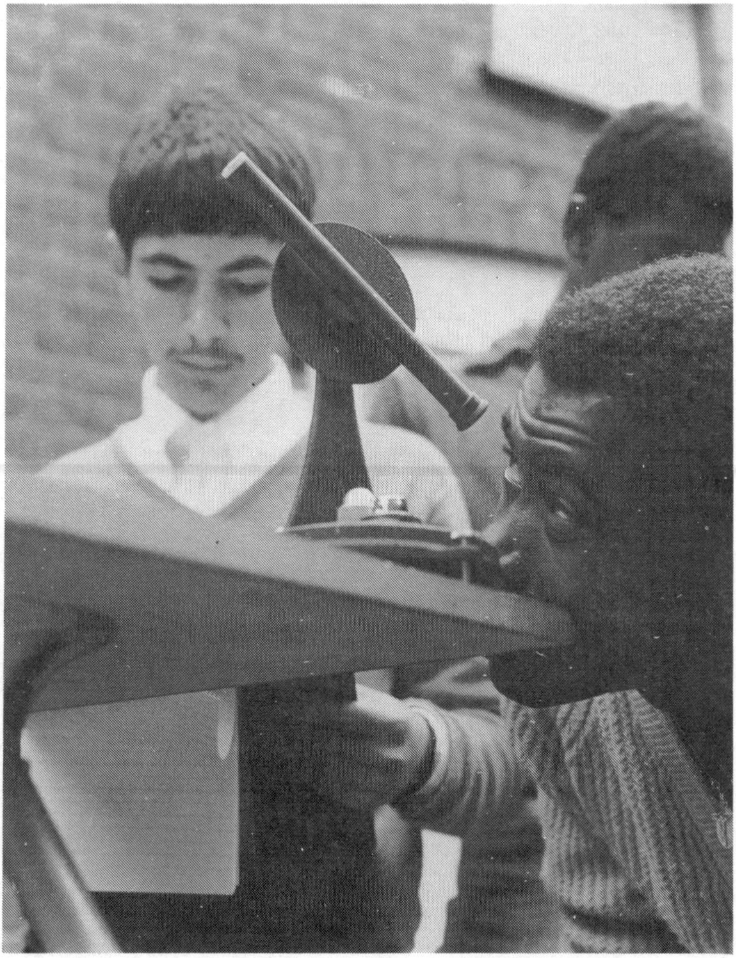

Discovering Heights.

Geometry

1 Shapes

Here are many different shapes and sizes! But if you look carefully, you will see that all the figures have two things in common. They all have straight sides and they all have corners. You could draw many more figures having any number of sides, but I wonder how many of the above figures you could name. What makes a square square? Make a list of the things around you that are square. I expect some of you will say that all the sides of a square are the same length. So they are, but look at these shapes. Are they all squares?

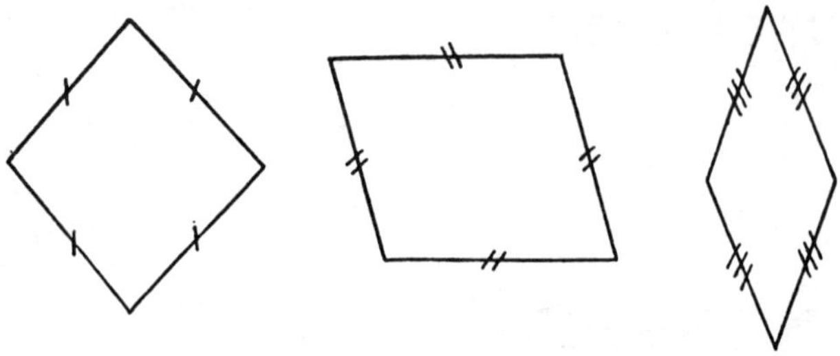

If you are lucky enough to own a Meccano set, try making a square out of four equal pieces but do not screw them together too tightly, and you will see that you can make other shapes like the ones above.

It is not only the length of the sides that makes a figure square but the angles between the sides have to be a certain size too.

Cut out two squares exactly the same size and label the corners like this.

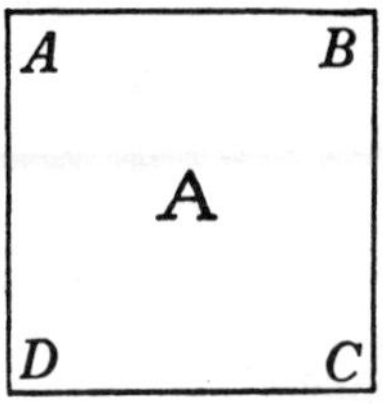

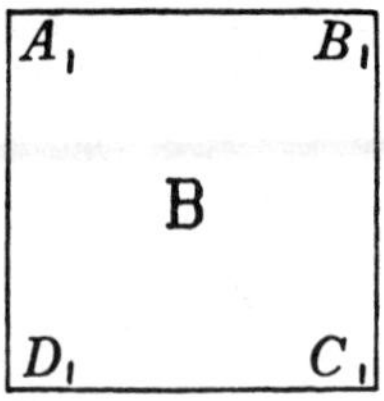

Now if you put one on top of the other, with *B*, according *A*, you will find that they fit no matter which way you turn them. The four angles are the same size: they are all **Right Angles**, all 90°.

Tear the corners off one of your squares and put them together at a point. They fill up all the space round the point and make a cross.

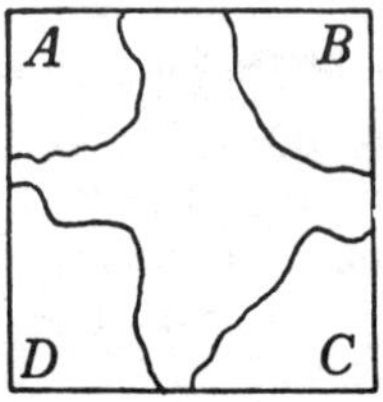

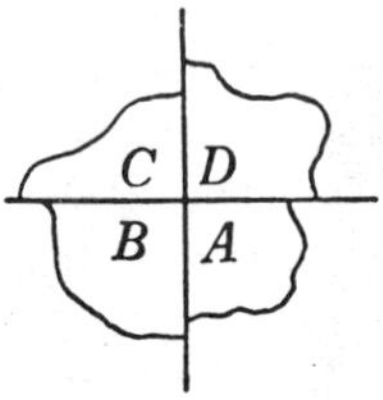

What other figure has four right angles like a square? The walls, the floor, your book and the desk on which it is lying are all . . . . . . ? . . . . . .

Can you imagine what a mess the place would look if all the rooms were of different shapes?

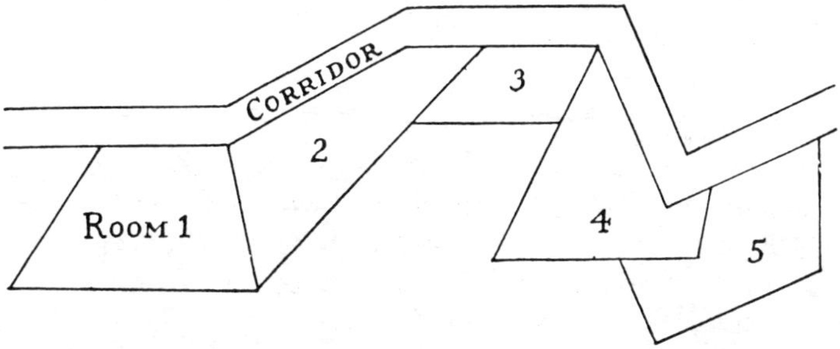

What would happen if your books were odd shapes? How would you like to sleep in a triangular bed?

The right angle is used more than any other angle. See if you can make a list of twenty places where you can find one.

The Eygptians found a useful way of constructing right angles, when they were building the Pyramids, a method which is still used today.

To construct a right angle they used a piece of rope divided into three. One division was three units long, one was four and the other was five. They would peg out *AB* four units long, then lay *BC* (three units) roughly at right angles to *AB*, then swing round the five unit length until it just touched *C* and they had a right angle at *B*.

To construct a right-angled triangle with a compass and ruler.

Draw base *AB* three units long.

With compass open to four units and point *A* construct an arc.

With compass open to five units and point at *B* construct a second arc.

The two arcs intersect at *C*.

Join *AC* and *BC* with straight lines we now have the required right-angled triangle. Draw these right-angled triangles.

1	3 cm	4 cm	5 cm	**3**	12 mm	16 mm	20 mm
2	6 cm	8 cm	10 cm	**4**	9 cm	12 cm	15 cm

5 In a piece of rope 250 cm long make knots at intervals of 60 cm, 80 cm and 100 cm. Use it to construct right angles, squares and rectangles.

We have thought a lot about right angles but what about the angles in between? Earlier we stated that a right angle was equal to 90° and that four right angles filled up the space round a point. So that there are 360° round a point.

Each hour the minute hand of a clock moves through 360°. The hour hand takes 12 hours to move through 360°.

1 Through how many degrees does the minute hand move in 5 minutes, 30 minutes, 20 minutes and 50 minutes?

2 Through how many degrees does the hour hand move in 3 hours, 5 hours, 8 hours and 11 hours?

3 What is the angle between the hands of these clocks?

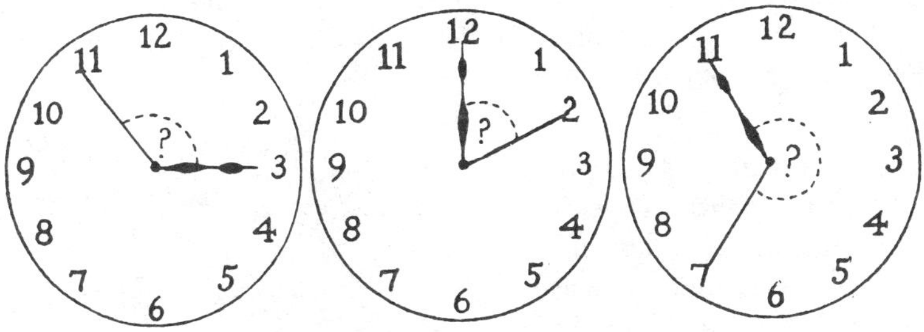

The instrument used for drawing and measuring these angles is called a protractor. It is in fact part of a circle marked in degrees.

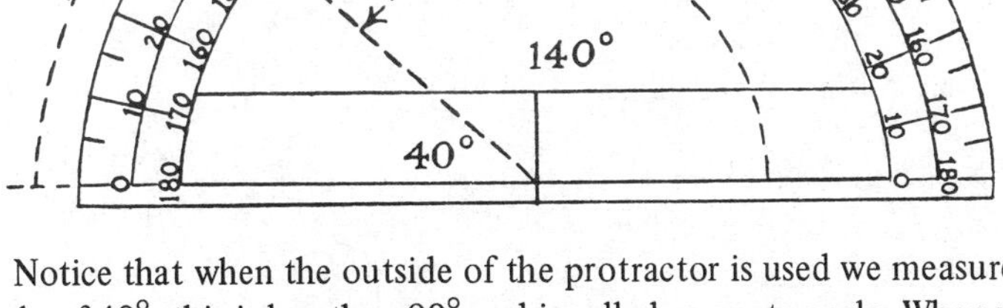

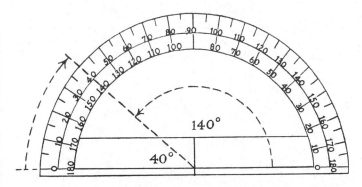

Notice that when the outside of the protractor is used we measure an angle of 40°, this is less than 90° and is called an acute angle. When using the inside of the protractor the angle becomes 140°. This angle is greater than 90° but less than 180° and is called obtuse.

What is the sum of the angles?

2 Angles at a Point

How many of you are Girl Guides or Boy Scouts? I expect you have learnt how to find the directions from the sun. Here are several interesting ways of finding North, for once you know North it is easy to find the other directions of the compass.

Point the hour hand of your watch to the sun; halve the angle between it and 12 and this will point South, turn round through two right angles and you will be facing North.

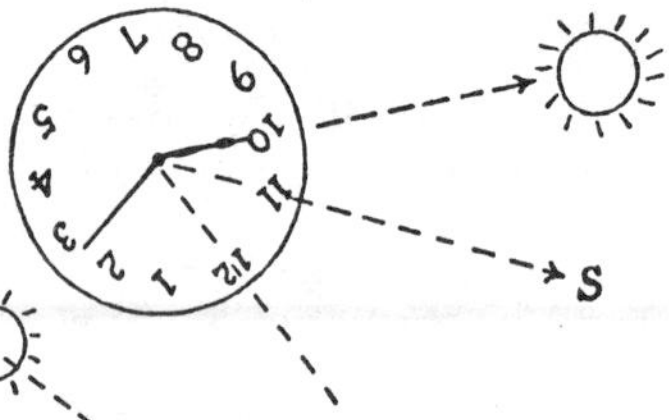

If you place a straight stick in the ground its shadow will point to North at noon G.M.T.

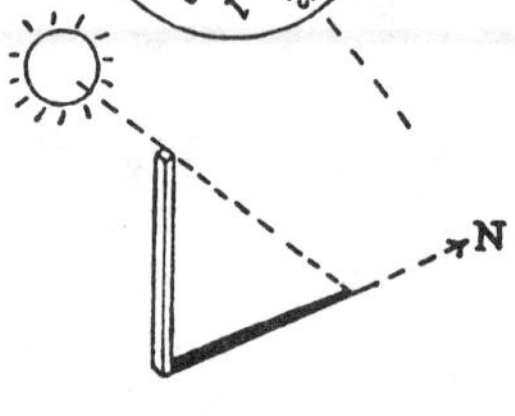

On a clear night you can find North from the stars. Pick out the Plough and follow the direction of its pointers to the Pole Star. Its distance from the Plough is approximately 5 times the distance between the pointers.

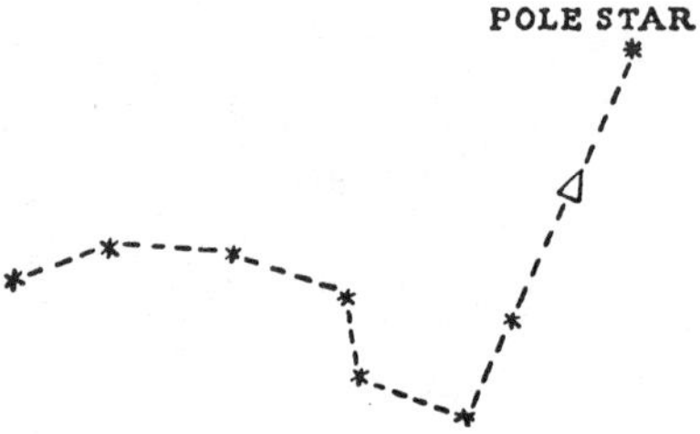

If you are near a town or village a church will help you for most of them are built in an East-West line, the altar being the East.

Points of the Compass

To find other directions, cut out a square of paper, and fold it into four making four small squares. Then fold it diagonally. Open it out and mark it like this.

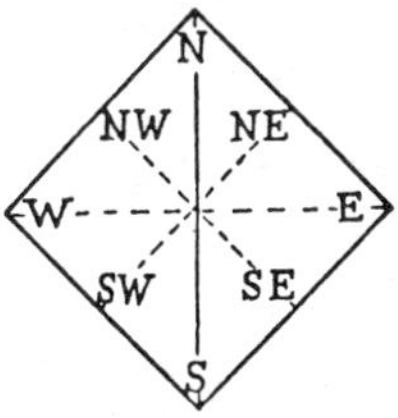

Place the paper so that your North really points North and now practise finding other directions.

Try following these directions

1 Face South and take two paces, turn East and take four paces then take two paces North and four paces West. How far from your starting point are you now?

2 In which direction must you walk to reach the door of your classroom?

3 If you take two paces South and two paces East in which direction must you go to get back to your place?

4 Find a clear spot on the playing field and draw a plan showing what you can see if you look in each direction.

EXAMPLE

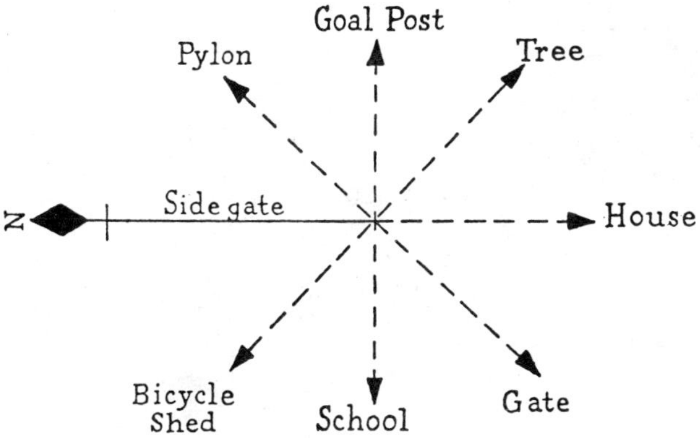

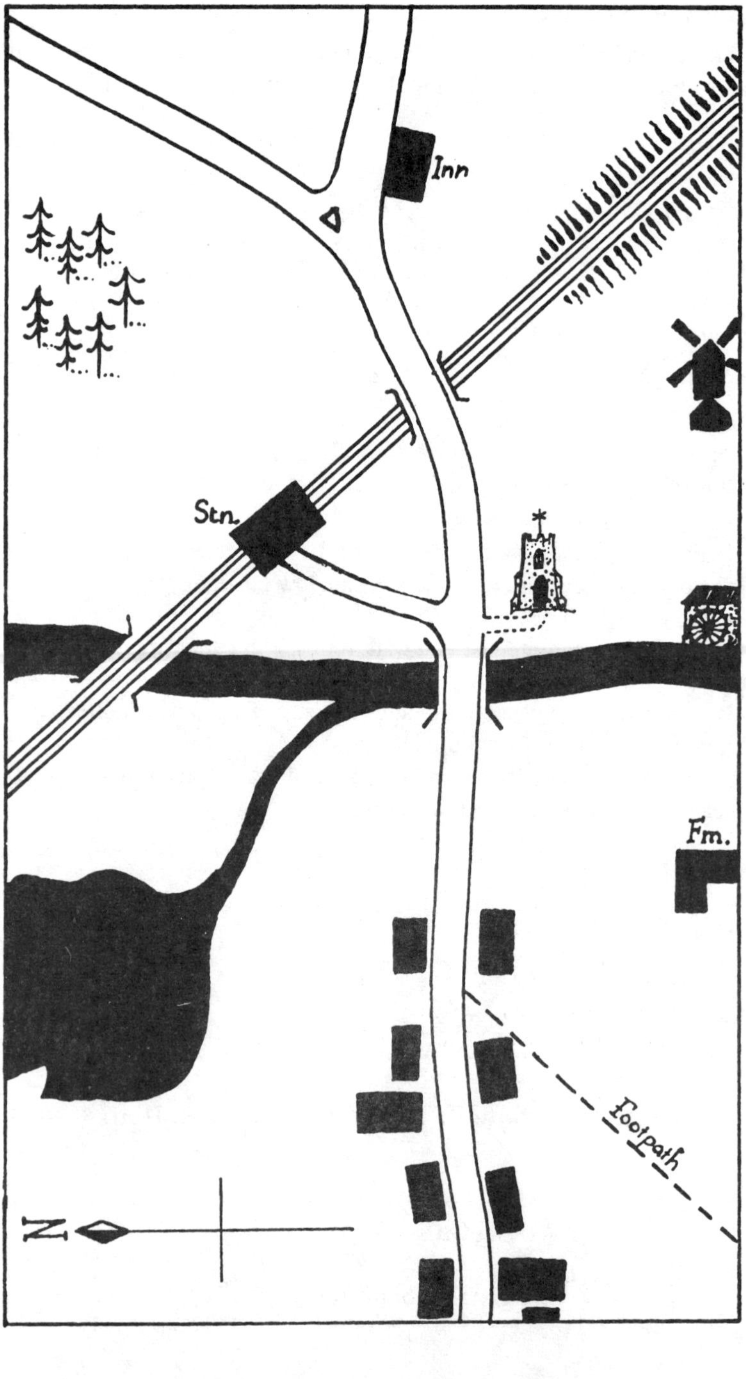
Inn
Stn.
Fm.
Footpath
N

EXERCISE 1

Study the map on the opposite page and answer these questions.

1. In which direction does the main road run?
2. In which direction does the river flow?
3. In which direction does the railway run?
4. Coming along the footpath on to the main road in the village, in which direction are you travelling?

Fill in the blanks in these sentences.

5. The embankment lies .... of the station.
6. The railway bridge over the river is .... of the pond.
7. The church is .... of the place where the railway crosses the road.

Now imagine that you are standing on the road bridge over the river, in which direction must you look to see the following landmarks?

8. The church.
9. The village.
10. The station.
11. The pond.
12. The clump of pine trees.
13. The windmill.
14. The inn.
15. The railway bridge across the river.
16. The farm.
17. The old water mill.

In what directions from the inn are the following

18. The windmill.
19. The village.
20. The railway embankment.

A man built a house with windows on all four sides. All the windows faced south. Where was the house?

What is the angle between each of the sixteen points shown on the above diagram of a compass?

EXERCISE 2

Calculate the angles (in a clock-wise direction) between these following directions.

1. N. and N.E.
2. W. and N.W.
3. W. and N.N.E.
4. N.W. and S.
5. S. and N.W.
6. W. and N.E.
7. N.N.E. and W.
8. N.W. and S.W.
9. S.E. and S.S.W.
10. E. and S.S.W.
11. S.E. and W.N.W.
12. S.S.E. and S.W.
13. W.S.W. amd S.E.
14. W.N.W. and S.S.E.
15. S.E. and W.S.W.
16. E.S.E. and S.S.E.
17. N.N.W. and S.S.E.
18. E.S.E. and S.W.
19. E.N.E. and S.S.W.
20. W.S.W. and N.N.E.

Say which of these are reflex, obtuse and acute angles.

So far we have only used the sixteen points of the compass. We might need to give a position that is more exact than these directions.

Bearings

Here is the method used by an airman in describing the course along which he is flying. If he is heading his aircraft in a direction 120°, the aircraft would be flying along a line 120° from North measured in a clockwise direction.

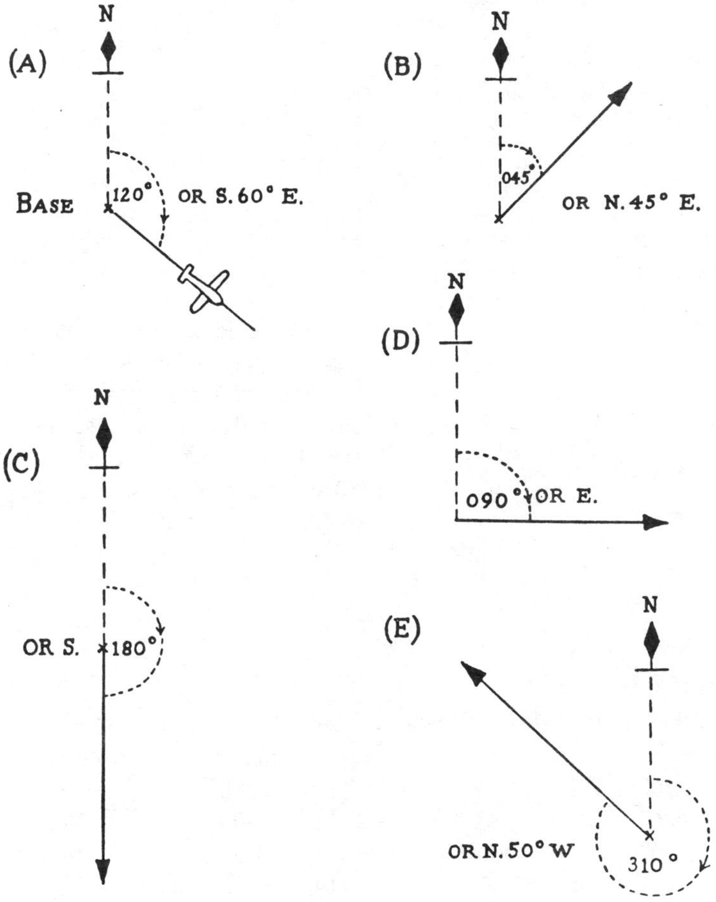

For the present assume that there is no wind and that the aircraft is flying in a true direction.

EXAMPLE 1

If the aircraft is flying at 300 km/h in the direction of 120° mark its position after 1 hour's flying.

On squared paper mark Home base, London Airport, and draw a dotted line pointing North. For convenience use the top of your page as North. With your protractor mark off 120° in a clockwise direction.

Using a scale of 1 cm to 50 km the aircraft will be 300 km (6 cm) from L after 1 hour.

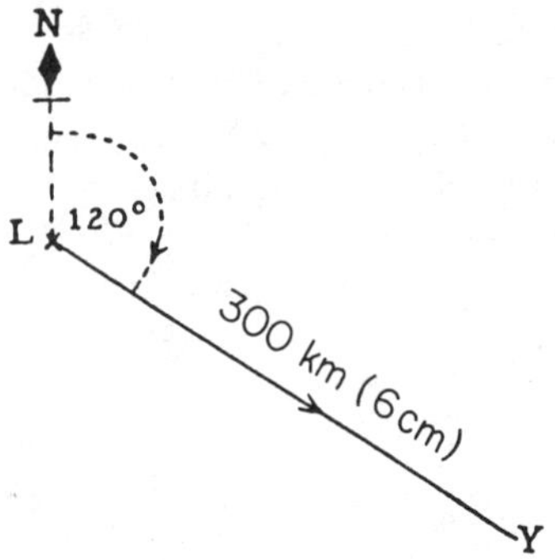

EXAMPLE 2

If the aircraft is flying at 200 km/h in the direction of 280° mark its position after 1 hour's flying.

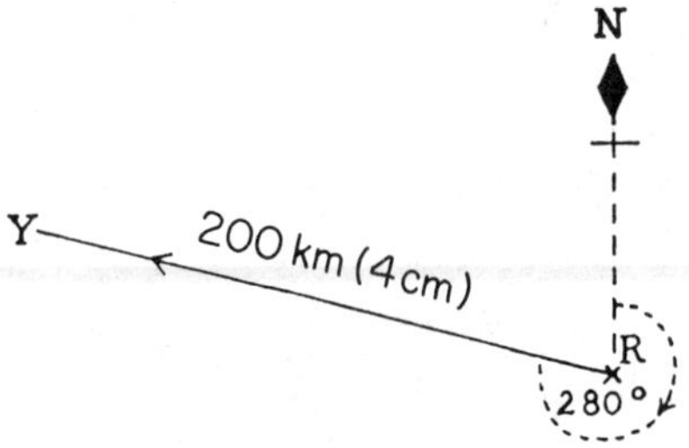

On squared paper mark Home base Ringway and draw a dotted line pointing north. Mark off an angle of 280° in a clockwise direction. Using a scale of 1 cm to 50 km, the aircraft will be 200 km (4 cm) from R after 1 hour.

EXERCISE 3

Plot the positions of the following aircraft after 1 hour's flying.

	BEARING	SPEED km/h		BEARING	SPEED km/h
1.	010°	350	11.	280°	190
2.	080°	150	12.	300°	240
3.	090°	200	13.	359°	140
4.	110°	200	14.	015°	350
5.	130°	150	15.	086°	340
6.	170°	250	16.	095°	280
7.	190°	180	17.	115°	200
8.	210°	320	18.	132°	160
9.	240°	230	19.	174°	180
10.	270°	140	20.	248°	110

When you have completed your drawings write the bearings by the side as 010° = N. 10° E. or 240° = S. 60° W.

EXAMPLE 3

An aircraft is flying at 200 km/h on a course of 090°. It changes course after 1 hour and flies for 1 hour on a course of 220°. How far is it from base at this point?

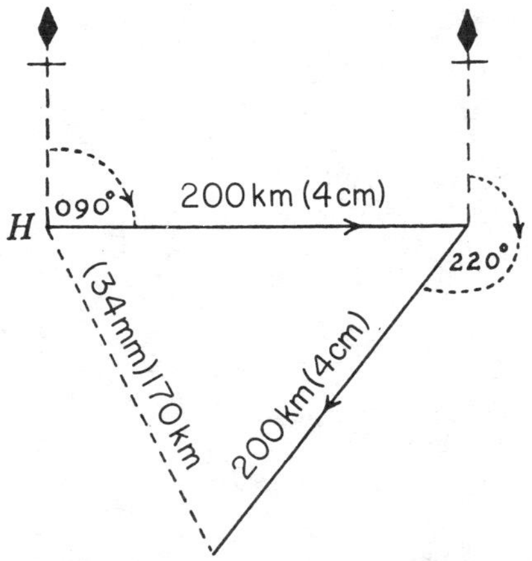

Mark home base *H* and indicate north with a dotted line. With your protractor mark off 090°. After 1 hour's flying the aircraft will be 200 km (4 cm) from *H*. At this point draw in another dotted line indicating north. Draw a line at 220° and mark position of aircraft after 1 hour's flying, 200 km (4 cm) from turning point.

Measure distance of aircraft from *H* in cm and convert back to km.

EXERCISE 4

Plot the positions of the following aircraft after 2 hours' flying, 1 hour on each heading, the speed being constant. Use a scale of 1 cm to represent 50 km.

	SPEED km/h	COURSE FROM BASE	COURSE FROM TURNING POINT
1.	300	080°	180°
2.	150	120°	270°
3.	200	180°	270°
4.	200	220°	090°
5.	150	270°	000°
6.	180	280°	045°
7.	140	000°	090°
8.	190	110°	080°
9.	240	080°	290°
10.	250	050°	300°

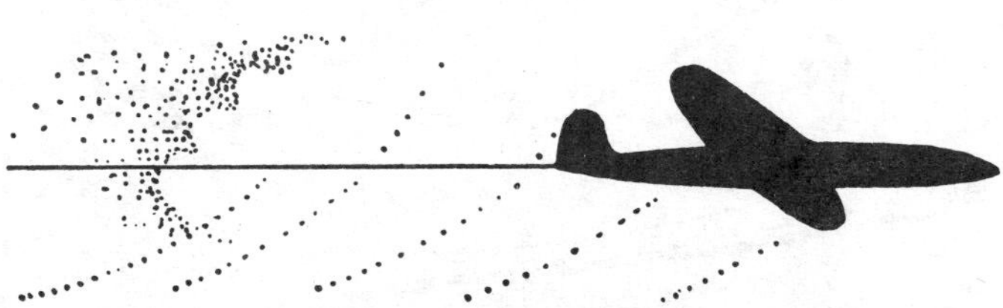

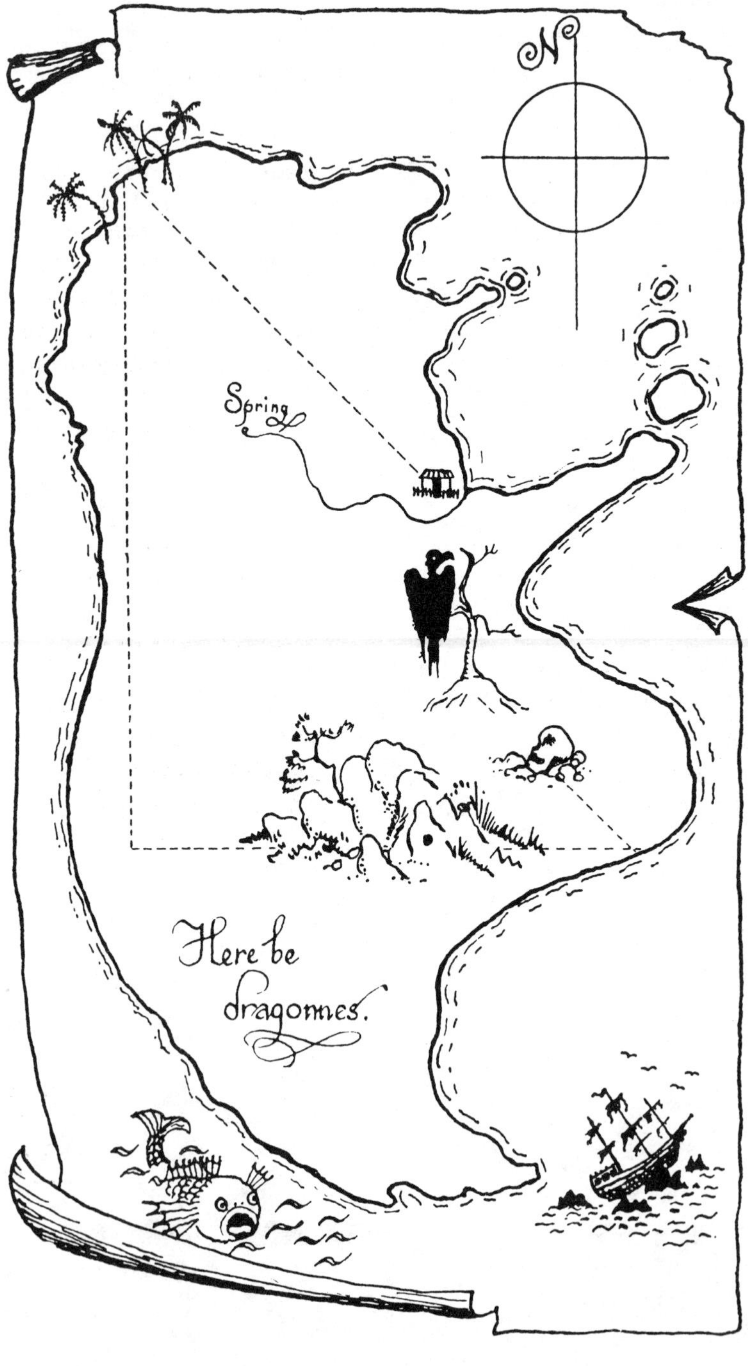
N
Spring
Here be
dragonnes.

Treasure Hunt

On the opposite page you see an old treasure map and the directions have been drawn in. They were:

Starting from the settlement at the mouth of the stream, walk to the North-West until you strike the coast near a clump of palm trees. Then turn due South and walk 400 paces. At this point turn towards the rising sun and keep very carefully to this direction over difficult ground until you come to the cliff edge to the north of the ill-fated *Santa Maria.* Turning away from the coast you will see to the North-West an isolated hill with a solitary dead tree upon it. Ninety paces towards this tree will bring you to a skull-shaped rock. Beneath this rock is the treasure.

A treasure has been buried in a field and the following directions have been found for locating it:

Start in the South-West corner of the field and walk 50 metres North, face N.30° E. and walk 15 metres, then 10 metres due South and 30 metres on a bearing of 120°. After this walk 50 metres North and 35 metres S. 15° W. Now dig.

One cm to 5 metres is a suitable scale for this map.

How far from the start was the treasure as the crow flies?

In what direction was it from the starting point?

You can make up more treasure hunts like these for each other to solve.

As you will see we have so far only used whole degrees. It is not easy to measure more accurately by drawing but more exact directions are often required. A navigator measures the angles of stars to seconds of a degree. The degree is divided into 60 minutes and each minute into 60 seconds. So we now have the following table:

60 seconds = 1 minute
60 minutes = 1 degree
90 degrees = 1 right-angle
4 right-angles or
360 degrees = one whole turn

3 Triangles

In the last chapter we plotted the course of an aircraft flying when there was no wind. What happens when the wind is blowing?

EXAMPLE 4

We are flying on a course of 110° at 300 km/h and the wind is blowing from 020° at 40 km/h. Plot the actual path of the aircraft.

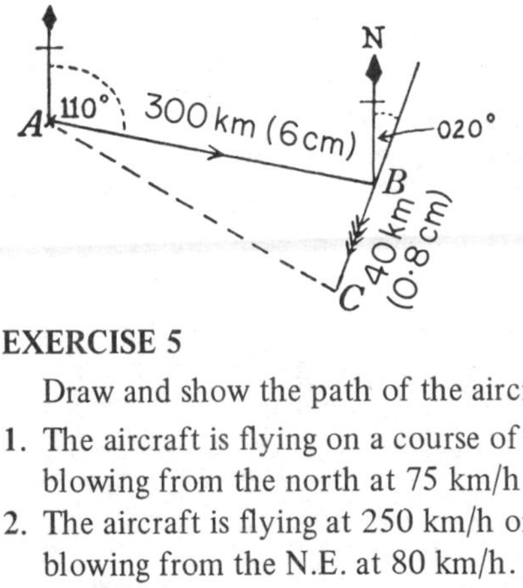

In 1 hour the aircraft would move from *A* to *B* a distance of 300 km, but the wind has been blowing on the aircraft for 1 hour so it will be at *C*. The actual path that it has travelled is *AC*.

EXERCISE 5

Draw and show the path of the aircraft.

1. The aircraft is flying on a course of 090° at 200 km/h with a wind blowing from the north at 75 km/h.
2. The aircraft is flying at 250 km/h on a course of 150°, the wind is blowing from the N.E. at 80 km/h.
3. A south wind is blowing at 100 km/h on an aircraft flying at 350 km/h on a course of 050°.
4. A plane is flying on a course of 270° at 380 km/h, the wind is blowing from 300° at 80 km/h.
5. Flying at 400 km/h on a course of 060° a plane's course if affected by a wind blowing at 120 km/h from 130°.

Now measure the three angles of each of your triangles and add them together. Draw other triangles and find the sum of their angles. If you are measuring them accurately, they should all add up to 180°, that is 2 right angles.

Cut out a triangle, label the corners and tear them off. Put them together like this:

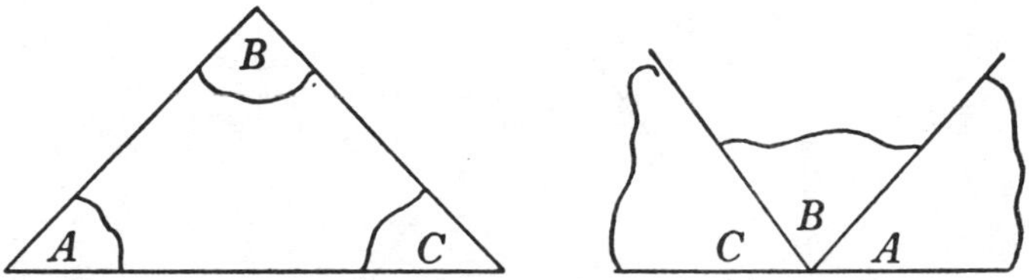

The angles round a point, you will remember, together make 360°. The angles on a straight line are just half of this, and the angles of a triangle are the same.

EXERCISE 6

Find x in the following examples.

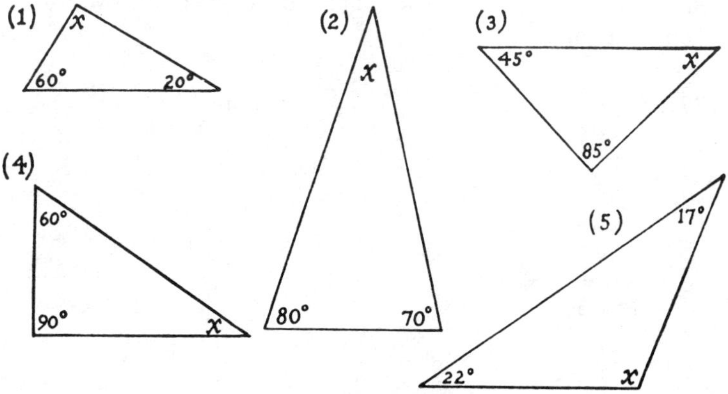

EXERCISE 7

You are given two angles of these triangles, what is the third?

1.	55°	80°	**11.**	78°	43°
2.	45°	70°	**12.**	18°	74°
3.	20°	80°	**13.**	13°	90°
4.	70°	35°	**14.**	9°	81°
5.	15°	27°	**15.**	25° 24′	38° 40′
6.	140°	30°	**16.**	68° 16′	72° 14′
7.	17°	91°	**17.**	112° 14′	30° 12′
8.	34°	72°	**18.**	58° 6′	71° 14′
9.	$24\frac{1}{2}°$	$57\frac{1}{2}°$	**19.**	89° 24′	14° 7′
10.	$18\frac{1}{2}°$	$74\frac{1}{2}°$	**20.**	25° 13′	14° 16′

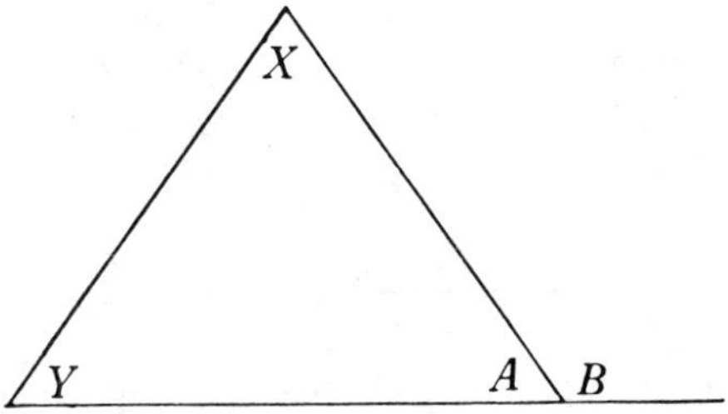

Angles A and B together make 180° because they are adjacent angles on a straight line.

Angles A, X, and Y also make 180° because they are the three angles of a triangle.

Therefore angles X and Y together must be equal to angle B. In other words the exterior angle of a triangle is equal to the sum of the two interior opposite angles.

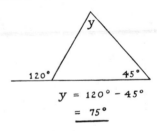

EXAMPLE 5

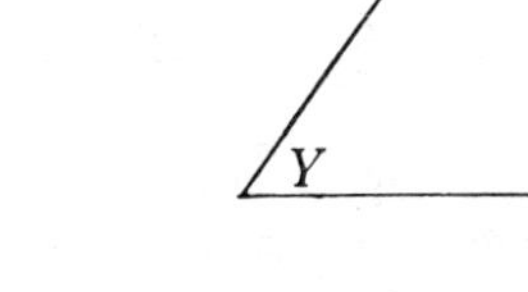

$$x = 70° + 30°$$
$$= \underline{100°}$$

EXAMPLE 6

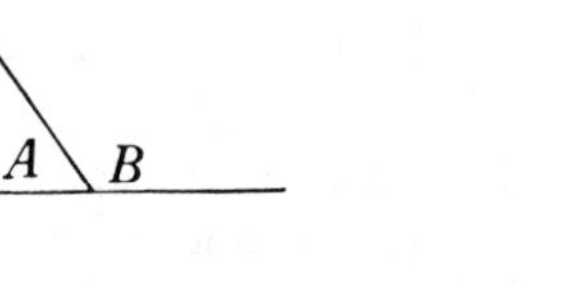

$$y = 120° - 45°$$
$$= \underline{75°}$$

EXERCISE 8

Find angle x in these diagrams.

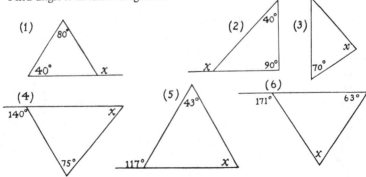

(1)

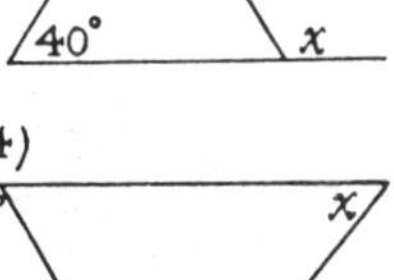

(2)

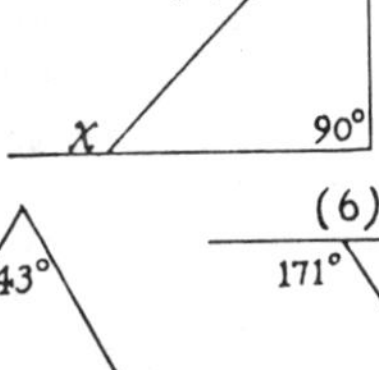

(3)

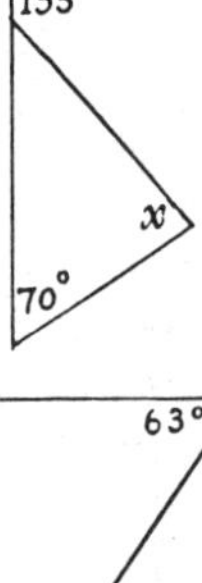

(4)

(5)

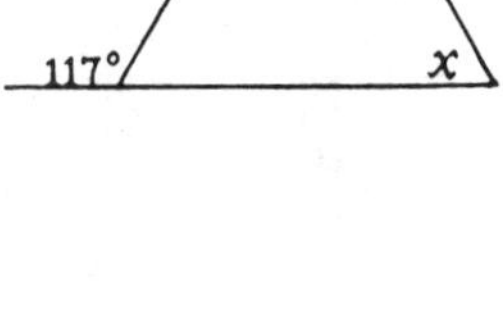

(6)

Some uses of Triangles

Make yourself a table with a piece of hardboard nailed to a stick about 1 metre long. Place a piece of drawing paper on the board and pin a card-board arrow to the middle of it.

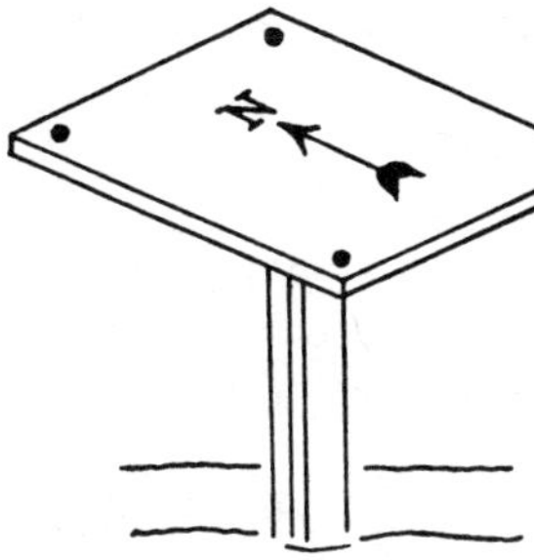

Erect your table in the middle of the grounds and with the aid of a compass, point your arrow to the North and mark this point on your paper. Now sight your arrow on prominent buildings and mark each on your paper. This can then be taken from the board and the bearings measured with your protractor.

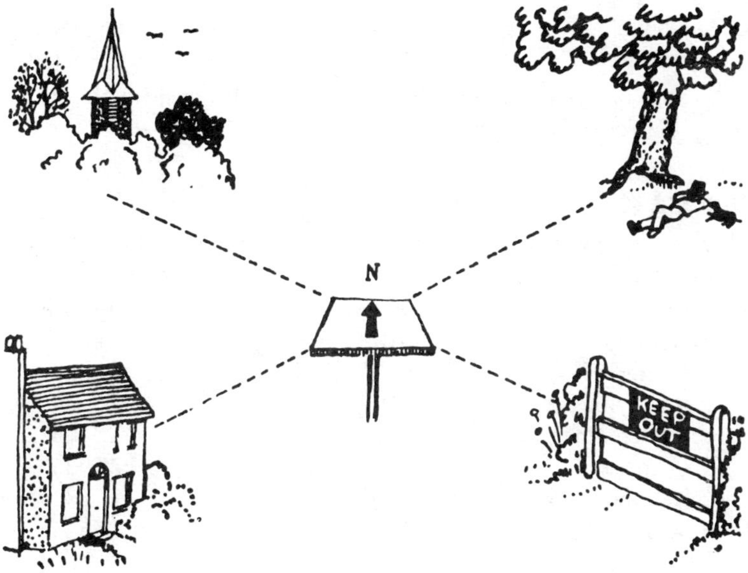

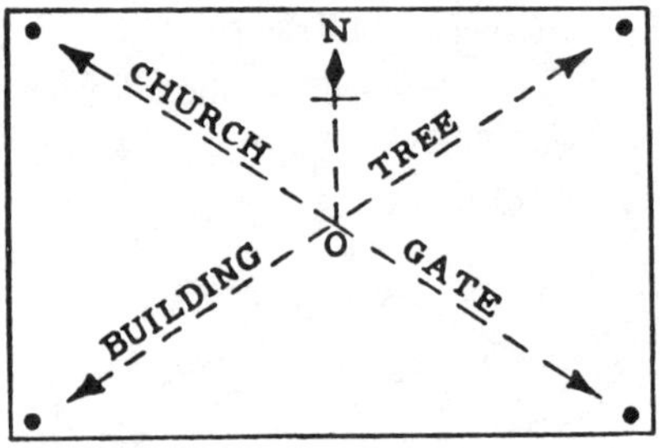

If the tree is 50 metres away we could show this distance on our drawing by using a scale of 1 cm to 10 metres. Then the tree would be 5 cm (50 metres) from *O*.

EXERCISE 9

The following table shows bearings and distances of objects from the centre of a school pi: ying field.

OBJECT	BEARING	DISTANCE
1. Cycle shed	020°	60 metres
2. Goal posts	120°	40 metres
3. Hockey posts	220°	30 metres
4. Main building	290°	65 metres
5. School gate	180°	60 metres

To a scale of 1 centimetre to 10 metres make a plan of these objects showing their actual position.

EXERCISE 10

The following were observed from a rise in the ground. Using a scale of 2 cm to 1 km plot their position.

OBJECT	BEARING	DISTANCE
1. Church spire	010°	3 km
2. Poplar trees	045°	3 km
3. Electric pylon	110°	3·5 km
4. Factory chimney	180°	1·5 km
5. Large house	220°	2·5 km
6. Bridge over river	270°	1·25 km
7. Signal box	290°	1·75 km
8. Water tower	120°	2 km
9. Flag pole	200°	0·75 km
10. Television mast	230°	1 km

To find the distance across a river.

Measure base line *AB* any convenient distance. Draw a similar base line on your plane table, to scale. Place table at *A* and sight point *B* along your base line. Sight *C* and draw direction line on table.

Take table to *B*, sight *A* along base line. Sight *C* and draw direction line on table. The diagram on your plane table should now look like this.

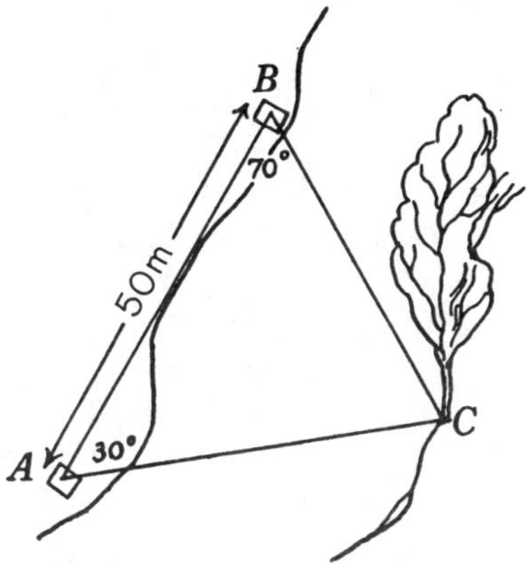

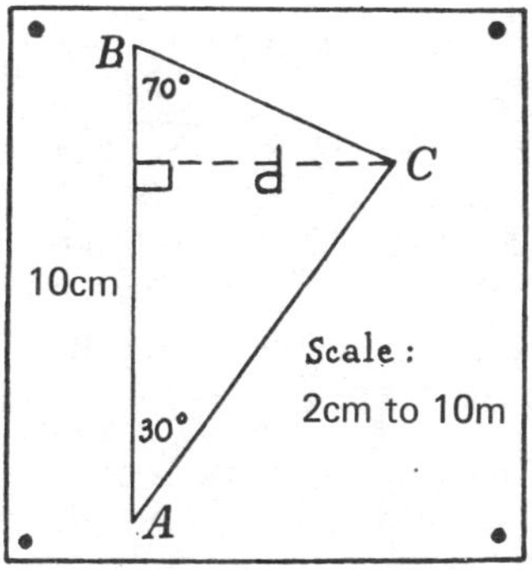

C will be the position of the tree. Draw a line at right angles to *AB* passing through *C* this is the distance across the river.

EXERCISE 11

Find by drawing the shortest distance from base line to point *C*. All angles in this exercise are between base line and object.

	LENGTH OF BASE LINE	ANGLE AT *A*	ANGLE AT *B*
1.	50 metres	45°	45°
2.	60 metres	45°	20°
3.	40 metres	50°	18°
4.	30 metres	60°	30°
5.	22 metres	70°	29°
6.	40 metres	44°	40°
7.	1 km	28°	30°
8.	14 m	30°	55°
9.	32 m	29°	61°
10.	2 km	35°	49°

The same problem can be solved taking all bearings from North.

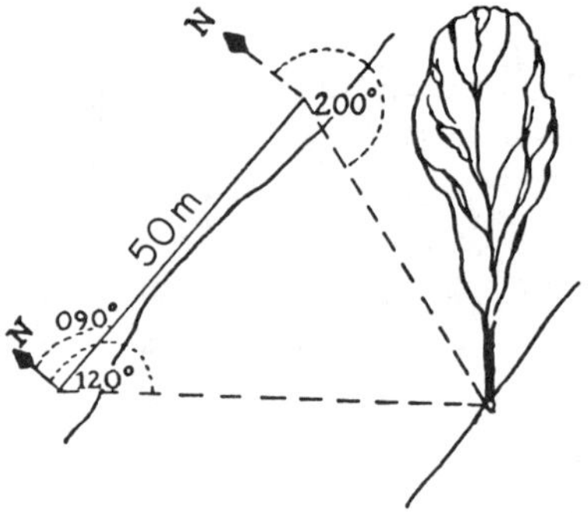

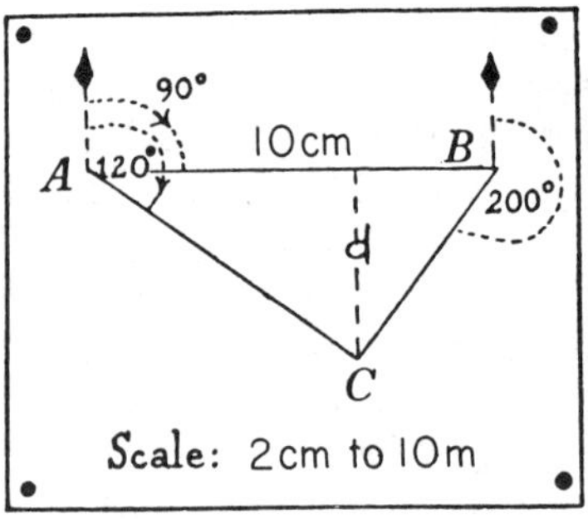

EXERCISE 12

Find the shortest distance of these objects from the base line.

1. From A the bearing of a church is 100°. B is 100 metres south of A. From B the bearing of the Church is 060°.
2. A is 50 metres west of B. From A the bearing of a boat is 030° and from B the bearing is 310°.
3. The bearing of a tree on an island is 215° from A. From B which is 440 metres south of A the bearing is 285°.
4. B is north-west of A and 200 metres away from it. The bearing of a hut on a mountain side is 240° from A and 175° from B.

From these exercises you will see that if the length of one side of a triangle and two angles are known, then we are able to construct that triangle. In what other ways are we able to construct triangles?

We saw how the Egyptians used their knowledge of constructing triangles to construct a right angle. Any triangle can be constructed if we know the lengths of its sides.

EXAMPLE 1 To construct a triangle with sides 4 cm, 6 cm and 8 cm.

Draw one of the sides, e.g. 6 cm side. Call it AB.

With compass open 4 cm and point at A describe an arc.

With compass open 8 cm and point at B describe an arc to cut first arc at C.

Join AC and BC thus making the required triangle.

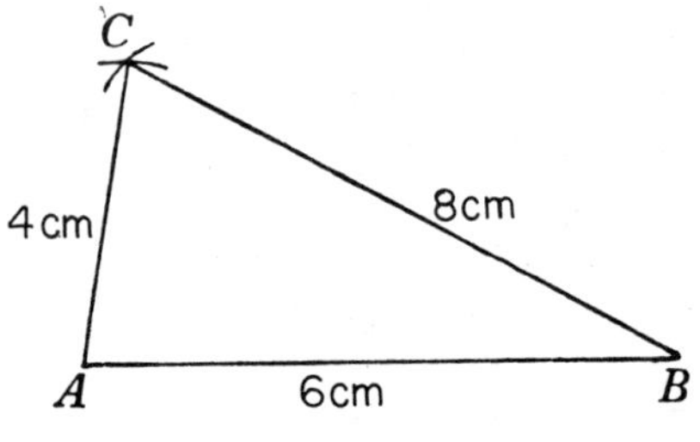

EXAMPLE 2

To construct a triangle with sides 6 cm., 4 cm. and included angle of 40°.

Draw one of the sides, e.g. 6 cm.
Label it *AB*.
Measure angle of 40° at *A*.
Measure 4 cm. along this line to *C*.
Join *BC* making the triangle *ABC*.

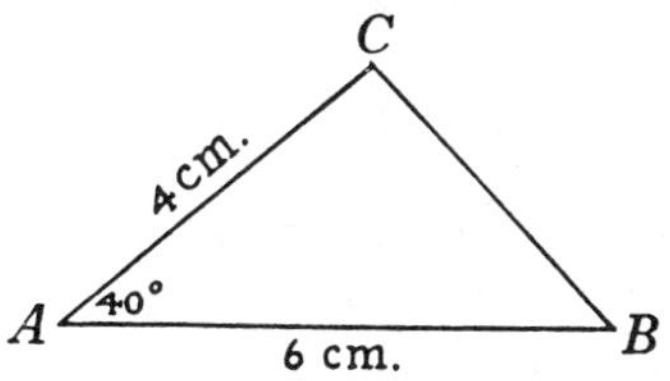

SUMMARY

A triangle can be constructed if the following facts are known.

1. The lengths of the three sides, S.S.S.
2. The lengths of two sides and the included angle, S.A.S.
3. The length of one side and two angles, S.A.A.

EXERCISE 13

Construct these triangles.

1. 4 cm.	45°	30°
2. 7 cm.	60°	45°
3. 10 cm.	40°	15°
4. 3 cm.	20°	24°
5. 2 cm.	24°	30°
6. 8 cm.	26°	36°
7. 6 cm.	28°	40°
8. 5 cm.	32°	47°
9. 4 cm.	34°	50°
10. 5 cm.	36°	56°
11. 8 cm.	70°	40°
12. 4 cm.	75°	30°
13. 7 cm.	80°	68°
14. 4 cm.	85°	28°
15. 35 mm.	90°	35°
16. 22mm.	95°	50°
17. 24 mm.	110°	32°
18. 47 mm.	34°	35°
19. 82 mm.	37°	51°
*20. 93 mm.	40°	40°

EXERCISE 14

1. 2 cm.	4 cm	30°
2. 2 cm.	3 cm	45°
3. 3 cm.	4 cm	60°
4. 4 cm.	2 cm.	90°
5. 3 cm.	1 cm.	120°
* 6. 4 cm.	4 cm.	35°
7. 5 cm.	7 cm.	40°
8. 35 mm.	20 mm.	55°
9. 45 mm.	30 mm.	65°
10. 20 mm.	25 mm.	70°
11. 10 cm.	7 cm.	35°
*12. 9 cm.	9 cm.	25°
13. 4 cm.	6 cm.	120°
14. 5 cm.	10 cm.	75°
15. 41 mm.	50 mm.	140°
16. 58 mm.	47 mm.	39°
17. 86 mm.	75 mm.	58°
18. 67 mm.	56 mm.	29°
19. 74 mm.	65 mm.	115°
20. 45 mm.	34 mm.	46°

EXERCISE 15

1.	4 cm.	3 cm.	2 cm.
* 2.	3 cm.	4 cm.	3 cm.
3.	2 cm.	3 cm.	15 mm.
† 4.	5 cm.	5 cm.	5 cm.
* 5.	3 cm.	5 cm.	3 cm.
6.	35 mm.	45 mm.	40 mm.
7.	45 mm.	28 mm.	24 mm.
† 8.	25 mm.	25 mm.	25 mm.
9.	28 mm.	33 mm.	15 mm.
*10.	35 mm.	43 mm.	35 mm.
†11.	55 mm.	55 mm.	55 mm.
12.	70 mm.	55 mm.	65 mm.
13.	3 cm.	4 cm.	5 cm.
14.	4 cm.	6 cm.	3 cm.
*15.	5 cm.	3 cm.	5 cm.
†16.	6 cm.	6 cm.	6 cm.
17.	7 cm.	5 cm.	3 cm.
*18.	7 cm.	4 cm.	4 cm.
19.	34 mm.	45 mm.	28 mm.
20.	58 mm.	35 mm.	45 mm.

EXERCISE 16

1.	4 cm.	50°	35°
2.	5 cm.	4 cm.	2 cm.
3.	6 cm.	35°	4 cm.
4.	35 mm.	48 mm.	28 mm.
5.	7 cm.	66°	25°
6.	5 cm.	25°	120°
7.	6 cm.	45°	10 cm
8.	10 cm.	90°	35°
9.	34 mm.	45°	25 mm.
†10.	45 mm.	60°	45 mm.
*11.	24 mm.	24 mm.	32 mm.
12.	5 cm.	95°	4 cm
13.	4 cm.	30°	65°
14.	5 cm.	135°	4 cm.
15.	5 cm.	20°	140°
16.	6 cm.	4 cm.	3 cm.
17.	4 cm.	3 cm.	5 cm.
*18.	55 mm.	45°	45°
19.	50 mm.	145°	21°
20.	10 cm.	45 mm.	66 mm.

When you have drawn some of these triangles measure the other sides and angles. What do you notice about the triangles marked * and †?

Can you give examples of triangles around you that have three equal sides? These are called Equilateral Triangles.

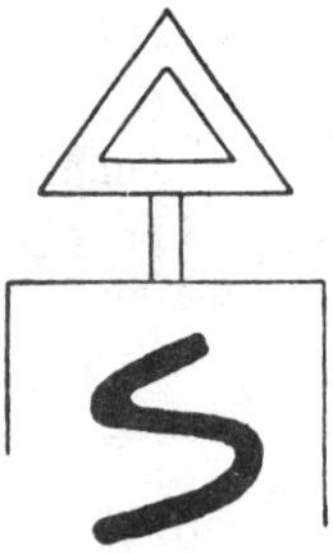

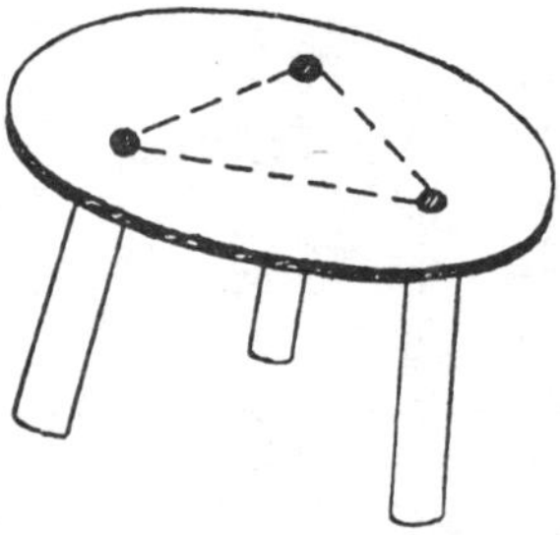

Many patterns can be made from equilateral triangles.

Have you ever seen a honeycomb? The cells in which the bees make their honey are shaped like the following diagram. Each cell can be formed by six equal triangles.

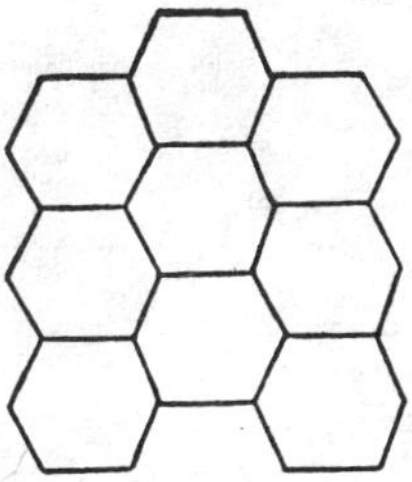

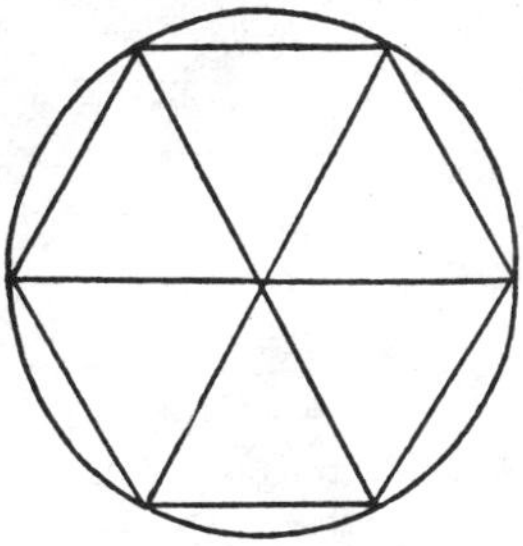

Triangles with two equal sides are called Isosceles Triangles.
Construct these patterns which are made up of equilateral and isosceles triangles.

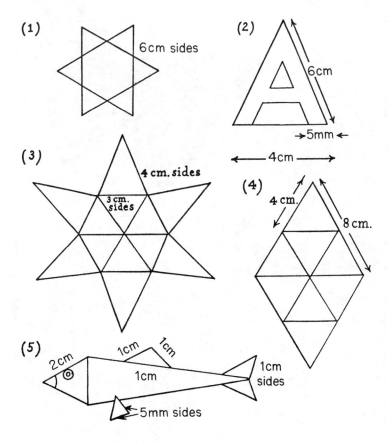

(1)

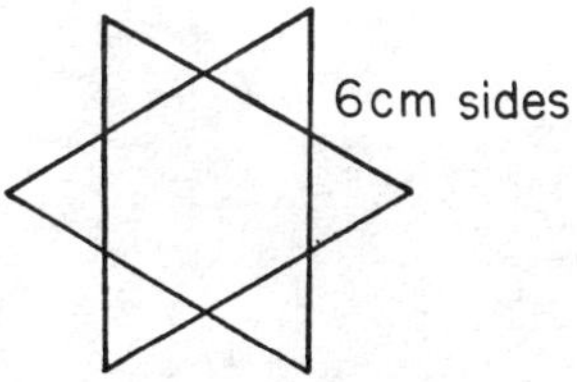

(2)

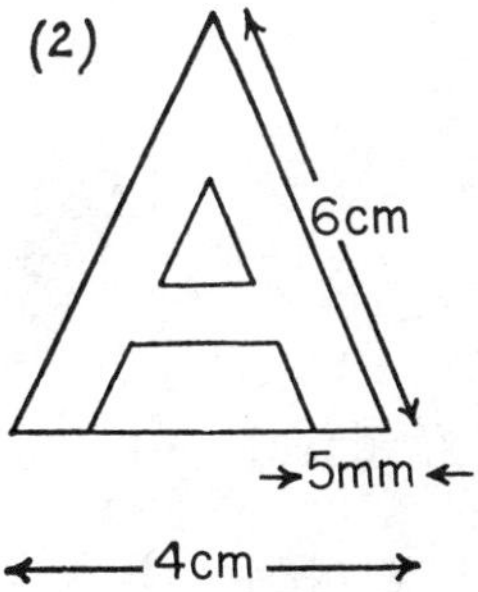

(3)

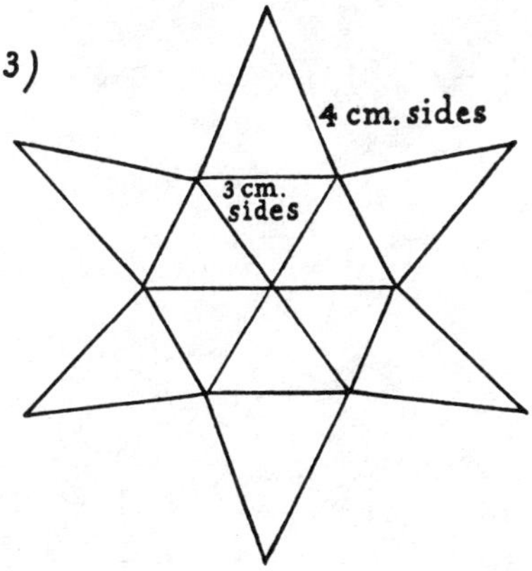

(4)

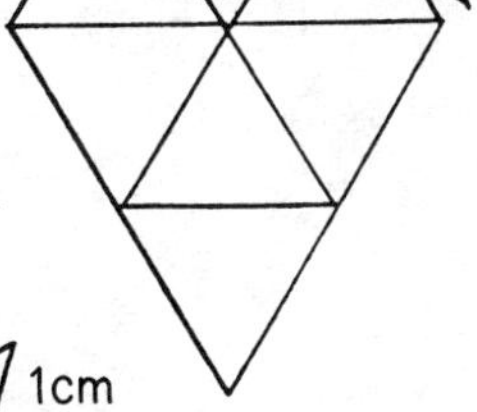

(5)

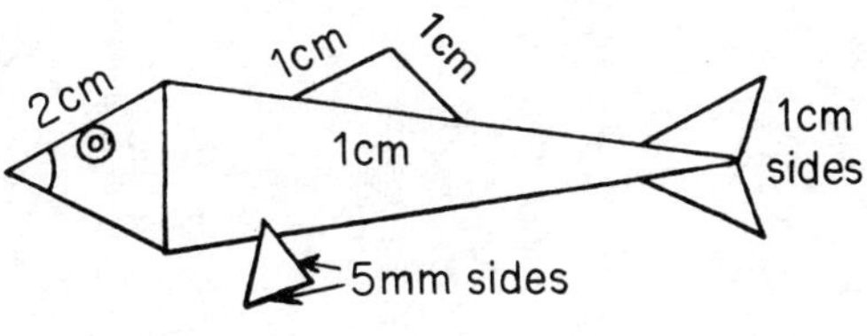

4 Ideas of Symmetry and Pattern

The illustrations on the page opposite give some ideas of symmetry which we find around us. The bat, the butterfly and the man are all symmetrical about a line which passes vertically through them.

Place a mirror, on its edge, along the dotted lines and you will see that the image in the mirror is the same as the half which is not reflected. When objects contain symmetry such as this, we say that they have bilateral symmetry. What about the snowflake? Does that have bilateral symmetry?

By placing your mirror along each of the axes in the snowflake you will see that it is symmetrical about six axes. Examine the doily in the same way. How many axes of symmetry does it contain? How many axes of symmetry does the circle have?

Fold a piece of paper and cut out some shape while it is still folded, then open up the paper and see the complete shape you have made.

The picture on the right hand side of the fold is the mirror image of the picture on the left hand side; check this is true by using your mirror. Can we say that the line X-Y is an axis.
of symmetry?

How many axes of symmetry does this shape have?

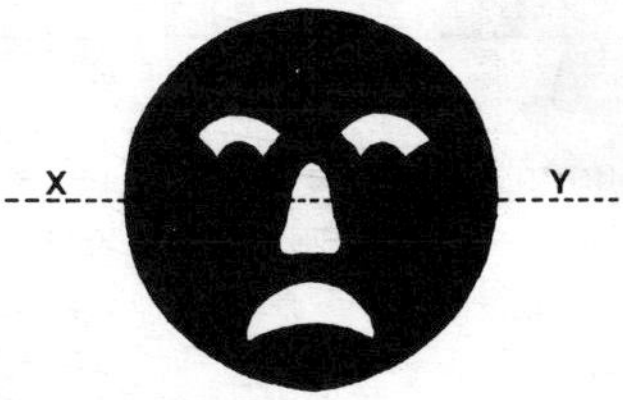

Use some circular pieces of coloured tissue to make some doily patterns. See if you can create some examples which will illustrate 1, 2, 3, 4, 5, 6, 7, and 8 axes of symmetry, by folding and cutting your paper. Is it possible to show all of these in this way?

Here are some quadrilaterals which we frequently meet in our work in mathematics. Use your mirror to find how many axes of symmetry they contain.

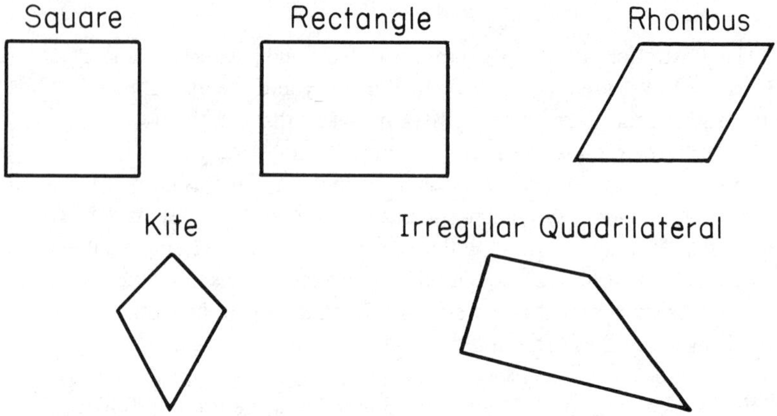

You will find that the square has four axes of symmetry. Does this tell you anything about he properties of the square?

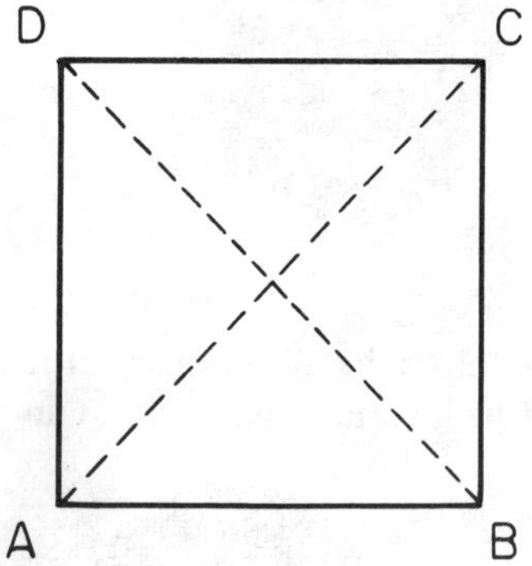

If I place a mirror along AC(reflecting D) the image will be the same as the half which is not reflected. From this we can deduce that DC = BC and AD = AB. Similarly if I place the mirror along BD (reflecting C), we can deduce that AD = DC and AB = BC.

Summarising the results : DC = BC
AD = AB
AD = DC
AB = BC

From these we can say that DC = BC = AB = AD. All the sides of the square are the same length.

By placing the mirror along the two other axes of symmetry, EG and FH, other properties of the square are revealed.

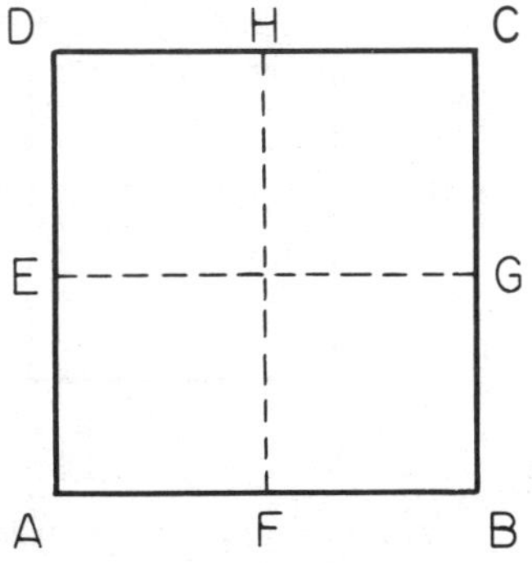

With the mirror along EG we can see that the ∠GCH = ∠FBG and ∠HDE = ∠EAF. Placing the mirror along HF reveals that the ∠HDE = ∠GCH and ∠EAF = ∠FBG.

Summarising then ∠GCH = ∠FBG
∠HDE = ∠EAF
∠HDE = ∠GCH
∠EAF = ∠FBG

Therefore ∠GCH=∠FBG=∠EAF=∠HDE.

All of the angles are the same, they are all 90° as we saw on page 8.

EXERCISE 17

Examine the set of quadrilaterals in this way and complete the table below by recording your results, as indicated by the square.

	Square	Rectangle	Rhombus	Kite	Irreg. Quad.
Axes of Symmetry	4				
Equal Sides	4				
Equal Angles	4				
Equal Diagonals	2				
Diagonals bisect each other	✓				

EXERCISE 18

Repeat your experiment on the set of triangles below.

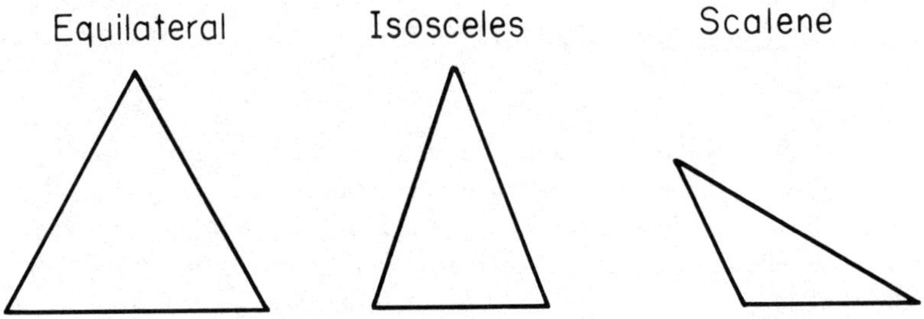

Another important characteristic of squares is that they will fit together around a point.

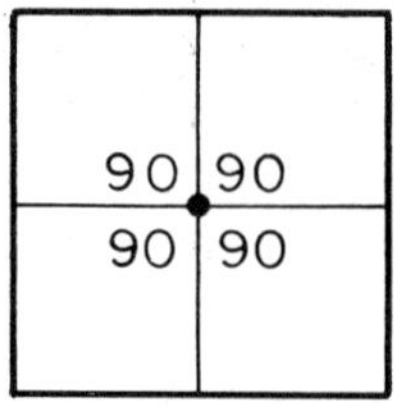

See how snugly these squares fit, leaving no gaps at all. In fact we could continue to add more squares so that this page was completely covered with them.

Are there any other shapes which have this property of covering space? What about the scalene triangle and the equilateral triangle?

Why will the square and equilateral triangle fit together in this way?

The angles of the square, as you discovered, are each 90°, and the angles of the equilateral triangle are each 60°. Is 360° a multiple of 90°? Is 360° a multiple of 60°? Here then is the answer to our questions.

We can build up other shapes from equilateral triangles which will fit together to cover space.

What shape is shown here? Cut out such a shape and see if it will fit round a point and so cover space without leaving gaps.

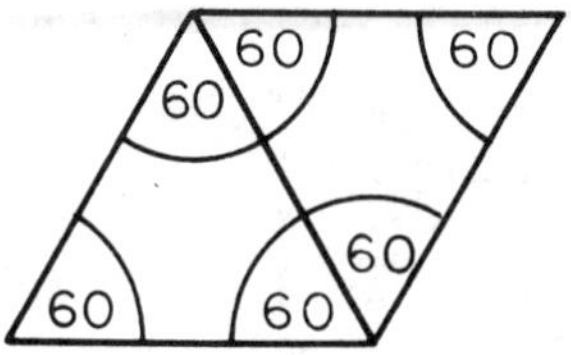

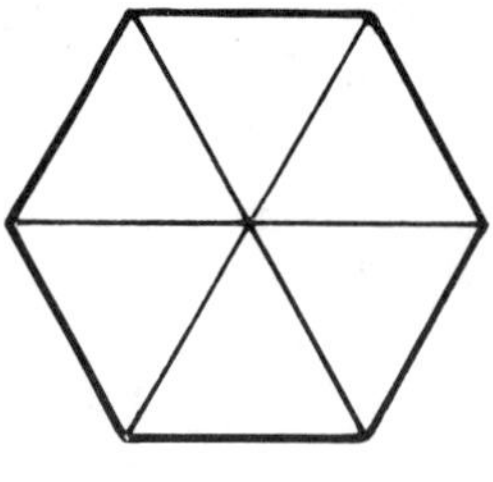

Here is another shape made up from equilateral triangles. What do we call this shape? Will it cover space without leaving gaps?

When a shape will cover space in this way and leave no gaps, we say that the shape tessellates. You will come across tessellations of squares again when you are shown how they are used to measure surfaces. We can also use tessellations for building up patterns for the decoration of material, wall paper and so on. Find out how many tessellations you can make by using the basic shapes which you know. Will the circle tessellate, or does it need some other shape combined with it in order to completely cover a surface?

The square and the equilateral triangle are regular figures. Can we change them in some way to make other shapes which will tessellate?

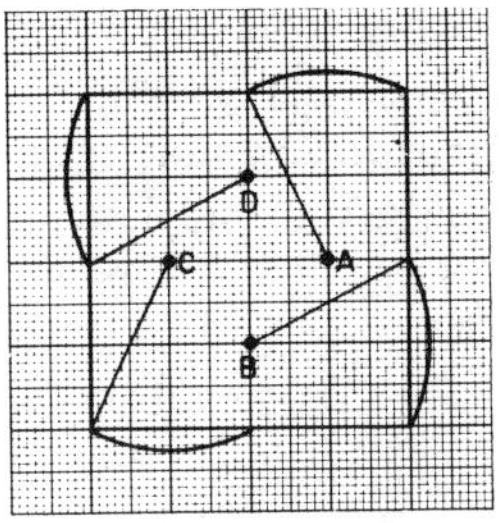

On cm squared paper, we have outlined a square with 4 cm sides. By placing the compass point on the marks A, B,C, & D, and keeping the same radius we have traced arcs as shown.

Using the same square and using E, F, G and H as centres we have drawn some more arcs.

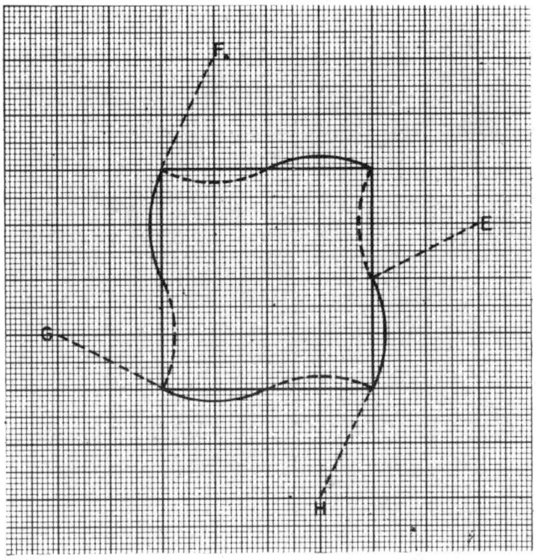

The completed shape which we have made from the square looks like this. Draw a similar shape in the way we have described. Cut out the shape and see if it will tessellate.

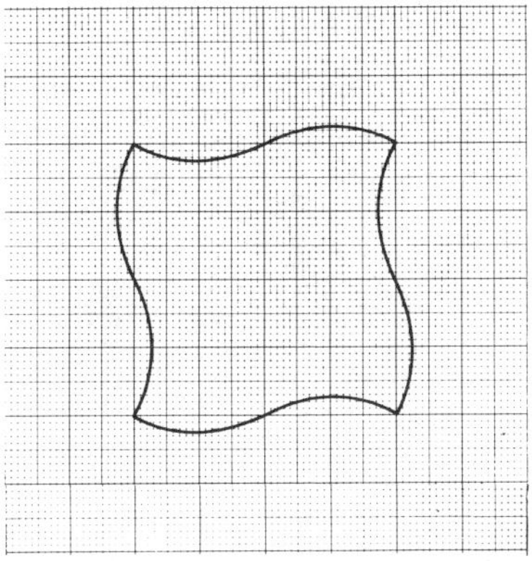

Did you notice that the shape will tessellate if we rotate it each time by 90° about one of its points? In making the shape we have cut a piece from the square and added a similar piece to the opposite side.

We can change the equilateral triangle in a similar way. It is an advantage to do this on paper which is covered with equilateral triangles.

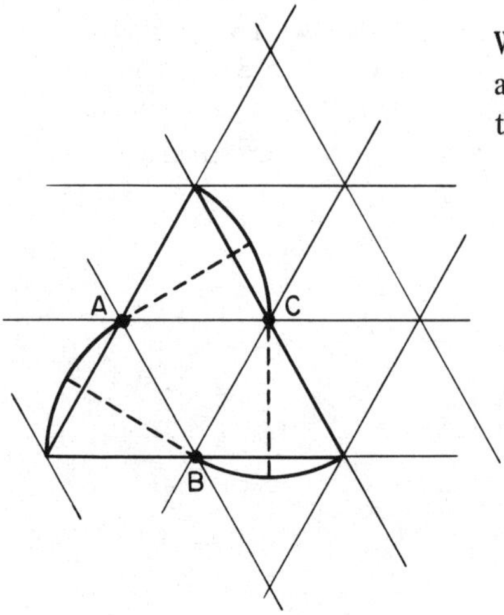

With the compass points at A,B, and C draw the curves shown. Keep the same radius.

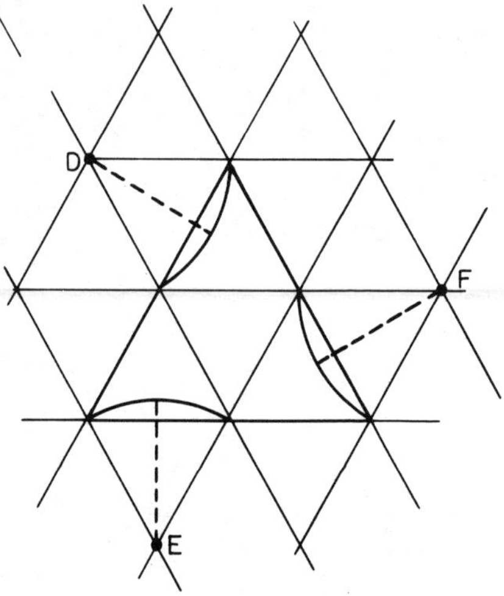

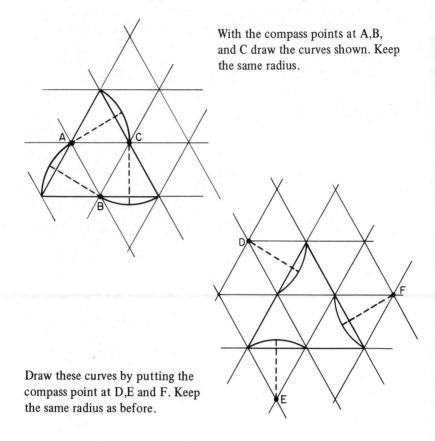

Draw these curves by putting the compass point at D,E and F. Keep the same radius as before.

The completed shape looks like this, and by rotating about any one of its points it will tessellate. Through what angle must you rotate it?

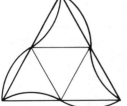

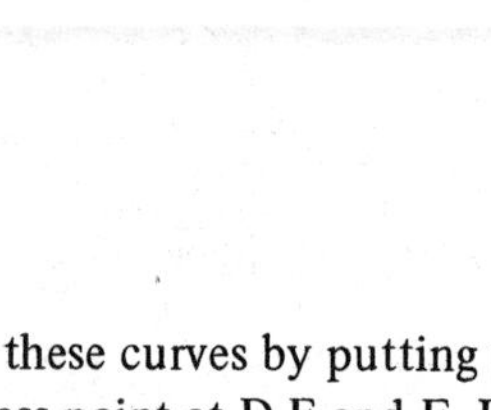

EXERCISE 19

1. Make a drawing of a Rhombus and a Hexagon and see if you can change their shapes to design other shapes which will tessellate.
2. Make copies of these shapes, cut out your drawings and see if they will tessellate. Construction lines have been elft in for your guidance.

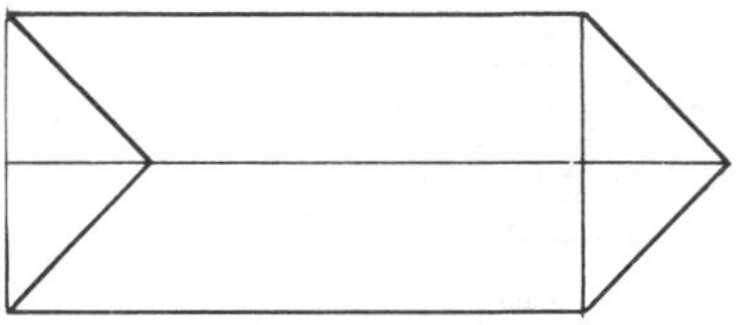

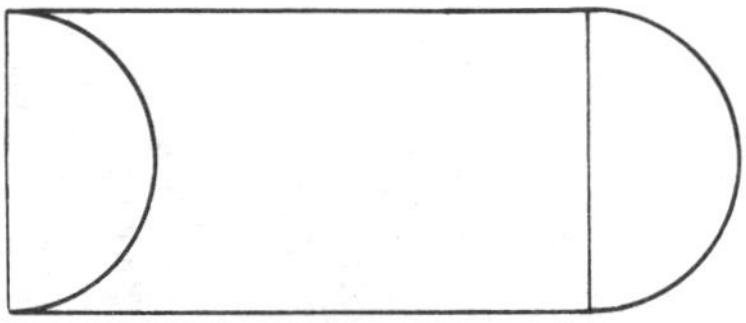

3. This shape tessellates as we have seen, but we have now added a design to it, and by making a line cut of it we can print designs on fabrics, and paper. Make up a design of your own and use it to cover a piece of A4 paper

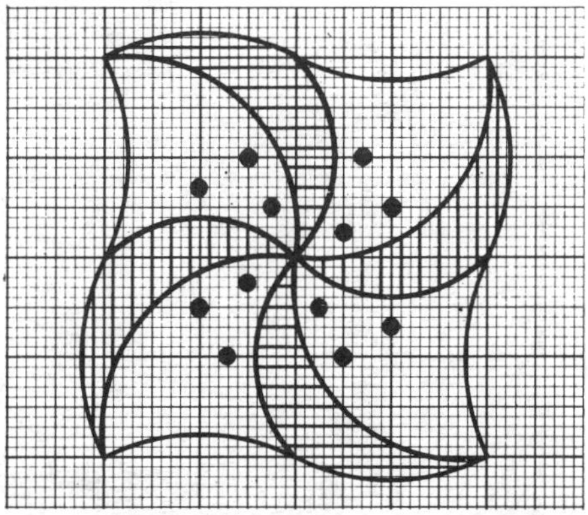

4. Can you see how these patterns have been made?

5 Parallels

On the illustration opposite are lines drawn for music. What do you notice about them? If you measure the distance between them you will see that it is the same all along their length.

When lines are an equal distance apart along the whole of their length we say they are PARALLEL. What other examples of parallel lines are there in the drawing?

If you make a piece of trellis with strips of card pinned together, you will find that you can stretch it.

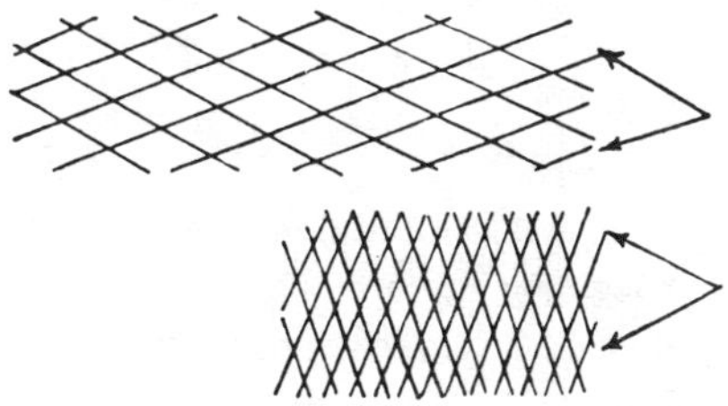

Press these ends together and the trellis will stretch.

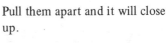

Pull them apart and it will close up.

You will see that in both of these cases the strips stay parallel. If you pin a piece of card across your model, it will be fixed and will neither open nor close.

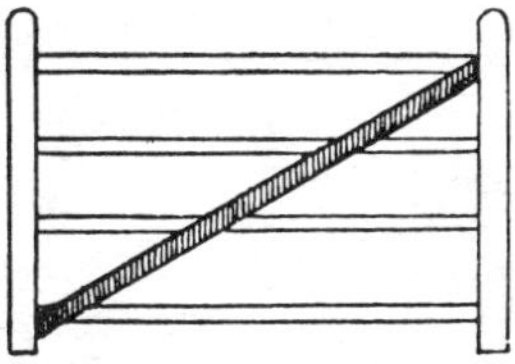

This is the reason for putting the strip diagonally across the gate. Without it the heavy gate would sag.

Cut four pieces of card, two pairs. One pair longer than the other. Pin them at the corners as shown.

The shape you now have is called a RECTANGLE.

What do you notice about the angles of the rectangle? What do you notice about the lengths of the sides?

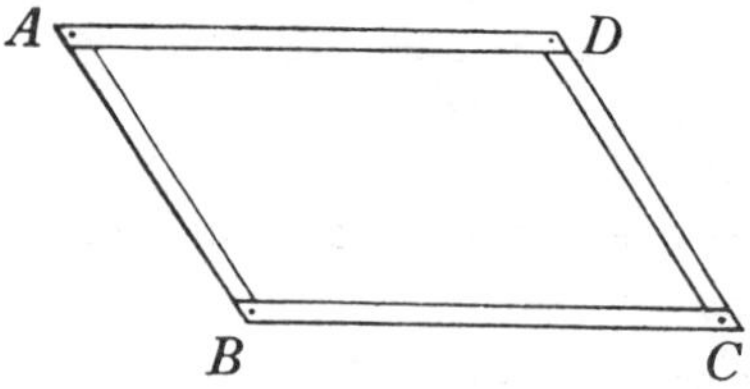

If you keep the side *BC* still and push *AD* over to the left your figure will look like this.

This is a PARALLELOGRAM.

Are the opposite sides still equal? Are they still parallel? What is the difference between this and the rectangle? Measure the angles and see if there is any definte pattern in this arrangement. You can make any number of parallelograms with your model. See if your discoveries are true for just one particular parallelogram or if they are true for ALL PARALLELOGRAMS.

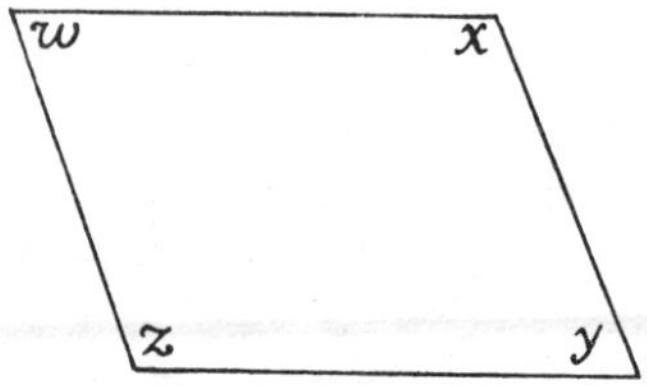

Add angles *w* and *x*, and *y* and *z*. You fill find that each pair adds up to 180°.

They are Interior Angles and since they add up to 180° they are called Supplementary Interior Angles.

We have now discovered these facts

In the RECTANGLE opposite sides are equal and parallel to each other.

Interior angles are all right angles.

In the PARALLELOGRAM opposite sides are equal and parallel to each other.

Opposite angles are equal to each other.

Adjacent angles are supplementary.

Fix a strip diagonally across your parallelogram.

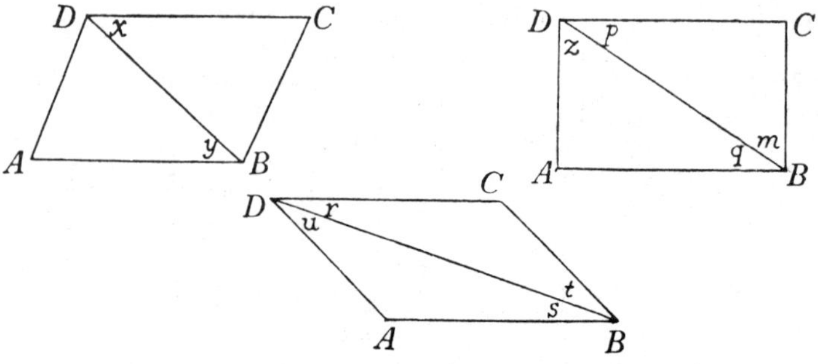

This fixes your model in one position. Measure the angles indicated.

What do you notice about the following pairs?

x and y	p and q
z and m	r and s
u and t	

These are called ALTERNATE angles because they are on alternate sides of BD.

In the letters *N* and *Z* there are pairs of alternate angles.

From your measurements you will see that the following pairs of angles are equal:

x and y, r and s, p and q — because DC is parallel to AB.

z and m, u and t — because AD is parallel to CD.

To add further strength to the model fix on the other diagonal.

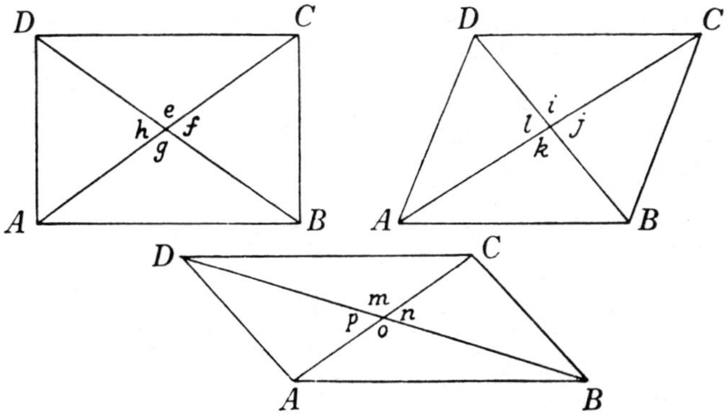

Measure the angles formed where the two diagonals cross. What do you notice about the following pairs of angles?

e and g	h and f	i and k
m and o	p and n	l and j

These are pairs of VERTICALLY OPPOSITE Angles.

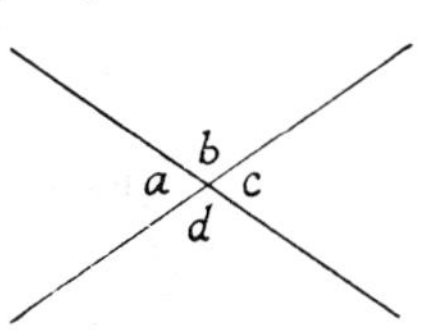

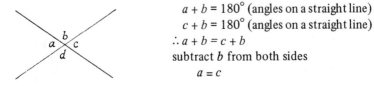

$a + b = 180°$ (angles on a straight line)

$c + b = 180°$ (angles on a straight line)

$\therefore a + b = c + b$

subtract b from both sides

$a = c$

Make a similar model with all four sides equal. This can either be a square or a RHOMBUS. Put in the diagonals and, as before, measure all the angles.

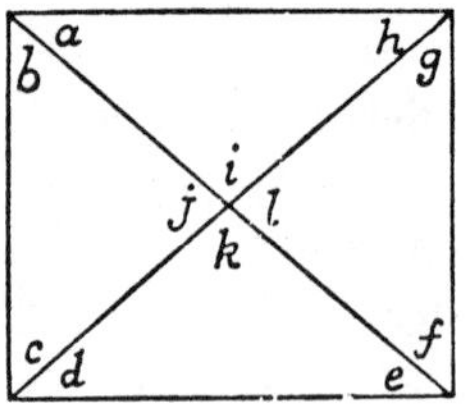

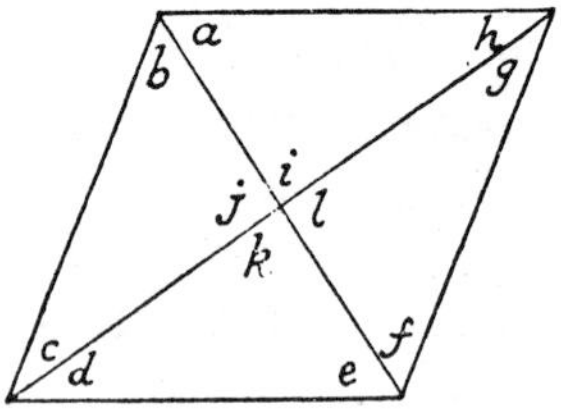

What can you say about these pairs of angles?

i and k	i and j	a and e
h and d	c and g	b and f
$(a+b)$ and $(e+f)$		$(h+g)$ and $(c+d)$
$(a+b)$ and $(c+d)$		$(h+g)$ and $(e+f)$

i, k, l and j are all . . . . . ? . . . . . .

In the Rhombus Opposite sides are . . . . . . and . . . . . . to each other.
The diagonals intersect at . . . . angles.
Opposite . . . . are equal to each other.

If you measure the lengths of the diagonals of the rhombus and the parallelogram you will find that they bisect each other.

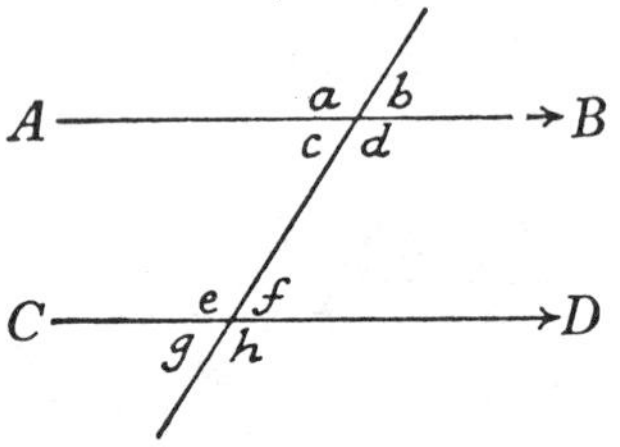

AB is parallel to *CD*.

$\therefore a = d$, $b = c$, $f = g$, $e = h$ } vertically opposite angles.

and $c = f$, $d = e$ } alternate angles.

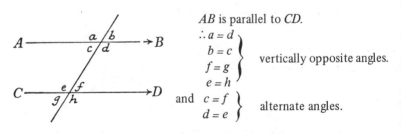

If you cut out a coloured piece of paper to cover the angle a you will see that it also covers the angle at e. Look at their position on the diagram They are both on the top left hand position in this group. They are in corresponding positions in their groups and are called CORRESPONDING Angles.

$a = e$, $b = f$, $c = g$, $d = h$ } corresponding angles

$c + e = 180°$, $d + f = 180°$ } supplementary interior angles

Summary

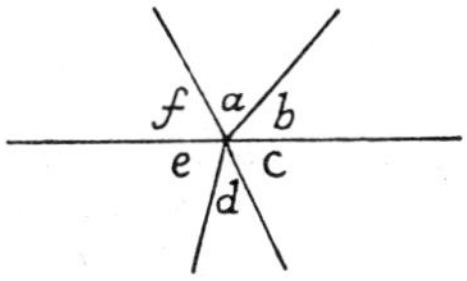

Angles at a point add up to 360°.

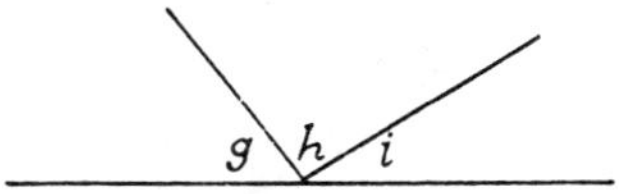

Angles on a straight line are supplementary.
$g + h + i = 180°$.

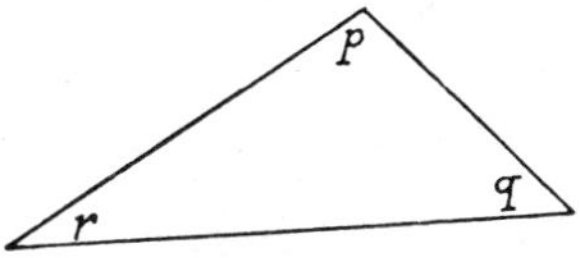

The angles of a triangle together make two right angles, i.e. 180°.

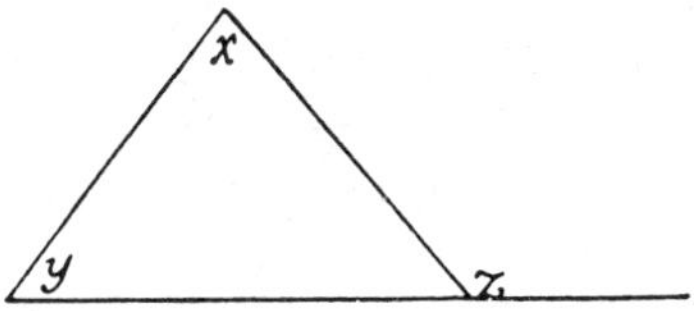

The exterior angle of a triangle is equal to the sum of the two interior opposite angles.

A Square has:

Four equal sides.
Four right angles.
Diagonals which are equal and bisect each other at right angles.

A Rectangle has:

Opposite sides which are equal and parallel.
Four right angles.
Diagonals which are equal and which bisect each other.

A Rhombus has:

Four equal sides.
Opposite angles which are equal.
Diagonals which bisect each other at right angles.

A Parallelogram has:

Opposite sides which are equal and parallel.
Opposite angles equal.
Adjacent angles supplementary.
Diagonals which bisect each other.

EXERCISE 20

What can you remember?

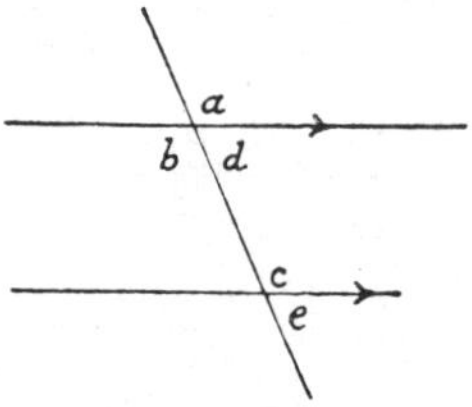

1. *a* and *b* are . . . . angles.
   a and *c* are . . . . angles.
   c and *e* are . . . . angles.
   b and *c* are . . . . angles.
2. If $a = 140°$ find angles *b, c, d* and *e*.

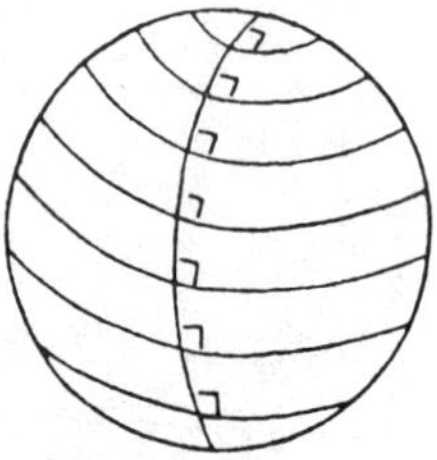

3. Lines of latitude are called parallels, e.g. 49th Parallel. What angle does a line of longitude make with a line of latitude?

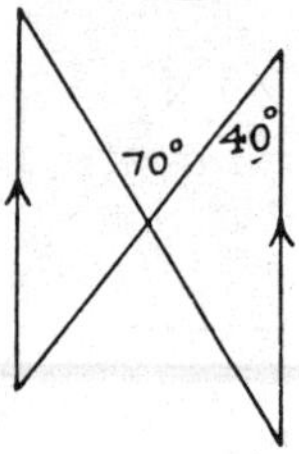

4. Copy this figure and put in all the missing angles.

5. Find angles *a* and *b*.

6. Find angles *a, b, c,* and *d.*

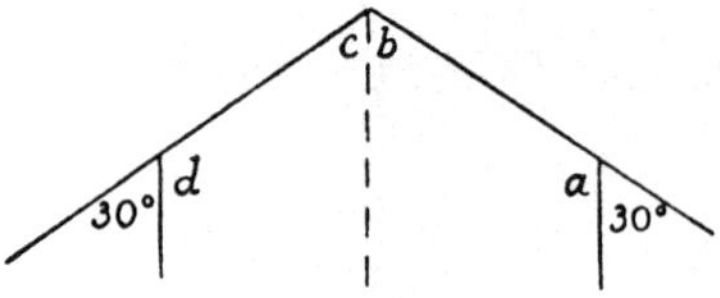

7. Find angles *a* and *b*.

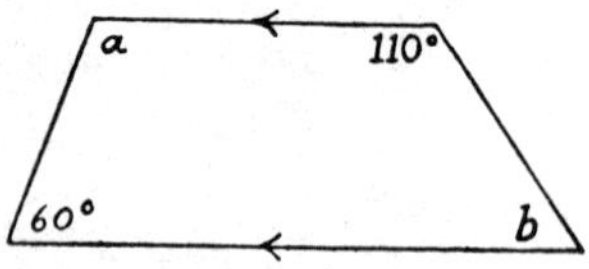

8. Put in the 3 angles of the triangle.

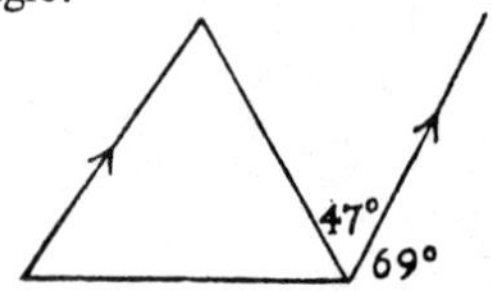

9. Find angle *p*.

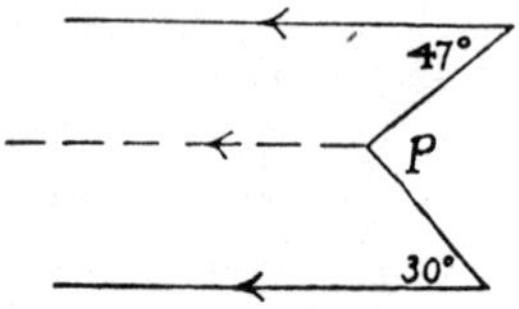

10. Put in the other 3 angles.

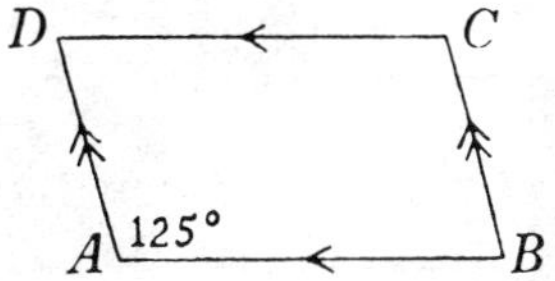

11. How big are *a*, *b*, *c* and *d*?

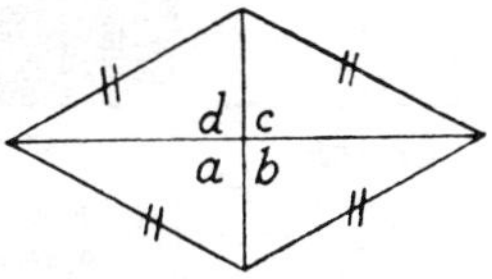

12. Name pairs of equal angles. Say why they are equal.

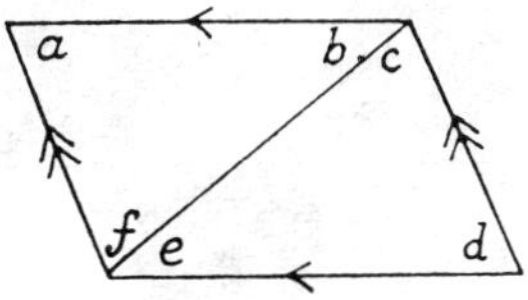

13. Find angle *a*.

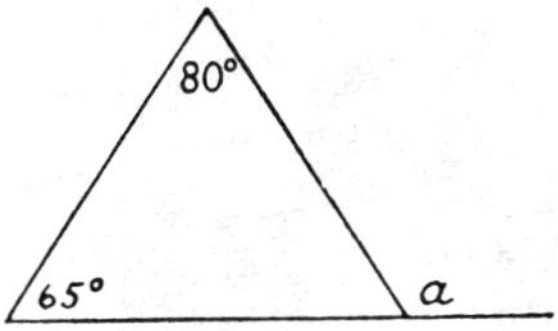

14. Find angle *b*.

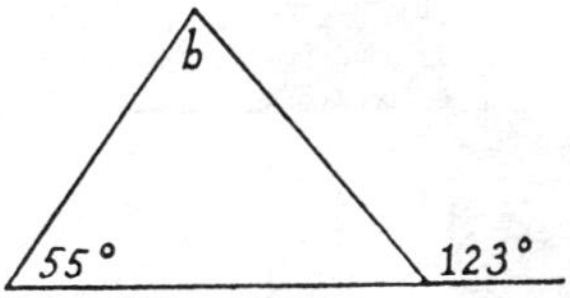

15. Find the value of *x*.

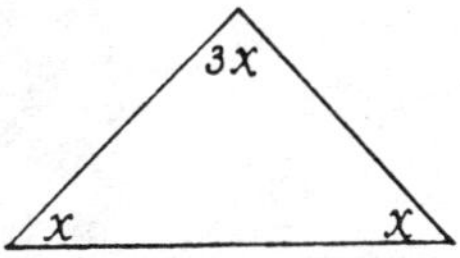

16. Find the angles of the triangle.

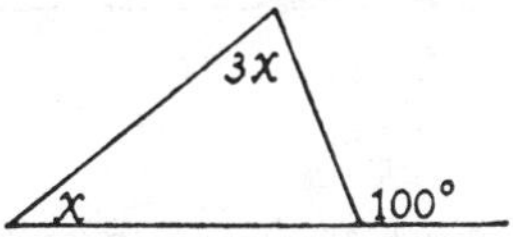

17. Copy this figure and put in all the missing angles.

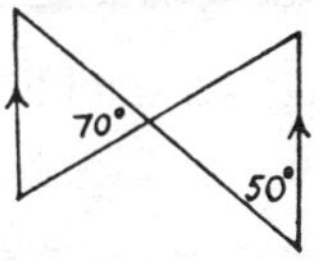

18. Find angles *x*, *a*, *b* and *c*.

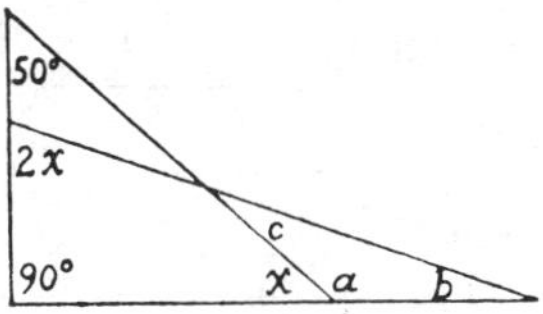

19. Find all the angles at this point.

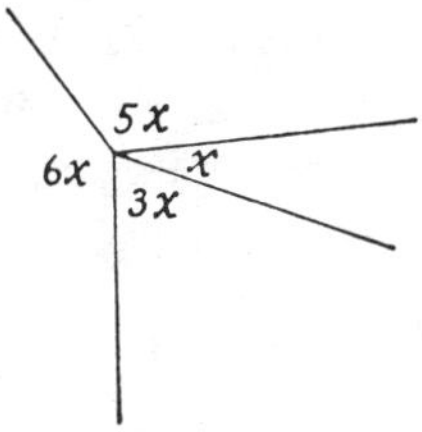

20. Find *x* 2*x* and 5*x*.

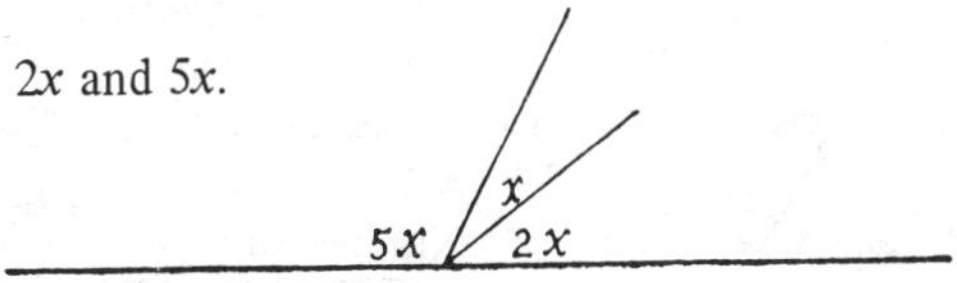

Copy this block and fill in the answers.

CLUES DOWN.

1. Find x.

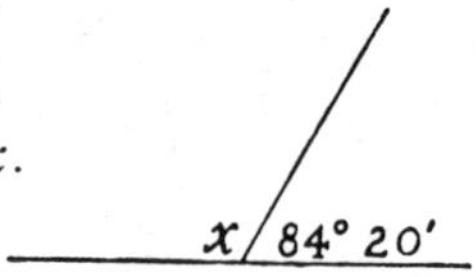

2. Course of plane flying N.E. (Bearing).
4. Course of plane flying N.25° E. (Bearing).
7. Find b.

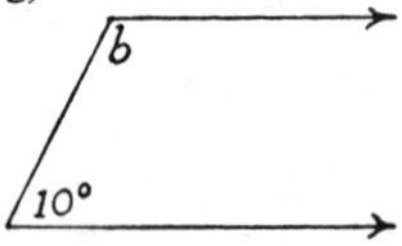

9. Find c.

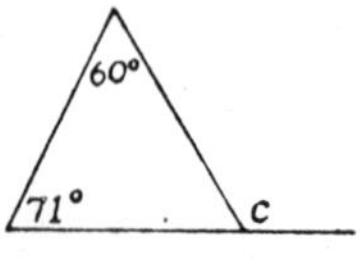

12. Sum of the angles at one corner of a box.
13. Bearing of a plane flying S. 35° W.
15. Number of seconds in 1°.

17. Find the sum of these angles.

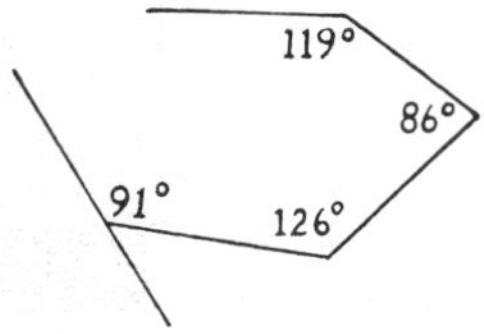

18. Find n.

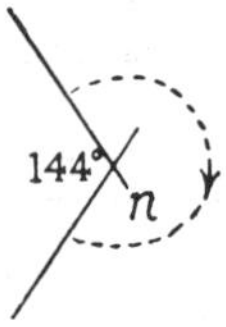

CLUES ACROSS.

1. Number of degrees in a right angle.
3. Sum of the angles of a square.
5. Angle through which a minute hand turns between 9 o'clock and 10.30.
6. Find x.

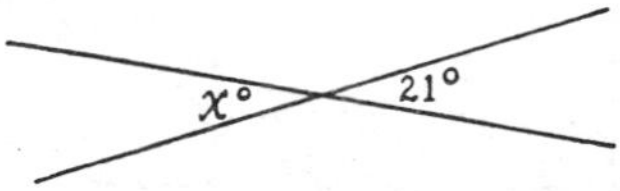

8. How many degrees between S. and S.W. measured in a clockwise direction?

10. Total number of degrees in the four angles of a square, three angles of a triangle + 36°.

11. Find a.

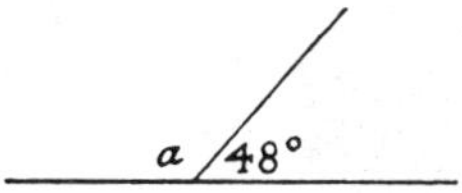

14. Find b.

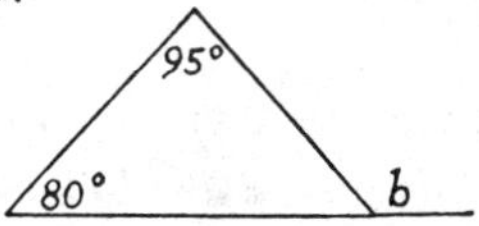

16. Find p.

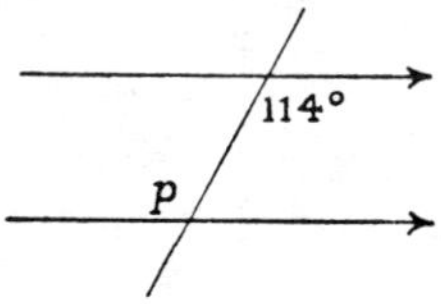

18. Find q.

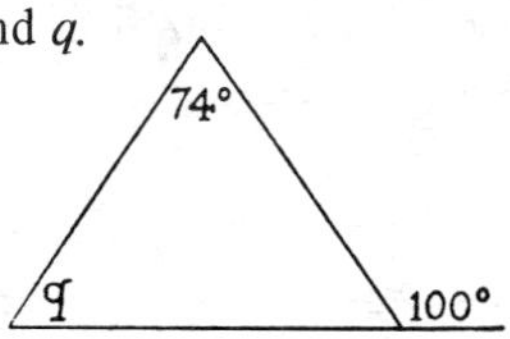

19. Find r.

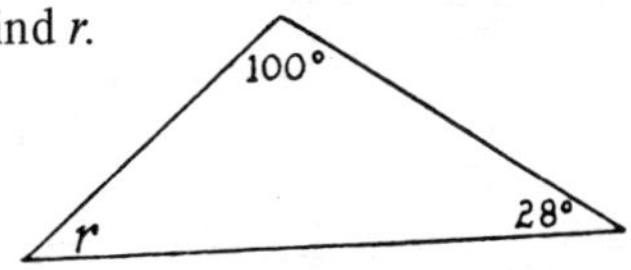

20. Angle through which the hour hand moves in seven hours.

21. Find s.

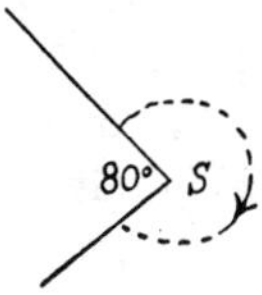

22. Find m.

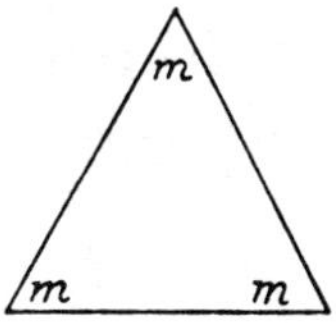

6 Scale

A drawing of your classroom might look like this. In order to draw it accurately in the space provided I have let each metre be represented by 1 cm.

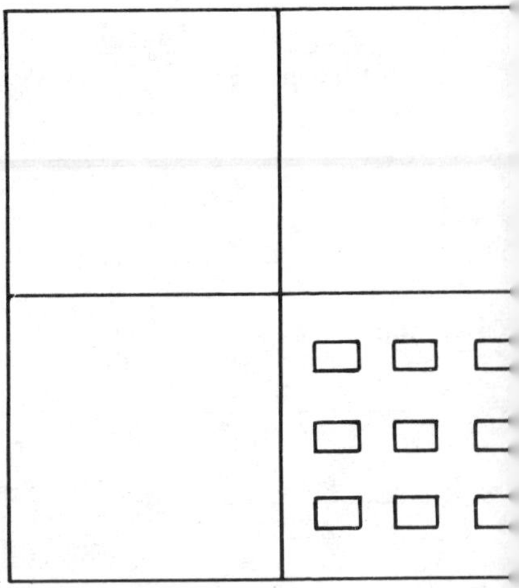

Yours could be one of four classrooms arranged like these. In this drawing I have let each 2 m be represented by 1 cm. Notice the desks are smaller.

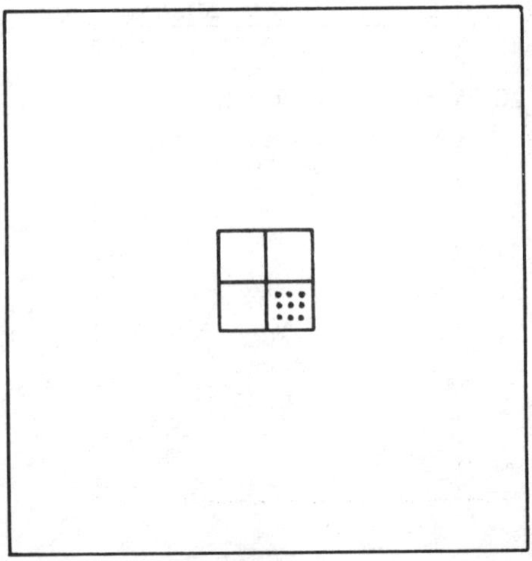

In this diagram I have shown the four classrooms in part of the school grounds. In this drawing I have let each 5 m be represented by 1 cm. In this plan the desks are merely dots.

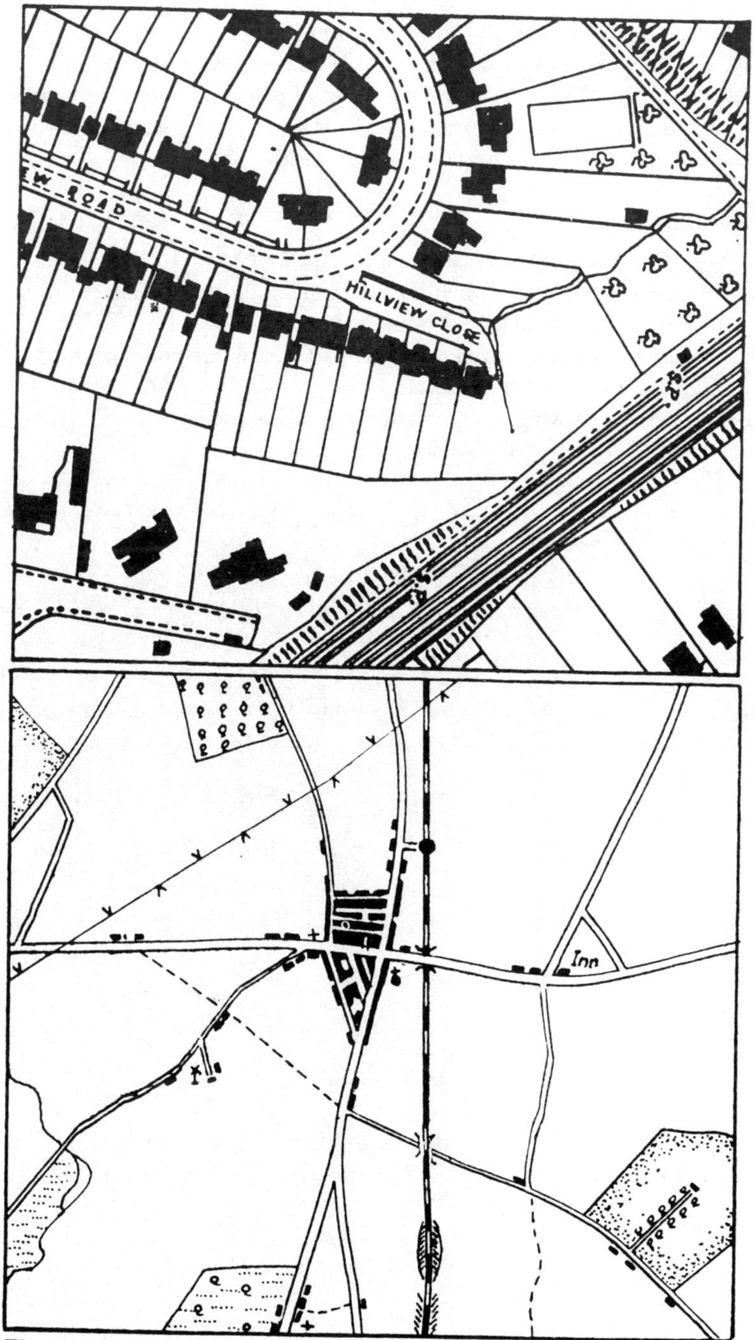

The top map is reproduced by permission of the Director General, The Ordnance Survey

Notice how the houses change from definite shapes on the large scale map to mere rectangles on the smaller scale map. It is very useful to be able to represent large areas on a small sheet of paper. The architect makes scale drawings of a house before it is built. Maps of large areas are drawn to scale so that you may study them conveniently.

EXERCISE 21

1. Make a drawing of your classroom to a scale of 1 cm. to 1 m.
2. Measure your desks and their position in relation to the walls, show these on your drawing.
3. Measure your school buildings or the part containing your classroom and draw it to a convenient scale.
4. Make a scale drawing of your football, hockey, rugby, netball pitches.
5. 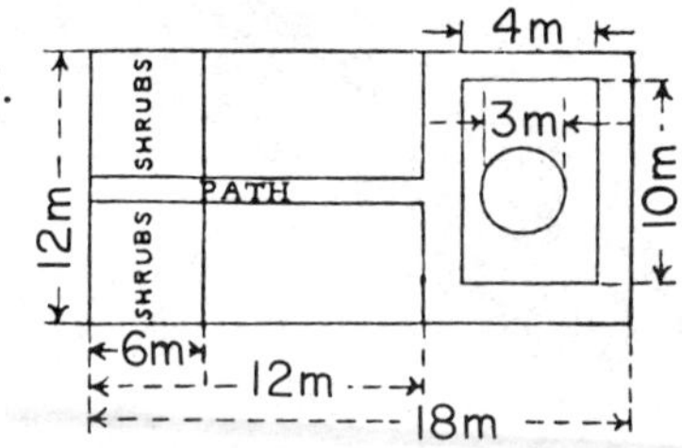

Width of all paths 1 m. Draw a plan of this garden to a scale where 1 cm represents 2 m.
6. Make a freehand sketch of a similar bookcase to the one shown in this diagram where all dimensions are a half of those indicated.

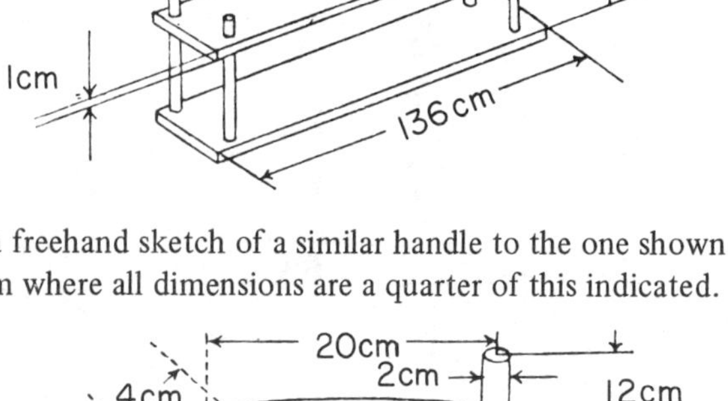

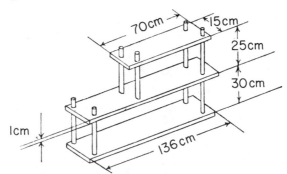

7. Make a freehand sketch of a similar handle to the one shown in this diagram where all dimensions are a quarter of this indicated.

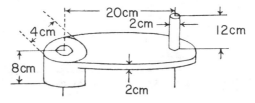

8. The following diagram shows the dimensions of articles which we find in the kitchen. Design your own kitchen and arrange these articles in suitable positions. The figures shown indicate the floor space required for each item. (Dimensions are in cm)

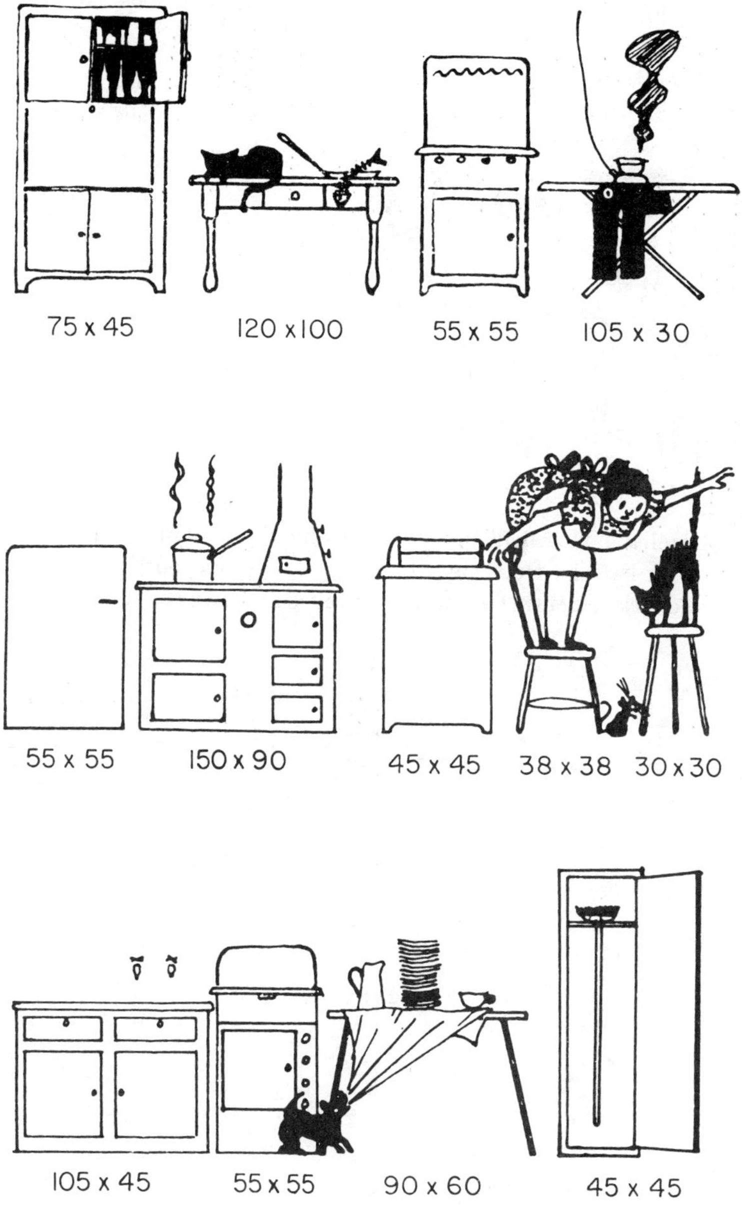

To make a survey of an irregular piece of ground.
Draw base line *AB*.
Draw offsets at right angles to *AB* to pass through objects being plotted.
Measure distance of offsets along base line from *A*.
Measure the length of each offset.
Surveying this field the entries in your field book would look like this.
Dimensions in metres.

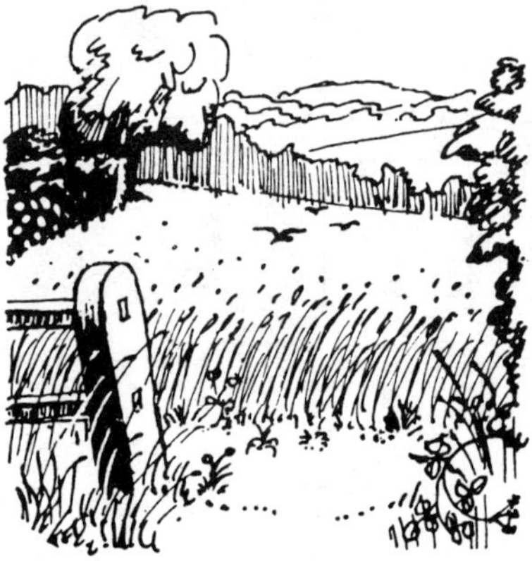

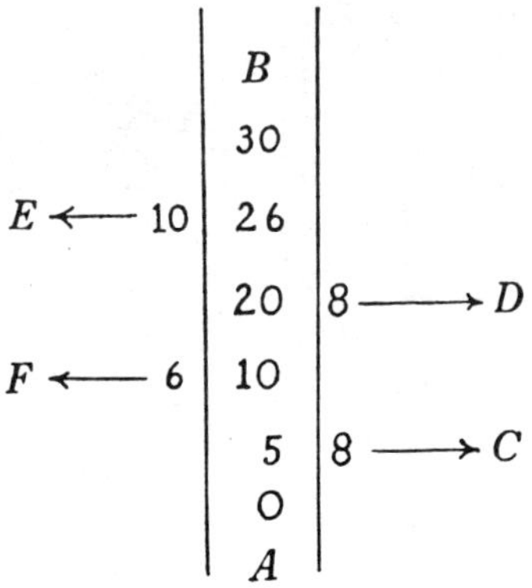

Notice that in your fieldbook you show the lengths of offsets and the point on the base line from which they are taken. From your fieldbook you could make a drawing of the area surveyed as shown below.

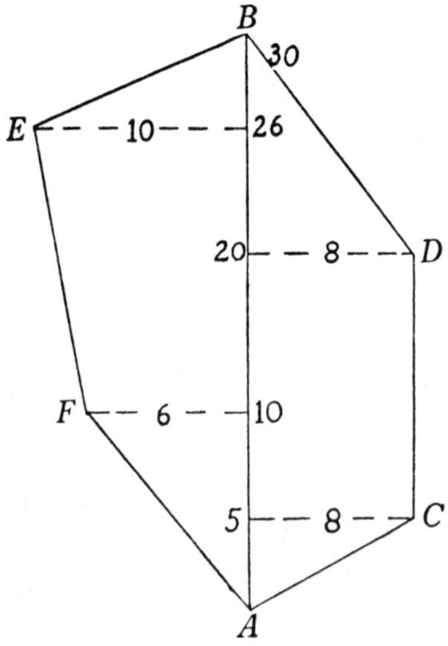

EXERCISE 22

Choose a suitable scale and draw the following. Find the areas of the first three questions. Dimensions are in metres.

(1)

	B	
	40	
Corner F 15	30	10 Corner E
Corner G 10	20	15 Corner D
	10	5 Corner C
	A	

(2)

	B	
20	35	
7	20	20
	16	15
10	5	
	A	

(3)

	B	
10	30	
	25	30
21	17	
	4	24
	A	

(4)

	B	
Pool 28	56	10 Gate
	48	19 Copse
Fox Hole 14	30	
	15	10 Sparrows' Nest
	A	

(5)

	B	
	30	
10	20	
	15	8
7	6	4
	A	

(6)

	B	
Hut 20	50	40 Cross Roads
	40	25 Bridge
Poplar Tree 30	20	
	10	15 Duck Pond
	A	

Heights and Distances

This instrument will help you to measure angles of elevation.

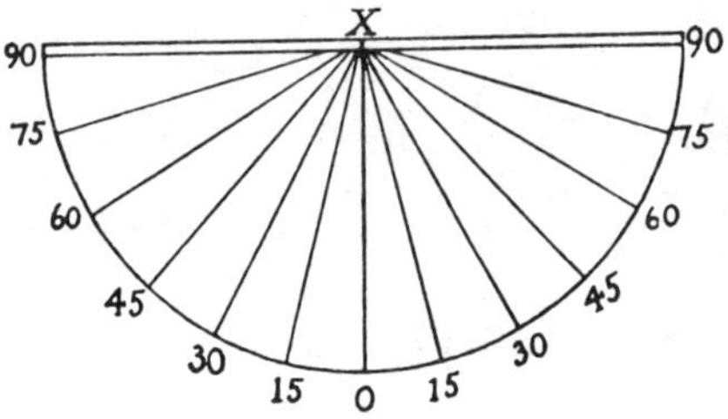

Cut a semicircle from a piece of card and mark it in degrees as shown.

Glue a narrow tube or milk straw to the straight edge.

Suspend a weighted thread from point *X*.

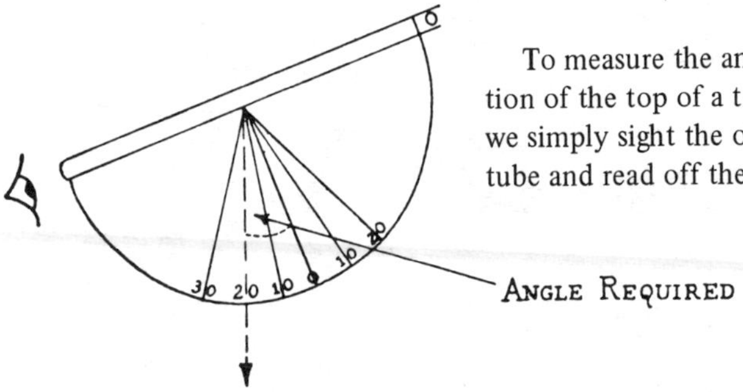

To measure the angle of elevation of the top of a tree or building, we simply sight the object along the tube and read off the angle as shown.

From the diagram you will see that the angle we find is the angle of elevation.

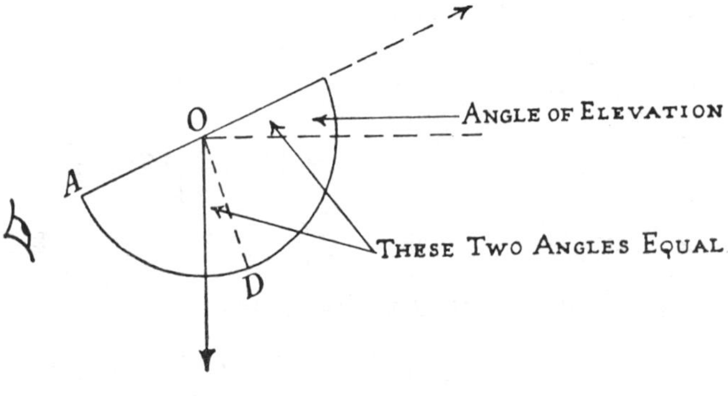

This instrument can be used for measuring angles in the vertical and horizontal planes.

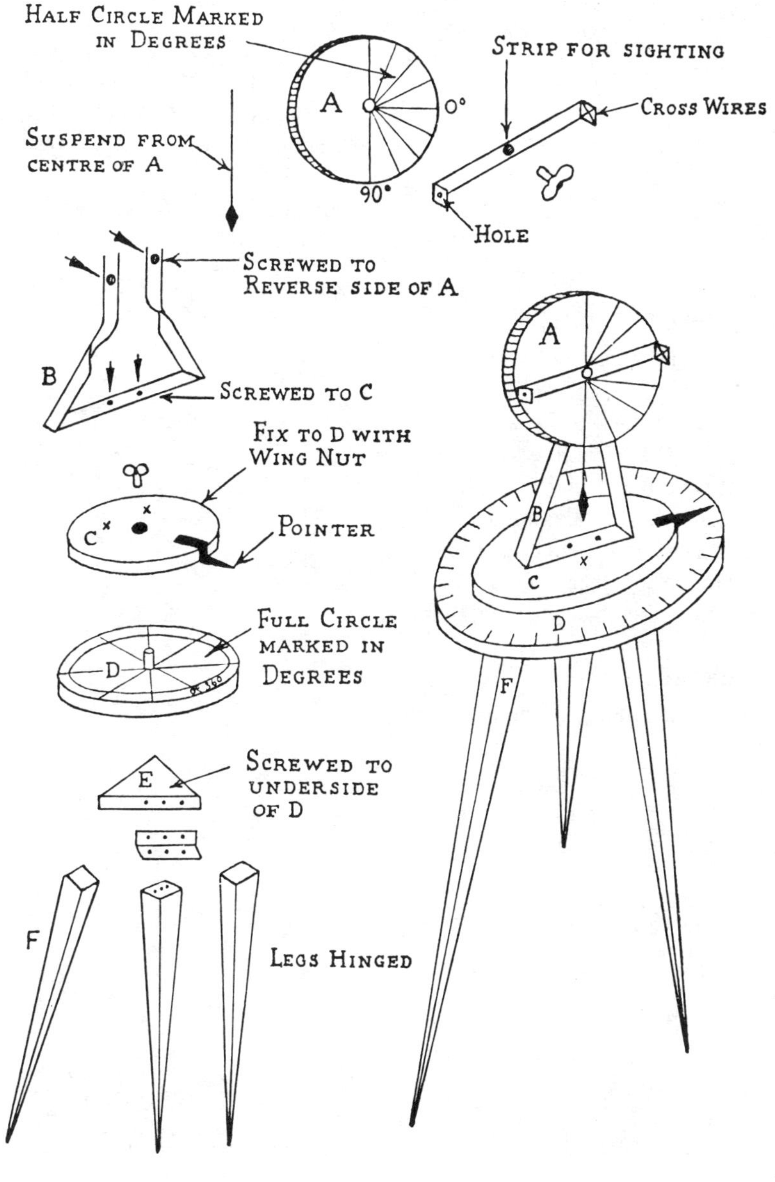

If we are able to measure our distance from the base of a tree and the angle which the top of the tree makes with the ground then we can find the height of the tree by drawing.

EXAMPLE 1

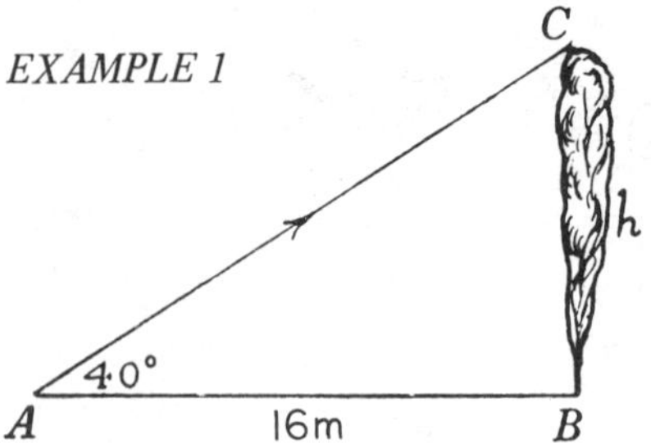

Let 1 cm. represent 2 m. Draw *AB* 8 cm. long. Mark off angle of 40° at *A*. Mark off angle at 90° at *B*. These lines will cut at *C*. Measure *BC* and convert back to metres. This is the height of the tree.

EXAMPLE 2

To find the distance of an object. Draw *AB* 5 cm. representing 50 m. Draw angle of 50° at *A* and 45° at *B*. These will intersect at *C*. Draw perpendicular from *AB* to pass through *C*. Measure *DC* and convert back to m. This is the shortest distance from the baseline to the object.

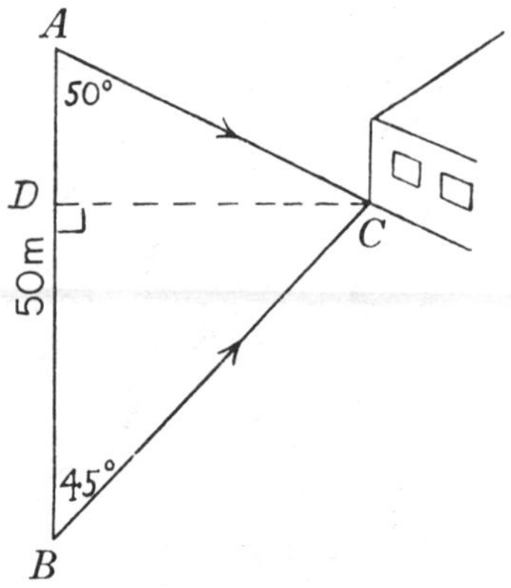

EXERCISE 23

1. Find the height of the tower.

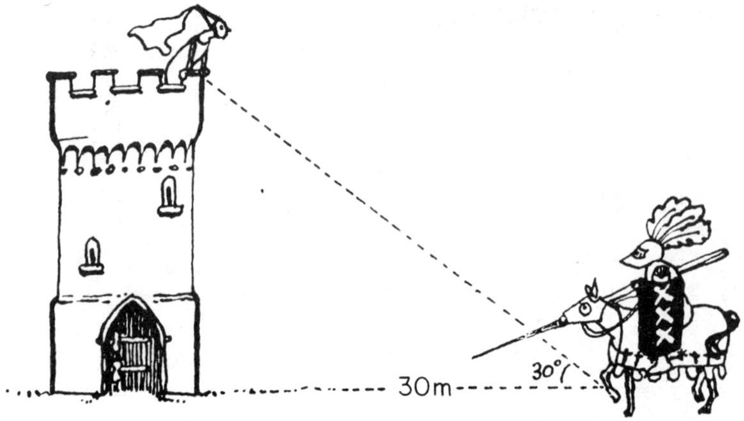

2. How far is the swimmer from the cliff?

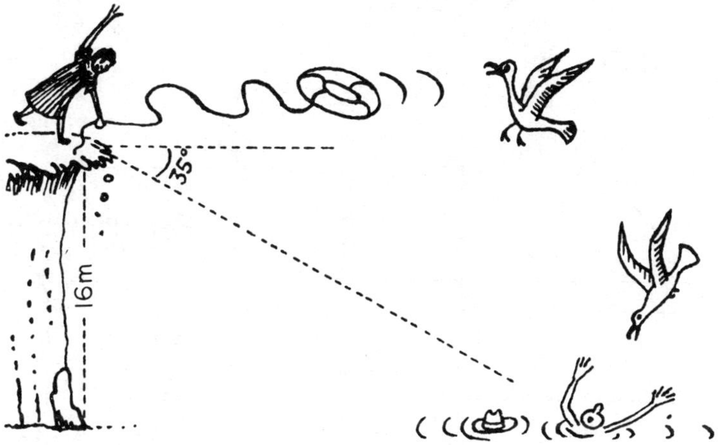

3.

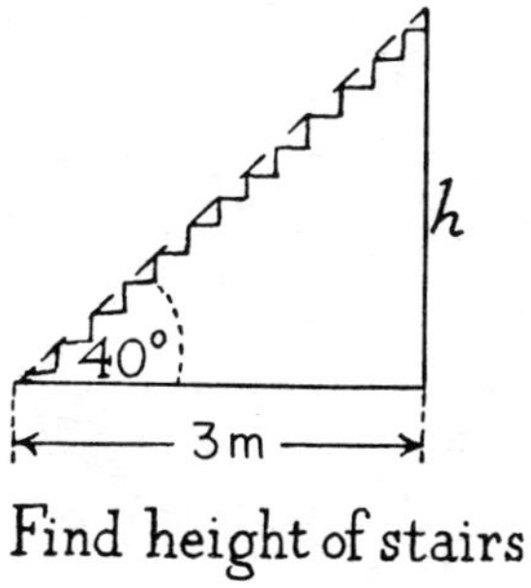

4.

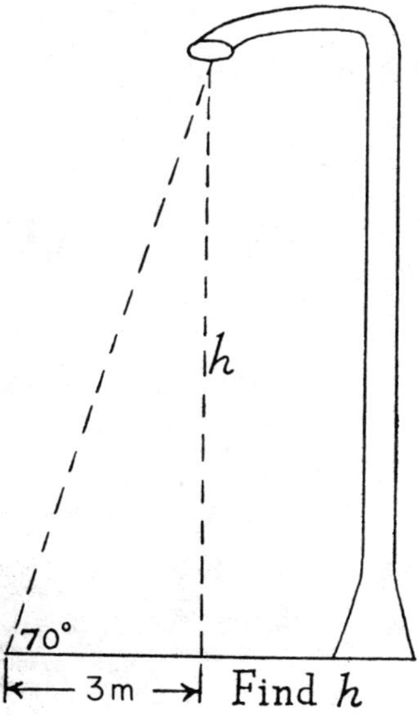

5.

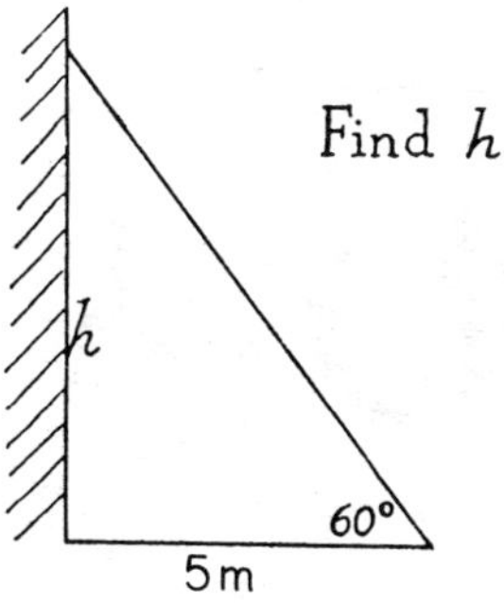

6.

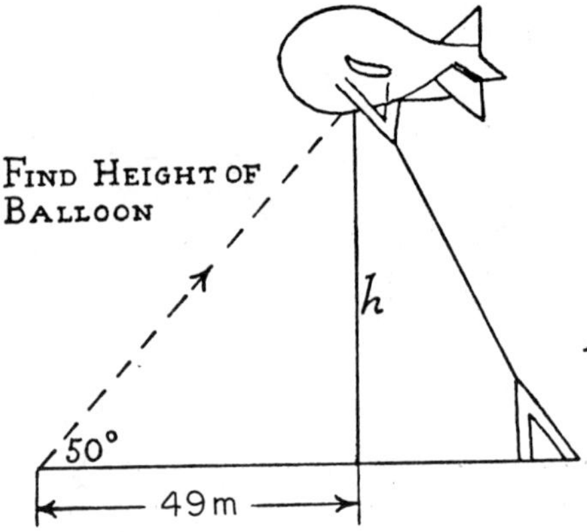

7.

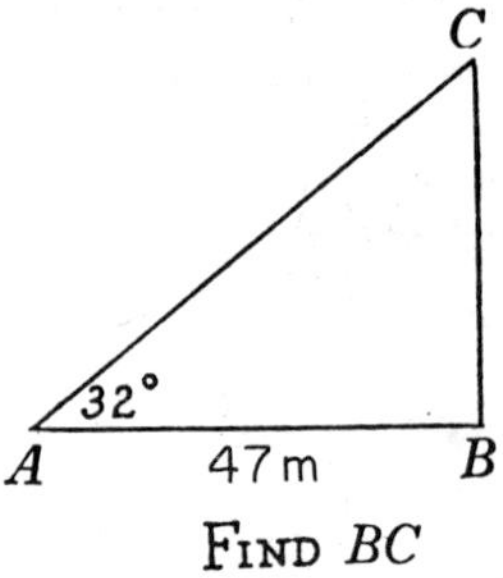

8.

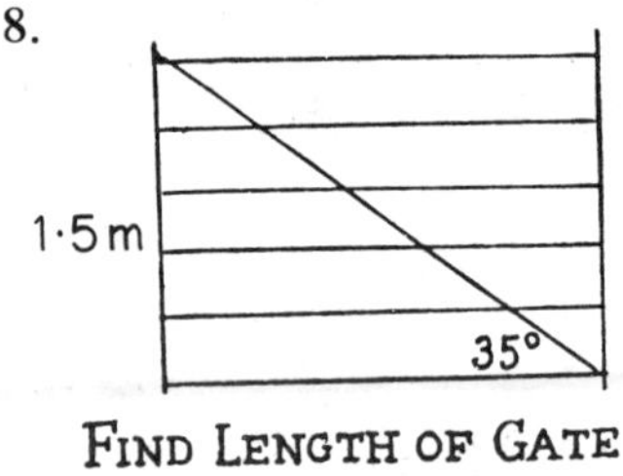

9.

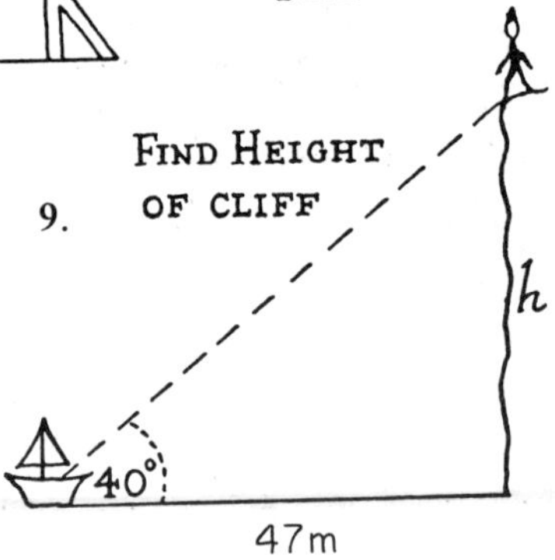

10. Which is the nearer gate?

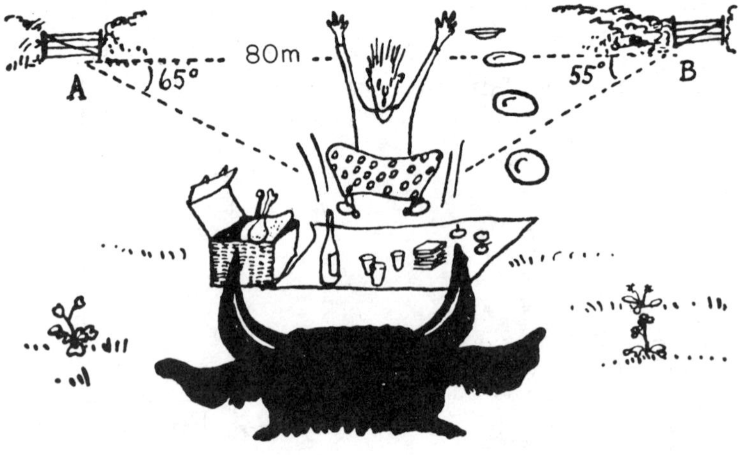

11. How far apart are the church, the pylon and the power station?

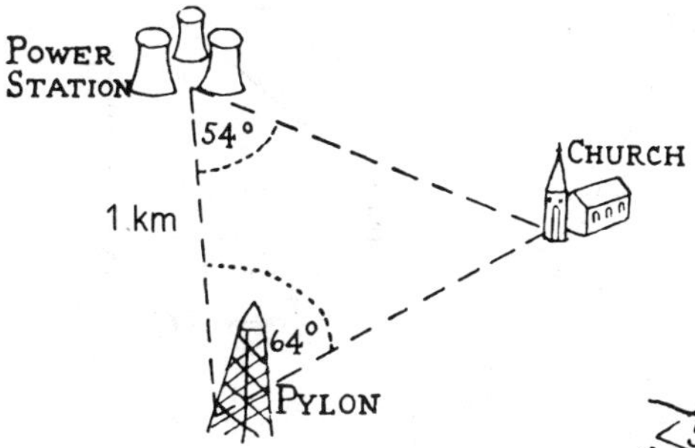

12. Find *(a) AC, (b) BC, (c) CD.*

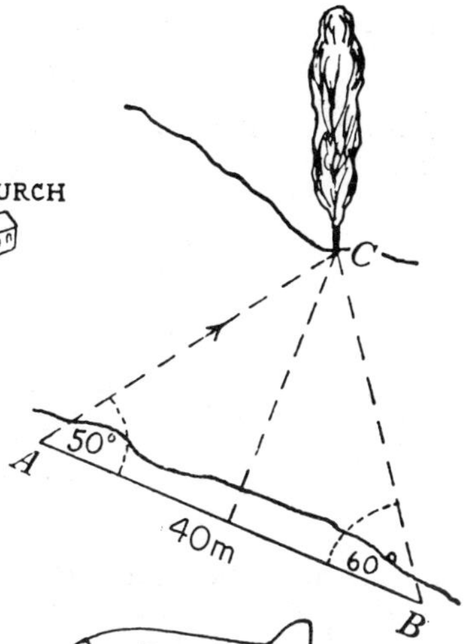

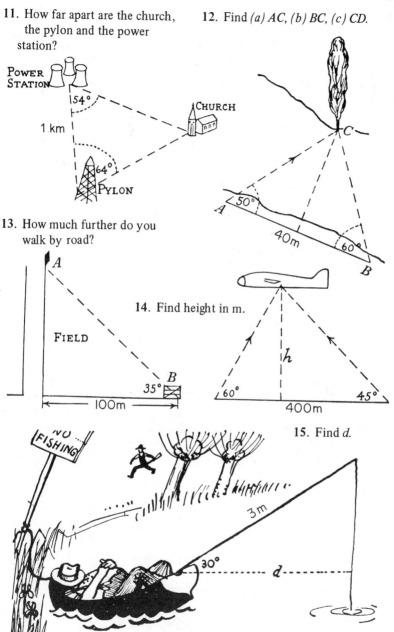

13. How much further do you walk by road?

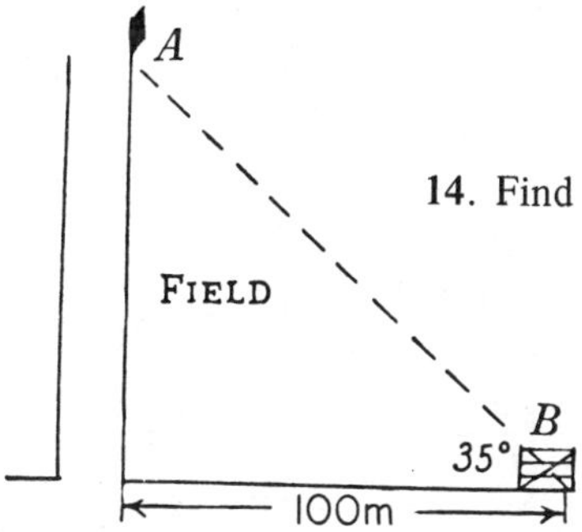

14. Find height in m.

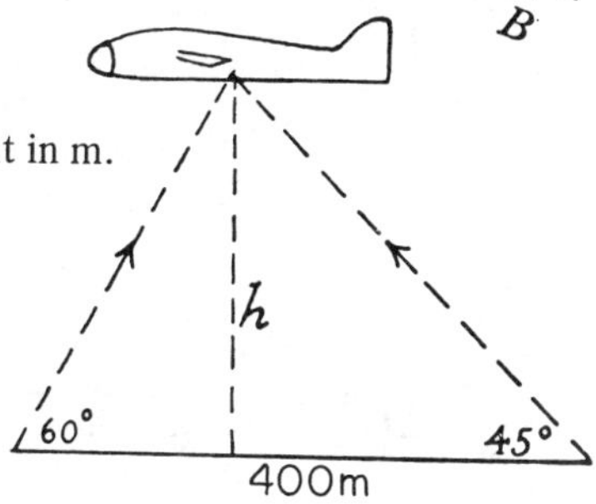

15. Find *d*.

Heights of inaccessible objects.

EXAMPLE 1

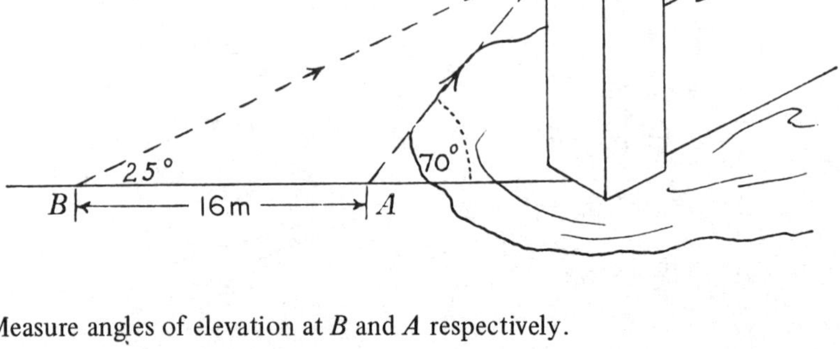

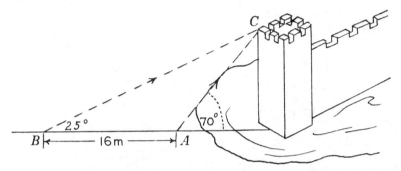

Measure angles of elevation at *B* and *A* respectively.

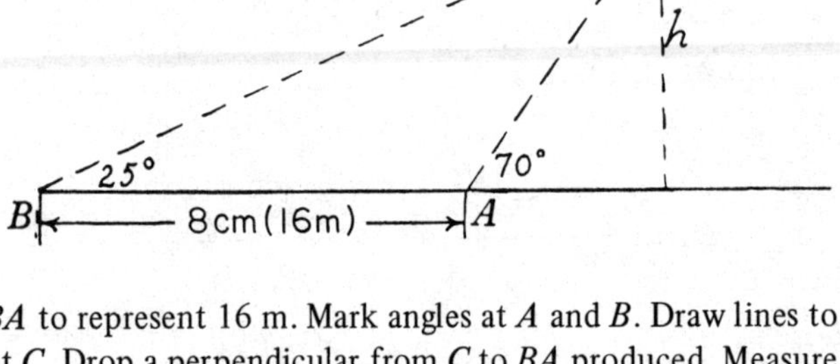

Draw *BA* to represent 16 m. Mark angles at *A* and *B*. Draw lines to intersect at *C*. Drop a perpendicular from *C* to *BA* produced. Measure *h*. This is the height of the castle.

EXERCISE 24

1.

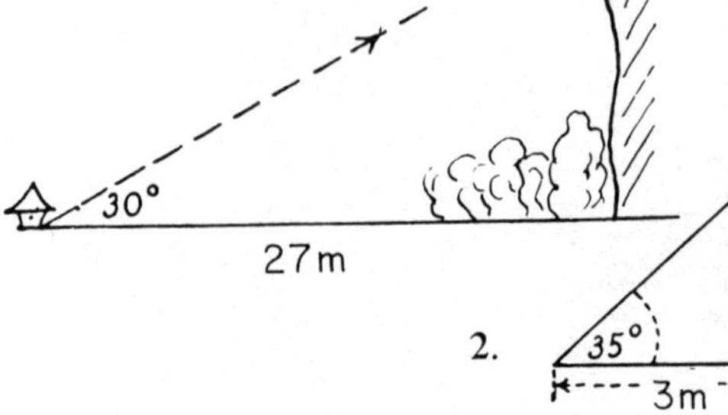

Find *h*.

2.

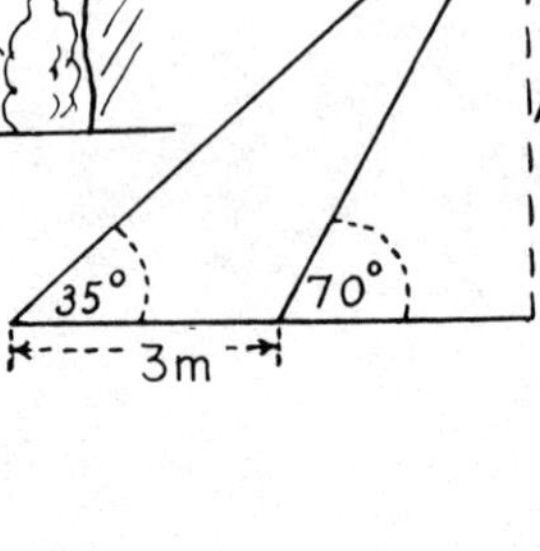

Find *h*.

3. What height is the aeroplane?

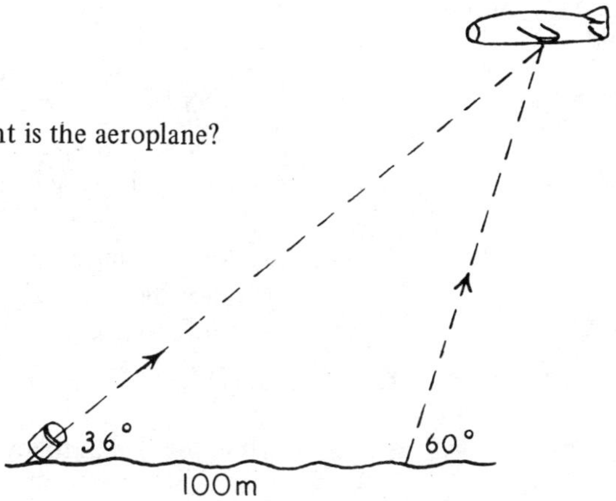

4. How far was ship from port when at *A*?

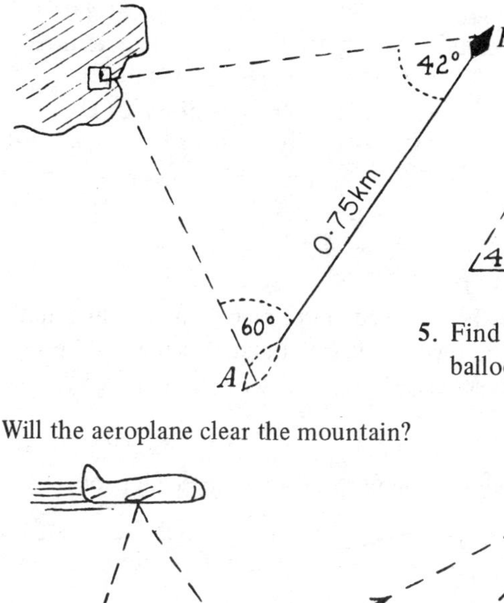

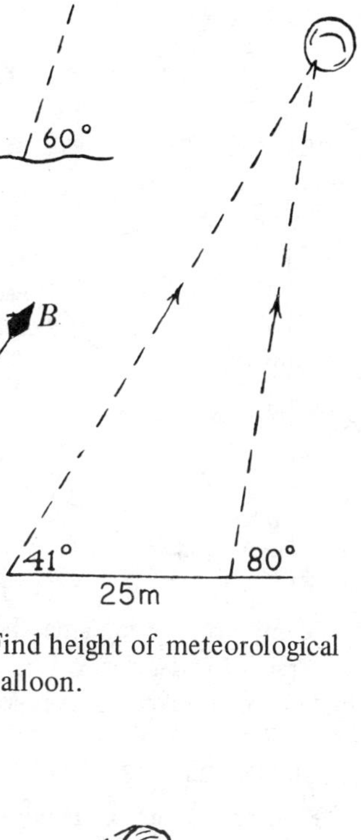

5. Find height of meteorological balloon.

6. Will the aeroplane clear the mountain?

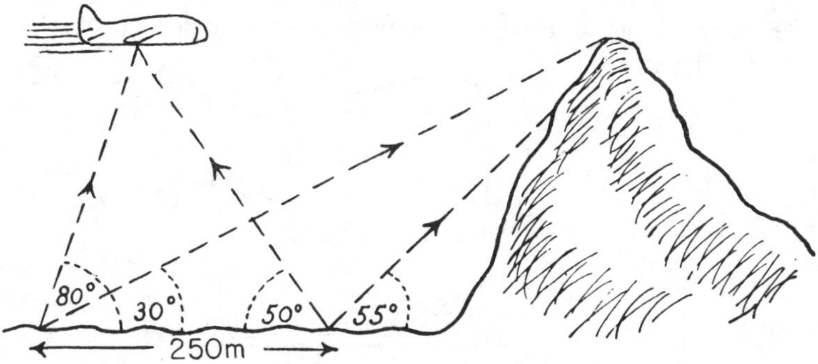

We can now use our knowledge of triangles and scale to survey the school grounds.

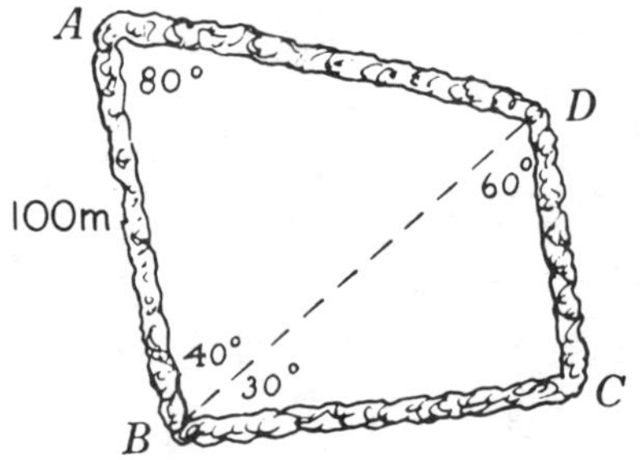

(*a*) Let the corners of the grounds be *A*, *B*, *C* and *D*.
(*b*) Measure *AB*, i.e. 100 m.
(*c*) Place poles along line *BD*.
(*d*) Measure angle *ABD*.
(*e*) Measure angle at *A*
(*f*) Measure angle *DBC*.
(*g*) Move to *D* and measure angle *BDC*.

On drawing paper to a scale of 1 cm. to 10 m proceed as follows.

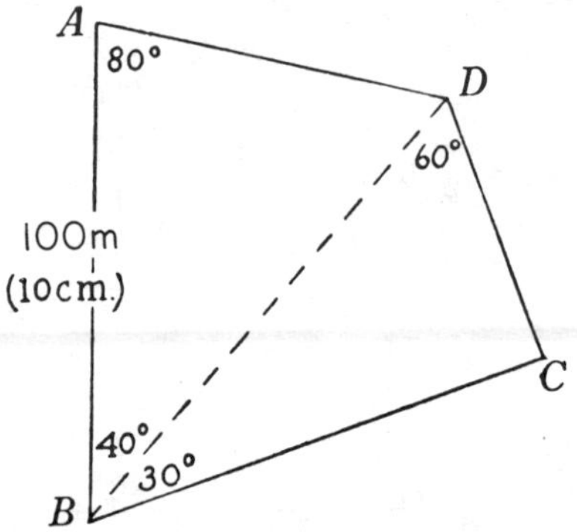

(A) Draw *AB* 10 cm. (100 m).
(B) Draw angle of 40° at *B*.
(C) Draw angle of 80° at *A*.
(D) Mark point where lines intersect and call it *D*.
(E) Draw angle of 60° at *D*.
(F) Draw angle of 30° on *BD* at *B*.
(G) Where these lines intersect mark *C*.

Many pieces of land may be surveyed in this manner by dividing up the the area into triangles. It is known as triangulation and is used in the making of Ordnance Survey Maps.

EXERCISE 25

Find the amount of fencing required to surround these plots of land to the nearest metre.

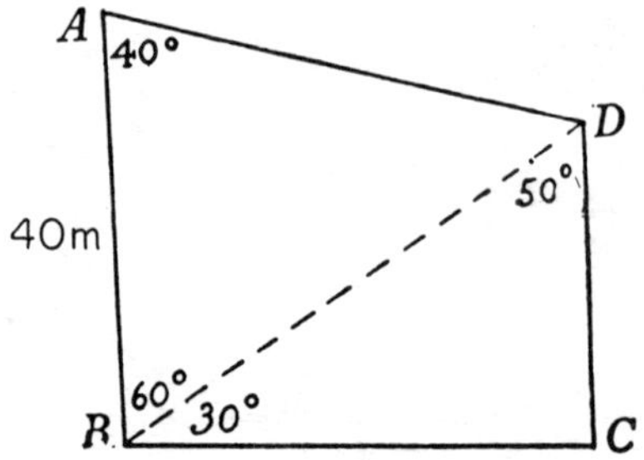

(2)

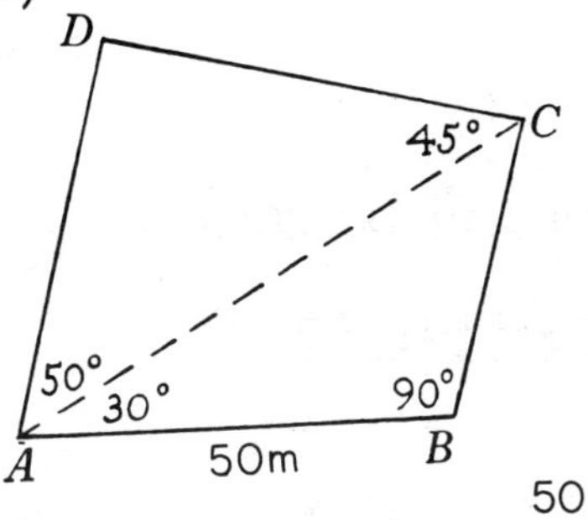

(3)

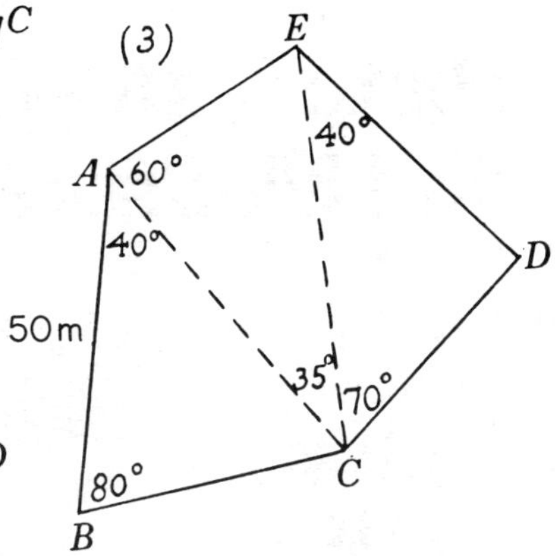

(4)

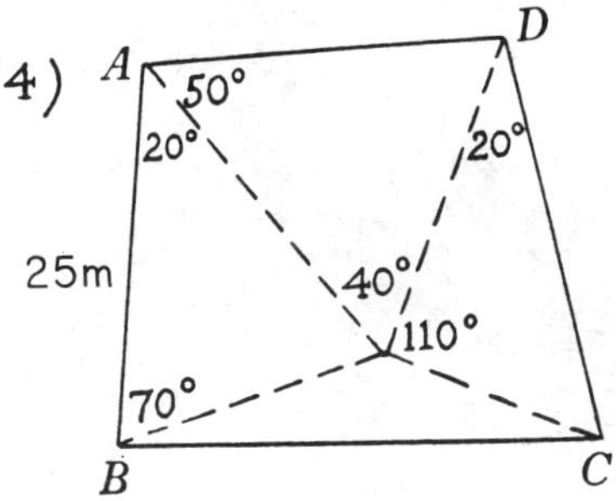

(5)

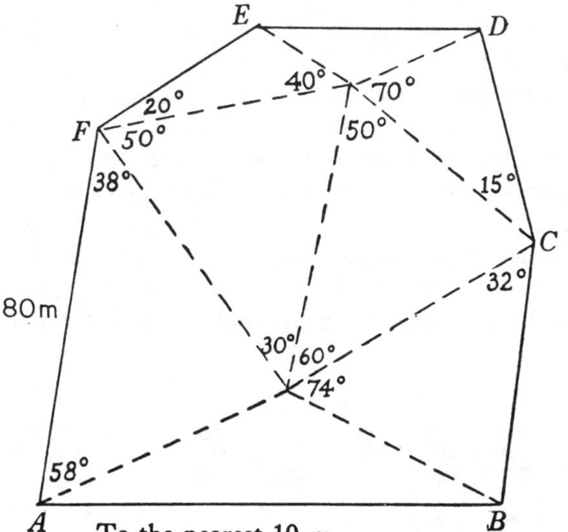

To the nearest 10 m

Arithmetic

7 History of Number and Measurement

Number

It was nearly dark. The old man looked at his son dragging the body of the bear they had killed.

"Not much further now," he said. "I can see the glow of the fire through the trees."

Over supper they told the stories of their adventures to the rest of the family. Then a younger son showed them the record he had kept while they had been away.

The nine notches which he had cut on the log with a sharp piece of stone represented the nine days that his father and brother had been away from home. Later on they cut a picture of their kill on the wall of their cave.

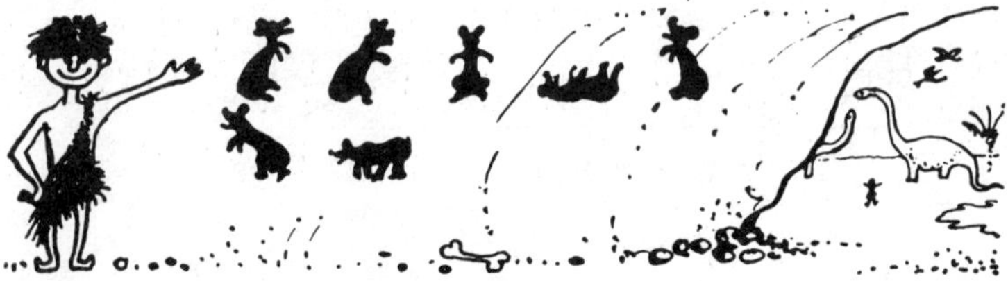

This made a total of seven kills since they moved into the cave after the flood went down. This was probably the earliest type of mathematics used in this country thousands of years ago.

At the same time, people in Egypt were keeping records on papyrus, a kind of paper made from the leaves of the papyrus plant. Joseph keeping records of the wheat stored in Pharaoh's barns would probably have used figures like these:

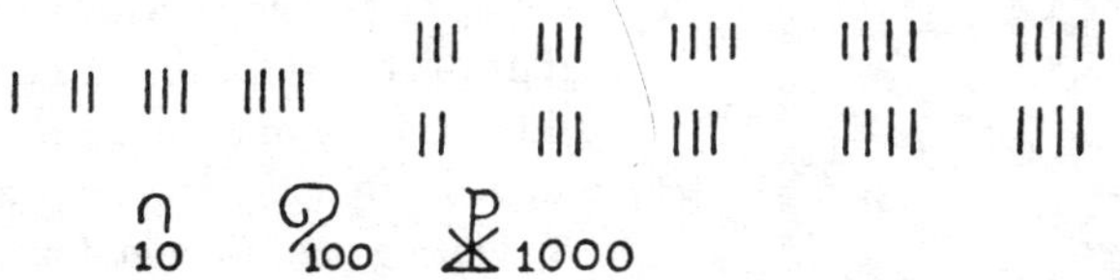

The people of Babylon wrote on clay tablets so their numbers looked like this:

In both countries they had special symbols for 10 and 100. The Egyptians wrote 132 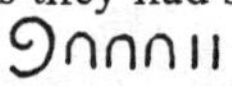 while the Babylonians wrote 

I expect most of you know the Roman Numerals which are sometimes on the face of a clock or numbering the chapters in the Bible.

I, II, III, IV, V, VI, VII, VIII, IX, X,

L = 50 C = 100 D = 500 M = 1000

Their symbols for 4 and 9 are found by subtraction. IV means 1 less than 5 and IX means 1 before 10 just as we say 20 minutes to 4 instead of 40 minutes past 3. So in Roman figures 1960 is MCMLX, that is M, 1000 + CM, 100 less than 1000 + L, 50 + X, 10.

The numerals we use in this country today are known as Arabic Numerals though they are not at all like the numerals used by the Arabs. They probably originated in India and were brought to Europe by traders through Baghdad about 1200 years ago. The main difference between these numerals and earlier systems was the symbol for an empty column, 0. We find the numbers which you will recognise, in a manuscript of the 13th century.

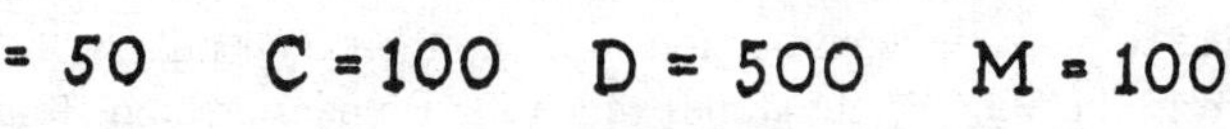

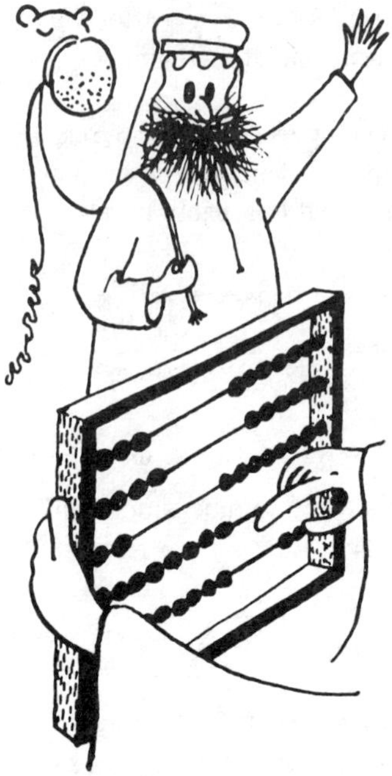

The abacus which you may think looks like a counting frame was used by many early people to do their calculating. Sometimes instead of putting beads on a frame they laid pebbles or shells in grooves in the sand or on a tray.

Japanese merchants still use a type of abacus for their calculations and can find the correct answer as quickly as Europeans can by writing the problem down.

Tally sticks were another aid to accounting which merchants used. Notches were cut in a stick then it was split down the length of it so that both buyer and seller had a record of the transaction. Their sticks tallied.

Length

Once man settled down to till the earth instead of moving from place to place hunting or tending his herds, he found that he needed to measure his land. At first distances were paced out but it was not a very satisfactory method as some paces would be longer than others. Our mile probably comes from the Mille Passus stones which the Romans erected along the roads which they built. The thousand paces would be double paces of the army on the march.

A furlong was the distance which could be ploughed by four yoke of oxen without a pause, a furrowlong. Shorter measurements were taken from parts of the body.

These units of measurement often varied according to who was measuring. When fairs and markets were held and traders travelled from place to place, disputes often arose over lengths, so standards had to be fixed. Today, standard units of measurements are kept in the local offices of weights and measures. Anyone can see the standard foot and yard which are set into the wall in Trafalgar Square. There are other standards which can be seen in the Guildhall in the City of London.

Weight

Pictures of simple balances on which weight was measured were found on Egyptian temple walls, and stones, which had been used as weights were found inside the pyramids. The Babylonians linked their weights to measurement of length by saying that one cubic foot of rain water weighed one talent. The talent was divided into 3000 shekels. In the Bible we read of Goliath who measured six cubits and a span, wore armour which weighed five thousand shekels of bronze and the head of his spear weighed six hundred shekels of iron.

The Roman unit of weight was the Libra, a pound. From this we get the abbreviation lb. for pound. This libra was divided into 12 unciae or ounces. In the fifteenth century uncia became onza, so we get the abbreviation oz. for ounce. After the Norman conquest of Britain, Troy weight was introduced from the town of Troyes in France. In this system 32 grains of wheat, dry and round make 1 pennyweight, 20 pence make 1 ounce and 12 ounces make 1 pound Troy. These weights are still used by goldsmiths today. Eventually, the Avoirdupois weights took the place of Troy weights. In this scale there are 16 ounces to the pound, there are 112 pounds to the hundredweight, that is 100 pounds plus 12 more for good measure.

Capacity

Liquids cannot easily be weighed though some measures which are used are based on measurements of weight and length. The Romans had a vessel which held a gallon of wine which was equal to a cubic foot of water. In Britain at this time, the wine gallon was only one eighth of that size, that is, slightly smaller than the 20th century gallon.

Saxon Ale measures were also in use, then the Normans brought over their wine measures which were different again so it was really essential that some standard be agreed. The standard quart and pint used in the days of Elizabeth I are very near to those recently in use.

The imperial Gallon was the amount of water that weighed 10 pounds.

Time

The movement of shadows during the day gave man a method of measuring time. The Egyptians had a wooden shadow clock which had to be turned round at midday. They also erected stone columns like Cleopatra's Needle which cast shadows, but these shadows are not very exact on account of their thickness. To tell the time during the night or in a cloudy spell other methods had to be found. Water clocks, burning tapers and hour glasses were used. In early times these primitive clocks were kept by priests who rang a bell at hourly intervals. It seems from this that the word clock is probably derived from the French word

cloche meaning bell, and the custom of having a clock on the church tower was started.

The Romans had clocks driven by wheels, and a clock driven by weights and wheels was set up in 1379. Galileo's study of the pendulum led to more accurate clocks being made though some of these would be very poor timekeepers by present-day standards.

Calendar

The Babylonians divided the year into 360 days, and the Ancient Egyptians were a little more accurate for they had 365 days in their year. Even this was not exact and so Julius Caesar put matters right by including an extra day every fourth year, a leap year. He also decided how many days there should be in each month. The names of our months are the same as Caesar's except for the months of July and August which were named after Julius and Augustus Caesar.

In the 14th Century it was discovered that the year was actually 11 minutes 14 seconds shorter than Caesar believed, so another correction had to be made. Centennial years are not leap years unless they are divisible by 400. The year 1900 was not a leap year but the year 2000 will be. This Gregorian Calendar, introduced by Pope Gregory 13th, was adopted in different countries at different times. In England it was not used until 1752 when it was necessary to miss out 11 days to correct the date. There was quite an uproar in some places because the people did not like the idea of losing 11 days.

Thirty days hath September, April, June and November,
All the rest have thirty-one–except February alone
Which has twenty-eight days clear and twenty-nine each Leap year.

Money

All boys know what bartering means, many will have learnt to their cost that exchange can sometimes be robbery. Primitive races all over the the world get things they need by bartering. Explorers in Africa used to carry brightly coloured materials and beads to exchange with the natives for food. The Maoris bartered large areas of New Zealand for blankets and guns.

In Europe as in Egypt, Babylon and China the difficulties of fair exchange were realised very early on and pieces of metal were used as coins. The size of the pieces was often determined by weight. So the Talent and Pound, units of weight, were also units of value.

Coins were made from bronze, silver and gold, and were often stamped with pictures. A coin stamped with the picture of an ox would probably at some time have been worth enough to buy an ox. When the Romans came to Britain they found that the natives already had pieces of metal which served as coins. However, a system based on Roman coins was soon introduced and the letters £ s d were reminders of Libra, Sestertii and Denarius used in those days.

1 gall. = 4·55 litres   1 litre = 1·76 pints

1 kg = 2·2 lb.   1 km = 0·6214 miles

The Tower pound, so called because it was minted at the Tower, was the value of a pound weight of silver coins. Until 1918 gold sovereigns were the pound coin, but after that date paper money came into use and in 1951 gold coins were taken out of circulation. Twenty years later, on 15th February, 1971, decimal currency was introduced into Britain. Instead of twelve pence to one shilling, twenty shillings to one pound, we have one hundred pence to one pound. The coins in use are the $\frac{1}{2}$p, 1p and 2p pieces which are made of bronze and the 5p, 10p and 50p pieces which are made of nickel. The 5p and 10p pieces are the same size and weight as the old shilling and two-shilling pieces which they replaced. The 50p piece is a rolleau heptagon.

The Metric System

In France at the end of the eighteenth century, many different standards of measurement were in use. As no one system was perfect, it was decided to work out a completely new set of standards. Scientists calculated the distance from the Equator to the North Pole; then this distance was divided by 10 000 000 to find the standard measure of length, the metre.

Multiples of a metre are given Greek prefixes while divisions are given Latin prefixes.

10 metres	=	1 decametre	dam	1 metre	=	10 decimetres	dm
10 dam	=	1 hectometre	hm	1 dm	=	10 centimetres	cm
10 hm	=	1 kilometre	km	1 cm	=	10 millimetres	mm

The measures of weight and capacity use the same prefixes as the metre and increase or decrease in multiples of 10. The unit of weight is the gram while the unit of capacity is the litre.

One cubic centimetre of water weighs one gram, and one cubic decimetre, which is 1000 cm^3, weighs one kilogram and has a capacity of one litre.

The use of the metric system was legalised in France in 1840. Since then it has been adopted in many countries. The change from Imperial to metric units is gradually taking place in Britain in the 1970s.

8 Number

1 2 3 4 5 6 7 8 9 10 11 12 13 14

These are the first 14 counting numbers. Each one of them can tell us how many objects there are in a certain set.

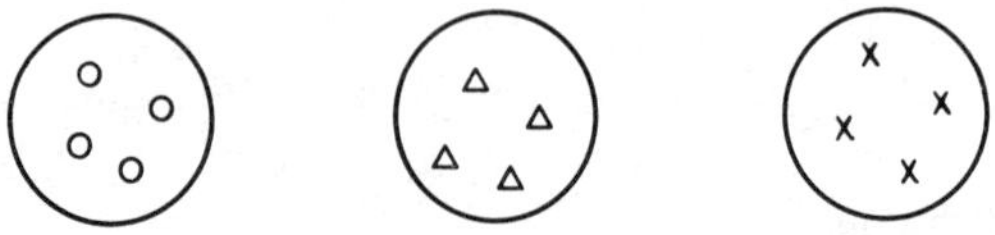

These sets each contain 4 things. If we add one object to each set, we have different sets represented by a different number, 5, and so on.

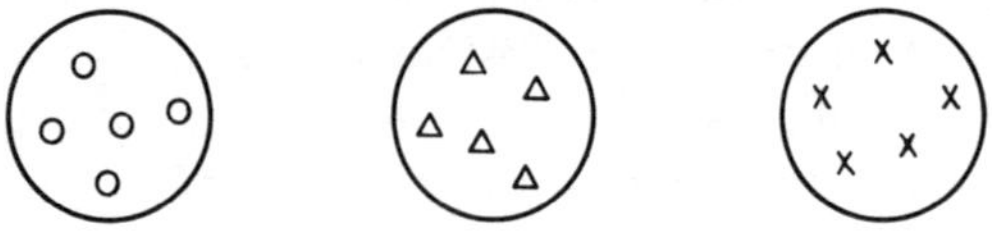

How many dots are there in each of these sets?

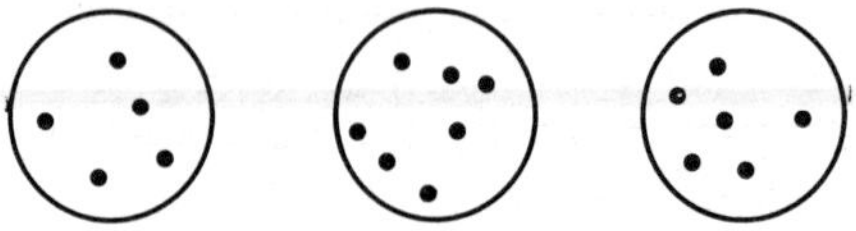

It is easier to answer this if the dots are arranged in patterns

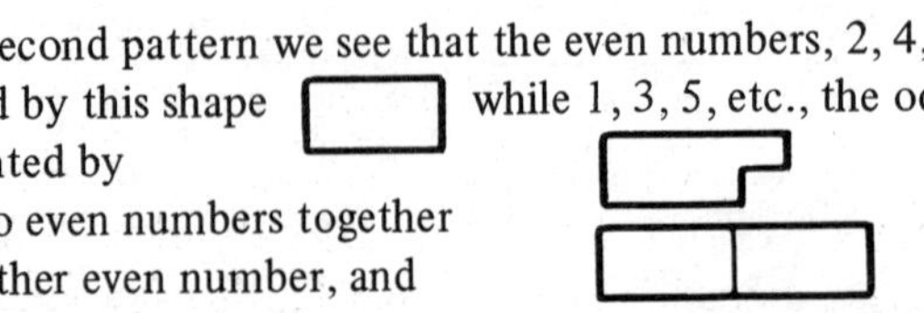

From the second pattern we see that the even numbers, 2, 4, 6, etc. can be represented by this shape 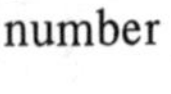 while 1, 3, 5, etc., the odd numbers, can be represented by 

Adding two even numbers together we get another even number, and adding two odd numbers also gives an even number. But adding an odd number to an even number gives us an odd number.

E stands for Even and O stands for Odd.

E + E = E

O + O = E

E + O = O

Since even numbers are represented by rectangles, we can take one even number and add it any number of times and still have an even number.

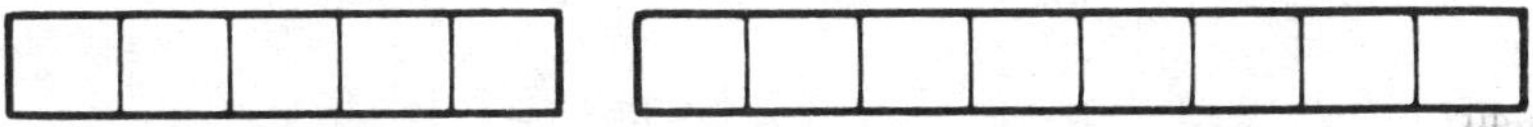

E x 5 = E   E x 8 = E

An odd number is represented by this shape  If we take an even number of these, we can pair them and so have a set of rectangles which results in an even number.

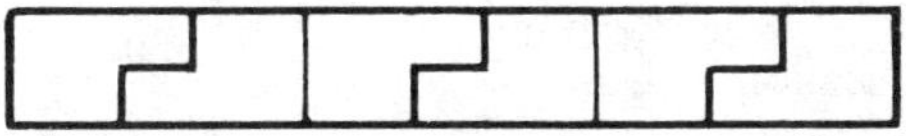

O x 6 = E

If we take this same odd number an odd number of times, the pairing leaves us with so many rectangles and one piece over.

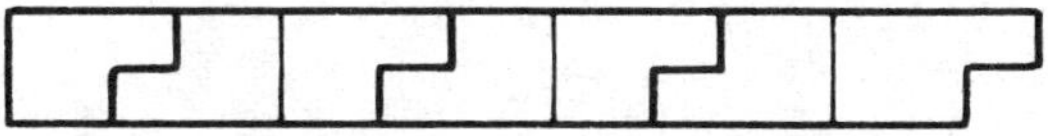

O x 7 = O

So we have E x E = E, E x O = E and O x O = O.
This can help us with our multiplication—to check whether our answer is possible or not. We know that 32 x 23 cannot be 55 and that 17 x 39 cannot be 204.
2, 4, 6, 8, 10. I expect many of you have used just the even numbers when counting. It is quicker to count things off two at a time. It is quicker still to count 3, 4 or 5 at a time if you can do it.

3,	6,	9,	12. . . . .
4,	8,	12,	16. . . . .
5,	10,	15,	20. . . . .

This way you can build up a multiplication table square. If we count 10 at a time we have 10, 20, 30, 40. . . . . Multiplying by 10, we appear to put a nought on the end of the number. What we actually do is to move the figures.

$$25 \times 10 = 250$$

The 5 units become 5 tens, 50, and the 2 tens become 2 hundreds, 200.
Multiply these numbers by 10 and write down the answers.

17, 32, 49, 61, 77, 235, 401, 376, 508, 450.

Now 30 = 10 x 3, so to multiply by 30, we first multiply by 10 and then by 3.

23 x 30 = 23 x 10 x 3 = 230 x 3 = 690.

EXAMPLE 1 23 x 15

```
 23
 15 x
230      23 x 10
115      23 x 5
345      23 x 15
```

EXAMPLE 2 27 x 32

```
 27
 32 x
810      27 x 30
 54      27 x  2
864      27 x 32
```

Division is the opposite of multiplication.

2 x 6 = 12 and 12 ÷ 6 = 2

If we divide 78 marbles between 6 boys, how many will they have each?
6 x 10 = 60 and 78 is bigger than 60, so they will have more than 10 each.
78 – 60 = 18, and 18 = 6 x 3, so they will have 10 + 3 each.

Usually we write

```
    13
6 ) 78
    60      6 x 10
    18
    18      6 x 3
    00
```

EXAMPLE 3

272 ÷ 16

```
     17    Ans.
16 )272
    160         16 x 10     = 160
    112         272 – 160 = 112
    112         16 x 7      = 112
    000
```

EXAMPLE 4

864 ÷ 32

```
     27    Ans.
32 )864
    640         32 x 20     = 640
    224         864 – 640 = 224
    224         32 x 7      = 224
    000
```

EXERCISE 26

1. 22 x 17	6. 54 x 21	11. 234 ÷ 13	16. 1701 ÷ 27
2. 36 x 41	7. 96 x 32	12. 513 ÷ 19	17. 1767 ÷ 31
3. 27 x 62	8. 65 x 26	13. 756 ÷ 21	18. 4752 ÷ 72
4. 48 x 34	9. 82 x 15	14. 1150 ÷ 25	19. 1395 ÷ 45
5. 19 x 56	10. 26 x 54	15. 2772 ÷ 33	20. 2408 ÷ 43

Factors

When we looked at number patterns earlier in this chapter, we saw that even numbers could be represented by rectangles. This was because even numbers can be arranged as pairs.

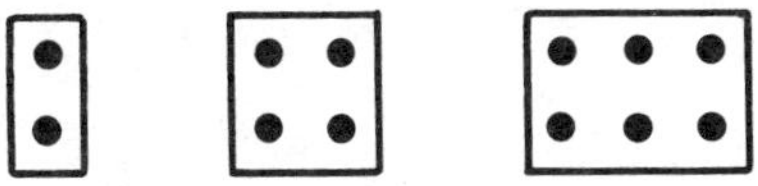

So even numbers can always be divided by 2. We say that 2 is a FACTOR of every even number. Most odd numbers have factors as well.

$15 = 3 \times 5$ $\qquad$ $77 = 7 \times 11$

Numbers like 2, 3, 5, 7, 11, 13, 17, 19, 23 and so on, which have no factors other than themselves and 1, are called PRIME NUMBERS.

All other numbers can be broken down into a product of primes.

$8 = 2 \times 2 \times 2 = 2^3$ $\qquad$ $105 = 3 \times 5 \times 7$

$36 = 2 \times 2 \times 3 \times 3 = 2^2 \times 3^2$ $\qquad$ $143 = 11 \times 13$

Write these numbers as products of primes.

48, 15, 135, 63, 78, 221.

The Highest Common Factor.

$12 = 2 \times 2 \times 3$ $\qquad$ $16 = 2 \times 2 \times 2 \times 2$

What factors do these two numbers have in common?
12 and 16 can be divided by 2 and also by $2 \times 2 = 4$, so 2 and 4 are the common factors of 12 and 16, 4 is the HIGHEST COMMON FACTOR.

EXAMPLE 5

Find the highest common factor of 30 and 45.

$30 = 2 \times 3 \times 5$ $\qquad$ $45 = 3 \times 3 \times 5$

Common factors are 3, 5 and $3 \times 5 = 15$.
15 is the highest common factor of 30 and 45.
Find the highest common factors of the following pairs of numbers.

24 and 16 $\qquad$ 45 and 75 $\qquad$ 100 and 80

Least Common Multiple

A common multiple of two numbers is any number of which they are both factors. A common multiple of 6 and 8 is 48, the product of the two numbers, but this is not the smallest one.

6 = 3 x 2   8 = 2 x 4

The LEAST COMMON MULTIPLE is $\overbrace{3 \times 2}^{6} \times 4 = 24$ with $\underbrace{2 \times 4}_{8}$

$$\overbrace{3 \times 2}^{6}\!\!\underbrace{\times 4}_{8} = 24$$

EXAMPLE 6

Find the Least Common Multiple (L.C.M.) of 10 and 15

10 = 2 x 5   15 = 3 x 5

L.C.M. = 2 x 3 x 5 = 30 Ans.

EXAMPLE 7

Find the H.C.F. and the L.C.M. of 24 and 32.

24 = 2 x 2 x 2 x 3

32 = 2 x 2 x 2 x 2 x 2

H.C.F. = 2 x 2 x 2 = 8 Ans.

L.C.M.= 2 x 2 x 2 x 2 x 2 x 3 = 96 Ans.

EXAMPLE 8

Find the H.C.F. and L.C.M. of 16, 24 and 36.

16 = 2 x 2 x 2 x 2

24 = 2 x 2 x 2 x 3

36 = 2 x 2 x 3 x 3

H.C.F.= 2 x 2 = 4 Ans.

L.C.M.= 2 x 2 x 2 x 2 x 3 x 3 = 144 Ans.

EXERCISE 27

Find the H.C.F. and L.C.M. of the following sets of numbers.

1. 9 and 15
2. 12 and 42
3. 24 and 40
4. 36 and 27
5. 52 and 78
6. 6, 9 and 15
7. 8, 12 and 24
8. 14, 35 and 49
9. 20, 35 and 100
10. 88, 110 and 132

9 Revision

EXERCISE 28

CLUES ACROSS

1. 3 x 8
3. 9 x 12
5. 13 x 20
7. 4 x 11
9. 32 x 70
11. 83 x 30
12. 6 x 9
16. 15 x 15
17. 20 x 20
18. 6 x 6

CLUES DOWN

1. 16 x 16
2. 13 x 13
4. 9 x 9
6. 12 x 12
7. 21 x 21
8. 7 x 7
10. 5 x 5
13. 11 x 11
14. 14 x 14
15. 8 x 8

EXERCISE 29

CLUES ACROSS

1. 11 x11
3. 25 x 7
6. 5 x 11
7. 6 x 6
8. 15 x 5
9. 3 x 4
11. 8 x 10
12. 9 x 12
14. 8 x 13
15. 6 x 4
17. 6 x 9
19. 9 x 7
21. 3 x 9
22. 10 x 5
23. 15 x 60
24. 8 x 45

CLUES DOWN

1. 12 x 13
2. 5 x 5
4. 7 x 11
5. 11 x 50
7. 8 x 4
9. 10 x 100
10. 128 x 8
11. 27 x 3
13. 7 x 12
16. 13 x 13
17. 3 x 19
18. 10 x 10
20. 3 x 10
22. 7 x 8

EXERCISE 30

CLUES ACROSS

1. 3×7
3. $2^5 \times 5 \times 11$
5. $3^2 \times 7^2$
7. 2×3^2
8. 5^2
10. $2 \times 3 \times 11$
12. $2^3 \times 3^2 \times 17$
14. $3 \times 5 \times 7 \times 11$
17. $2^3 \times 3$
18. $2^2 \times 13$
20. $2^2 \times 3 \times 5$
22. $2^3 \times 5 \times 11$
23. $2^6 \times 5 \times 7$
24. $2^3 \times 7$

CLUES DOWN

2. $2^2 \times 37$
3. $2^4 \times 7$
4. $2 \times 3 \times 11$
6. $2^7 \times 3^3 \times 5^2$
7. $2^3 \times 5 \times 11 \times 23$
9. 3×17
10. 2×31
11. 3×5
13. 3^3
15. 2×7
16. 5×11
19. $2^4 \times 3 \times 5$
20. 5×11^2
21. 2^5

EXERCISE 31

CLUES ACROSS

1. 24 x 59
4. 11^2
7. 20^2
9. 1176 ÷ 21
10. L.C.M. of 25, 35, 36
11. L.C.M. of 27, 60, 72
14. L.C.M. of 44, 46
16. $2^7 \times 3^3$
19. 816 ÷ 24
20. L.C.M. of 75, 125
22. Number of pence in £6·40
23. 89 x 35

CLUES DOWN

1. 11 x 13
2. HCF of 80, 120, 280
3. Number of metres in 1 km
5. Number of grams in $\frac{1}{4}$ kg
6. 40^2
8. Number of seconds in 1 hour
12. L.C.M. of 32, 51, 60
13. L.C.M. of 64, 66, 72
15. L.C.M. of 29, 39
17. 37 x 12
18. 37 x 15
21. 1207 ÷17.

Patterns in Number

Earlier we saw that the multiplication tables went up in equal steps.

3, 6, 9, ...

5, 10, 15, ...

You could easily continue these patterns, but what about the ones which follow?

1. 1, 3, 5, 7, ...

2. 2, 6, 10, ...

3. 3, –, 9, –, 15, ...

4. 100, 80, –, 40, ...

5. 75, 60, 45, ...

Each of these patterns depends on adding or subtracting the same number each time.

Can you continue this sequence 1, 3, 6, 10, ... ?

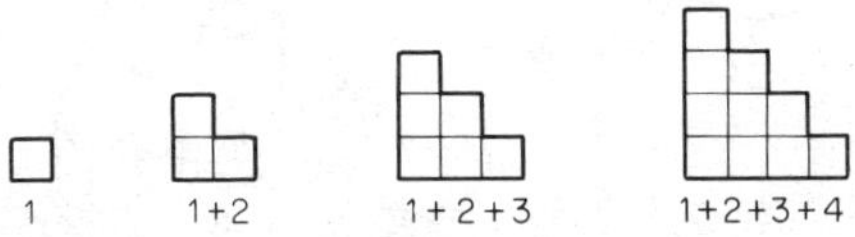

These are called triangular numbers, can you think why?

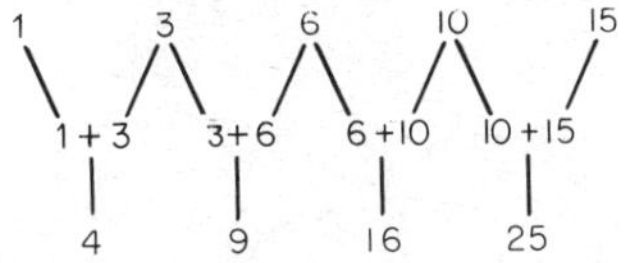

What can you say about this new sequence?

If we put the triangles together in pairs, we make squares.

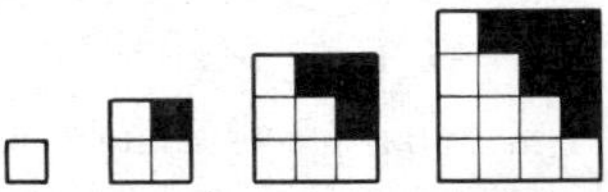

The square numbers can be made by multiplying

1 x 1, 2 x 2, 3 x 3, 4 x 4, ...

or by adding the odd numbers

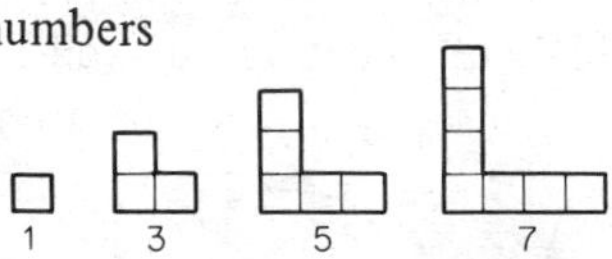

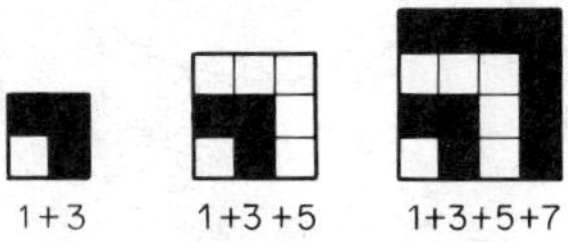

Without adding, find the answers to 1 + 3 + 5 + 7 + 9 + 11 and 1 + 3 + 5 + 7 + 9 + 11 + 13 + 15 + 17 + 19.

10 Fractions

Can you imagine what a fuss there would be if four boys had to share three cakes? There are several ways in which they could do it, some ways fairer than others.

Method 1

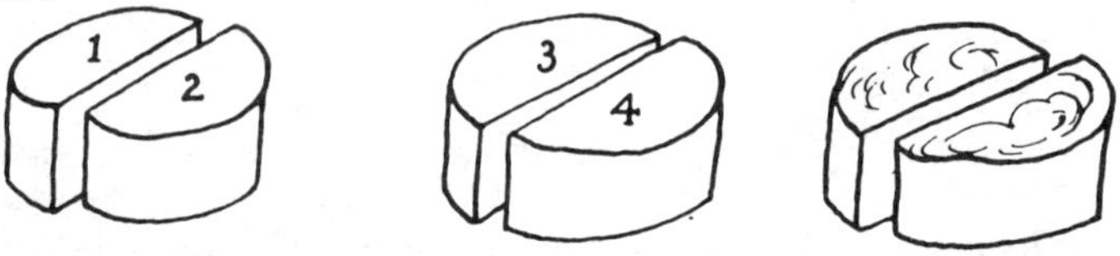

This group halved each cake having one half each and there were two halves left.

Halving these again they had one piece each. How big was each of these pieces? Altogether each boy had $\frac{1}{2} + \frac{1}{4}$ of a cake. How much is this altogether?

Method 2

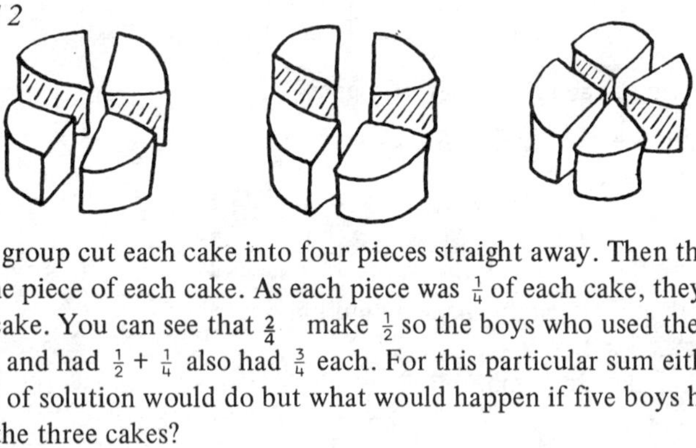

This group cut each cake into four pieces straight away. Then they all took one piece of each cake. As each piece was $\frac{1}{4}$ of each cake, they all had $\frac{3}{4}$ of a cake. You can see that $\frac{2}{4}$ make $\frac{1}{2}$ so the boys who used the first method and had $\frac{1}{2} + \frac{1}{4}$ also had $\frac{3}{4}$ each. For this particular sum either method of solution would do but what would happen if five boys had shared the three cakes?

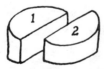

Using the first method and cutting them in half they would have half each and half over. I think we will leave the division of that portion to them while we use the other method.

Cut each cake into five pieces. If each boy takes one piece of each cake they will have the equivalent of $\frac{3}{5}$ of a cake. The 5 tells us that the cake was cut into 5 pieces and the 3 tells us that they had 3 pieces each.

Can you say what fraction each boy would have if 3 cakes were divided among 8 boys? If 5 cakes were divided among 6 boys? If 7 cakes were divided among 10 boys? If 1 loaf was divided among 3 boys?

What fraction of these figures is shaded?

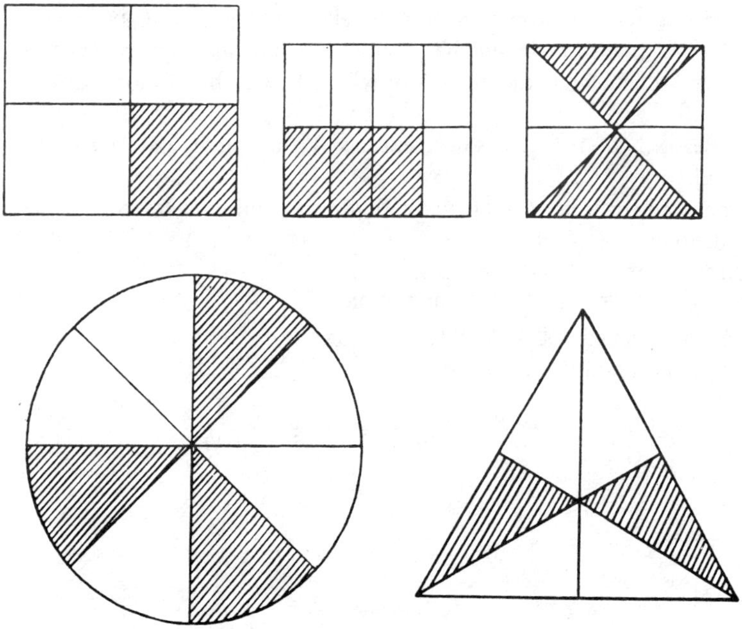

What fraction is not shaded.

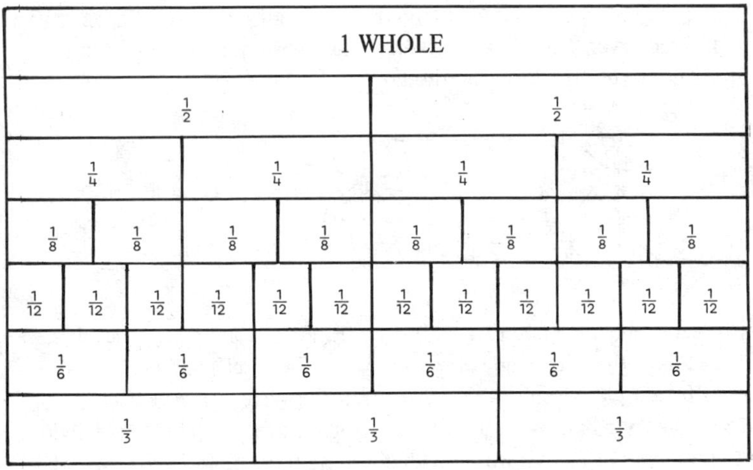

Using the fraction chart, copy and fill in the missing numbers.

1 whole = $\frac{\_}{2} = \frac{\_}{3} = \frac{\_}{4} = \frac{\_}{6} = \frac{\_}{8} = \frac{\_}{12}$

The missing numbers (top numbers) are called the NUMERATORS. The DENOMINATORS (bottom numbers) tell us into how many parts our object has been broken, while the numerator tells us how many of those parts we have.

$\frac{5}{8}$ means that our cake has been cut into 8 equal pieces and we have 5 of them.

The word fraction comes from the Latin word meaning broken, just as fracture does.

Again, using the chart, we can see that

$\frac{2}{8} = \frac{1}{4}$, $\frac{3}{6} = \frac{1}{2}$, $\frac{4}{12} = \frac{1}{3}$ and so on.

These are called EQUIVALENT FRACTIONS.

Copy and fill in the missing numbers

$$\frac{1}{2} = \frac{\_}{4} = \frac{3}{\_} = \frac{\_}{8} = \frac{6}{\_} = \frac{\_}{10} = \frac{10}{\_} = \frac{\_}{50} = \frac{\_}{100}$$

$$\frac{1}{4} = \frac{\_}{8} = \frac{3}{\_} = \frac{\_}{16} \qquad \frac{3}{4} = \frac{9}{\_} = \frac{\_}{8} = \frac{15}{\_}$$

$$\frac{1}{3} = \frac{\_}{6} = \frac{6}{\_} = \frac{\_}{21} \qquad \frac{2}{3} = \frac{4}{\_} = \frac{\_}{12} = \frac{18}{\_}$$

EXERCISE 32

Find the common denominators and arrange the following in order, greatest first.

1. $\frac{2}{3}$ $\frac{1}{4}$
2. $\frac{1}{2}$ $\frac{3}{4}$
3. $\frac{1}{6}$ $\frac{5}{12}$ $\frac{5}{6}$
4. $\frac{1}{3}$ $\frac{1}{4}$ $\frac{2}{3}$
5. $\frac{1}{5}$ $\frac{1}{10}$ $\frac{3}{10}$
6. $\frac{7}{20}$ $\frac{2}{5}$
7. $\frac{3}{8}$ $\frac{1}{2}$ $\frac{5}{16}$
8. $\frac{1}{4}$ $\frac{11}{16}$ $\frac{3}{4}$
9. $\frac{3}{4}$ $\frac{1}{2}$ $\frac{2}{3}$ $\frac{1}{4}$
10. $\frac{7}{20}$ $\frac{3}{5}$ $\frac{3}{10}$ $\frac{2}{5}$ $\frac{1}{10}$
11. $\frac{1}{2}$ $\frac{1}{4}$ $\frac{2}{3}$ $\frac{5}{6}$ $\frac{1}{6}$ $\frac{5}{12}$
12. $\frac{1}{5}$ $\frac{1}{3}$ $\frac{1}{4}$
13. $\frac{1}{16}$ $\frac{2}{3}$
14. $\frac{3}{5}$ $\frac{7}{10}$ $\frac{4}{15}$
15. $\frac{3}{4}$ $\frac{5}{6}$ $\frac{2}{3}$
16. $\frac{7}{16}$ $\frac{1}{2}$ $\frac{2}{9}$
17. $\frac{2}{3}$ $\frac{4}{5}$ $\frac{7}{30}$
18. $\frac{4}{11}$ $\frac{7}{22}$ $\frac{1}{2}$
19. $\frac{3}{7}$ $\frac{4}{21}$ $\frac{7}{14}$
20. $\frac{1}{7}$ $\frac{1}{6}$ $\frac{1}{5}$

Addition and Subtraction, Find the Common Denominator

The common denominator should always be the smallest number that all the denominators will go into, so that the sum may be kept as simple as possible.

EXERCISE 33

Use your fraction chart on p. 86 to help you to solve these.

1. $\frac{1}{2} + \frac{1}{4}$
2. $\frac{1}{16} + \frac{1}{8}$
3. $\frac{1}{12} + \frac{1}{4}$
4. $\frac{3}{4} + \frac{1}{2}$
5. $\frac{3}{8} + \frac{1}{4}$
6. $\frac{1}{2} + \frac{3}{10}$
7. $2\frac{1}{2} + \frac{3}{4}$
8. $1\frac{3}{4} + \frac{3}{8}$
9. $\frac{1}{4} + 1\frac{7}{8}$
10. $3\frac{1}{2} + \frac{3}{16}$
11. $2\frac{1}{5} + 1\frac{7}{10}$
12. $1\frac{5}{8} + 2\frac{7}{8}$
13. $\frac{1}{2} - \frac{3}{8}$
14. $\frac{11}{16} - \frac{1}{8}$
15. $2\frac{1}{12} - \frac{1}{4}$
16. $\frac{1}{2} - \frac{1}{4}$
17. $1\frac{3}{8} - 1\frac{1}{4}$
18. $\frac{1}{2} - \frac{3}{10}$
19. $2\frac{1}{2} - \frac{3}{4}$
20. $1\frac{3}{4} - \frac{3}{8}$
21. $2\frac{1}{4} - 1\frac{7}{8}$
22. $3\frac{1}{2} - \frac{3}{16}$
23. $2\frac{1}{5} - 1\frac{9}{10}$
24. $3\frac{5}{8} - 2\frac{7}{8}$
25. $3\frac{1}{2} - 1\frac{3}{16}$
26. $2\frac{1}{5} - 1\frac{7}{10}$
27. $4\frac{5}{8} - 2\frac{7}{8}$
28. $2\frac{1}{4} - \frac{5}{8}$
29. $2\frac{3}{16} - \frac{7}{8}$
30. $1\frac{1}{4} - \frac{7}{8}$

EXAMPLE 1

$\frac{3}{4} + \frac{5}{8}$		Find the common denominator of 4 and 8, that is 8.
$= \frac{6}{8} + \frac{5}{8}$		Change $\frac{3}{4}$ to $\frac{6}{8}$
$= \frac{11}{8}$		Add 6 and 5
$= \underline{1\frac{3}{8}}$	Ans.	Change $\frac{8}{8}$ to 1 whole so that $\frac{11}{8} = 1$ whole $+\frac{3}{8}$.

EXERCISE 34

Add these setting them down as above.

1. $\frac{2}{3} + \frac{1}{6}$
2. $\frac{1}{16} + \frac{1}{8}$
3. $\frac{1}{4} + \frac{1}{12}$
4. $\frac{1}{2} + \frac{3}{4}$
5. $\frac{1}{4} + \frac{3}{8}$
6. $\frac{6}{10} + \frac{1}{2}$

7. $\frac{3}{4} + \frac{1}{8}$
8. $\frac{3}{8} + \frac{3}{4}$
9. $\frac{7}{8} + \frac{1}{4}$
10. $\frac{1}{3} + \frac{2}{7}$
11. $\frac{3}{11} + \frac{1}{4}$
12. $\frac{2}{7} + \frac{3}{5}$
13. $\frac{4}{9} + \frac{2}{3}$
14. $\frac{1}{8} + \frac{2}{7}$
15. $\frac{3}{4} + \frac{1}{6}$
16. $\frac{1}{6} + \frac{5}{12} + \frac{5}{6}$
17. $\frac{2}{3} + \frac{1}{2} + \frac{1}{4}$
18. $\frac{1}{5} + \frac{1}{10} + \frac{3}{10}$
19. $\frac{7}{20} + \frac{2}{5} + \frac{1}{4}$
20. $\frac{3}{8} + \frac{1}{2} + \frac{5}{16}$

We subtract in the same way. First find the common denominator then change the fractions and subtract.

EXERCISE 35

Subtract these.

1. $\frac{2}{3} - \frac{1}{4}$
2. $\frac{1}{8} - \frac{1}{16}$
3. $\frac{1}{4} - \frac{1}{12}$
4. $\frac{3}{4} - \frac{1}{2}$
5. $\frac{3}{8} - \frac{1}{4}$
6. $\frac{1}{2} - \frac{3}{10}$
7. $\frac{3}{4} - \frac{3}{8}$
8. $\frac{3}{11} - \frac{1}{4}$
9. $\frac{3}{5} - \frac{2}{7}$
10. $\frac{2}{3} - \frac{4}{9}$
11. $\frac{2}{7} - \frac{1}{8}$
12. $\frac{7}{9} - \frac{3}{5}$
13. $\frac{5}{8} - \frac{3}{32}$
14. $\frac{2}{3} - \frac{1}{4} - \frac{1}{8}$
15. $\frac{3}{4} - \frac{1}{2} - \frac{1}{12}$
16. $\frac{7}{8} - \frac{1}{4} - \frac{1}{3}$
17. $\frac{3}{5} - \frac{2}{7} - \frac{1}{35}$
18. $\frac{7}{12} - \frac{1}{6} - \frac{1}{4}$
19. $\frac{4}{5} - \frac{1}{3} - \frac{1}{30}$
20. $\frac{7}{10} - \frac{1}{3} - \frac{2}{15}$

EXAMPLE 2

$2\frac{1}{4} + 1\frac{7}{8}$ — Add whole numbers first, then find common denominator, and equivalent fractions.

$= 3\frac{2}{8} + \frac{7}{8}$

$= 3\frac{9}{8}$ — Add 2 and 7, then change $\frac{9}{8}$ to $1\frac{1}{8}$ and add the 1 whole one to the 3 we already have

$= \underline{4\frac{1}{8}}$ Ans.

EXERCISE 36

Add these.

1. $2\frac{1}{3} + 1\frac{1}{7}$
2. $2\frac{1}{4} + 3\frac{1}{3}$
3. $2\frac{7}{8} + 1\frac{1}{6}$
4. $3\frac{2}{5} + 2\frac{1}{3}$
5. $4\frac{1}{2} + 2\frac{7}{9}$
6. $6\frac{1}{7} + 2\frac{1}{3}$
7. $1\frac{1}{7} + 2\frac{1}{4}$
8. $1\frac{2}{3} + 2\frac{3}{4}$
9. $4\frac{1}{4} + 2\frac{1}{8}$
10. $2\frac{3}{10} + 1\frac{1}{3}$
11. $3\frac{3}{16} + 2\frac{1}{4}$
12. $4\frac{1}{5} + 3\frac{2}{3}$
13. $2\frac{1}{4} + 1\frac{2}{3} + 1\frac{1}{2}$
14. $3\frac{3}{8} + 4\frac{1}{4} + 2\frac{1}{16}$
15. $2\frac{1}{2} + 1\frac{1}{3} + 1\frac{1}{4}$
16. $2\frac{7}{12} + 1\frac{1}{4} + 2\frac{7}{24}$
17. $5\frac{1}{9} + 2\frac{1}{18} + 3\frac{1}{36}$
18. $1\frac{3}{8} + 2\frac{1}{2} + 3\frac{3}{5}$
19. $3\frac{5}{9} + 1\frac{1}{4} + 2\frac{2}{3}$
20. $3\frac{3}{10} + 2\frac{4}{5} + 1\frac{1}{2}$

EXAMPLE 3

$2\frac{7}{12} - 1\frac{4}{5}$ Subtract whole numbers first, 2 – 1 = 1

$= 1\frac{35}{60} - \frac{48}{60}$ Common denominator is 60. 48 is too big to be taken from 35, so the whole 1 is changed

$= \frac{95}{60} - \frac{48}{60}$ to $\frac{60}{60}$ and added to the $\frac{35}{60}$ making $\frac{95}{60}$.

$= \frac{47}{60}$ Ans. 95 – 48 = 47

EXAMPLE 4

$2\frac{1}{3} - 1\frac{1}{2} + 2\frac{1}{4}$ Whole numbers first.

$= 3\frac{4}{12} - \frac{6}{12} + \frac{3}{12}$

Common denominator is 12.

$= 3\frac{1}{12}$ Ans. 4 + 3 – 6 = 1

EXERCISE 37

1. $2\frac{1}{3} - 1\frac{7}{8}$
2. $3\frac{1}{3} - 2\frac{1}{4}$
3. $2\frac{7}{8} - 1\frac{1}{6}$
4. $3\frac{2}{5} - 2\frac{1}{3}$
5. $4\frac{1}{2} - 2\frac{7}{9}$
6. $6\frac{1}{7} - 2\frac{1}{9}$
7. $5\frac{3}{8} - 3\frac{1}{2}$
8. $2\frac{3}{4} - 1\frac{2}{3}$
9. $4\frac{1}{4} - 2\frac{1}{8}$
10. $2\frac{3}{10} - 1\frac{1}{2}$
11. $3\frac{3}{16} - 2\frac{1}{4}$
12. $4\frac{1}{5} - 3\frac{2}{3}$

From the fraction chart on page 86 you can see that $\frac{1}{2}$ of $\frac{1}{2}$ is $\frac{1}{4}$, $\frac{1}{2}$ of $\frac{1}{4}$ is $\frac{1}{8}$, $\frac{1}{2}$ of $\frac{1}{3}$ is $\frac{1}{6}$ and so on.
We can use the multiplication sign instead of the 'of' since when we write 2 x 2 we mean 2 lots 'of' 2, so we have $\frac{1}{2} \times \frac{1}{4} = \frac{1}{8}$.

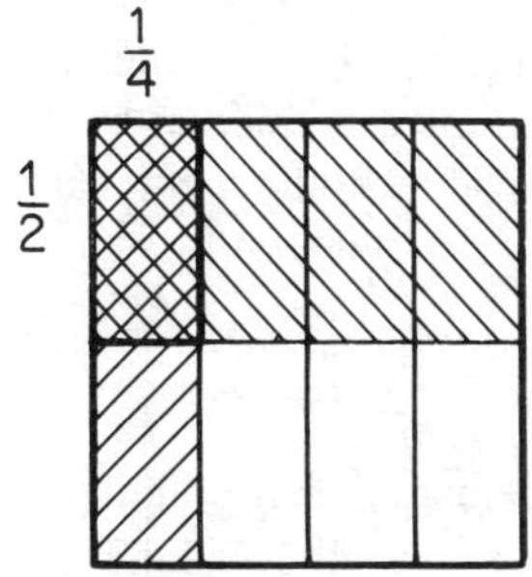

Each vertical strip is $\frac{1}{4}$ of the unit square. Each horizontal strip is $\frac{1}{2}$ of the unit square. The rectangle which is outlined represents $\frac{1}{2}$ of $\frac{1}{4}$ or $\frac{1}{4}$ of $\frac{1}{2}$. It is $\frac{1}{8}$ of the square.
So $\frac{1}{2} \times \frac{1}{4} = \frac{1}{8} = \frac{1}{4} \times \frac{1}{2}$.

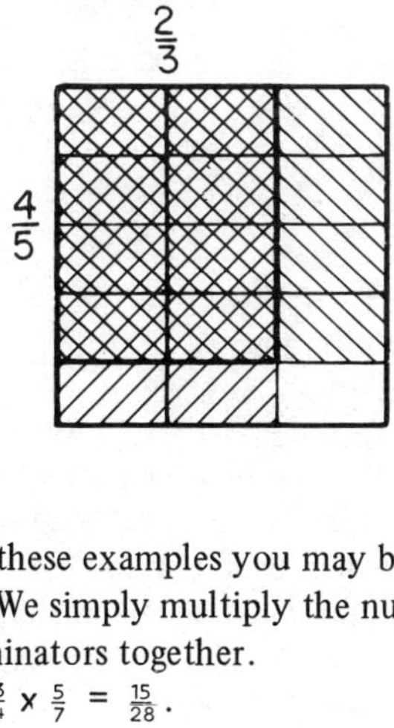

To find $\frac{2}{3} \times \frac{4}{5}$.
Each small rectangle is $\frac{1}{15}$ of the unit square. The outlined rectangle represents $\frac{2}{3}$ of $\frac{4}{5}$ and is $\frac{8}{15}$ of the unit square.
So $\frac{2}{3} \times \frac{4}{5} = \frac{8}{15}$.

From these examples you may be able to find a rule for multiplying fractions. We simply multiply the numerators together then multiply the denominators together.
Thus $\frac{3}{4} \times \frac{5}{7} = \frac{15}{28}$.

EXERCISE 38

1. $\frac{1}{2} \times \frac{1}{3}$	**6.** $\frac{1}{4} \times \frac{3}{5}$	**11.** $\frac{4}{5} \times \frac{1}{2}$
2. $\frac{2}{3} \times \frac{1}{5}$	**7.** $\frac{3}{8} \times \frac{5}{11}$	**12.** $\frac{1}{3} \times \frac{6}{7}$
3. $\frac{3}{4} \times \frac{1}{2}$	**8.** $\frac{2}{5} \times \frac{4}{7}$	**13.** $\frac{2}{3} \times \frac{3}{4}$
4. $\frac{5}{8} \times \frac{3}{4}$	**9.** $\frac{3}{7} \times \frac{2}{3}$	**14.** $\frac{1}{4} \times \frac{2}{9}$
5. $\frac{3}{10} \times \frac{2}{5}$	**10.** $\frac{4}{9} \times \frac{3}{5}$	**15.** $\frac{5}{8} \times \frac{3}{10}$

In this exercise you will have found that some of your answers could have been written more simply.
For instance $\frac{2}{3} \times \frac{3}{4} = \frac{6}{12}$ but $\frac{6}{12} = \frac{1}{2}$ (see fraction chart).
We could have reduced this before multiplying. This process is called cancelling or dividing out.

$$\frac{\cancel{2}^{\,1}}{\cancel{3}_{\,1}} \times \frac{\cancel{3}^{\,1}}{\cancel{4}_{\,2}}$$

$$= \frac{1}{1} \times \frac{1}{2} = \frac{1}{2}$$

Dividing the 2 in the numerator and the 4 in the denominator by 2, and then dividing both the numerator and the denominator by 3.

EXAMPLE 5

$$\frac{3}{16} \times \frac{4}{9}$$

Divide out, 4 and 16 will divide by 4
3 and 9 will divide by 3.

$$= \frac{\cancel{3}^{\,1}}{\cancel{16}_{\,4}} \times \frac{\cancel{4}^{\,1}}{\cancel{9}_{\,3}}$$

$$= \frac{1}{12} \quad \text{Ans.}$$

To multiply mixed numbers we draw another diagram.

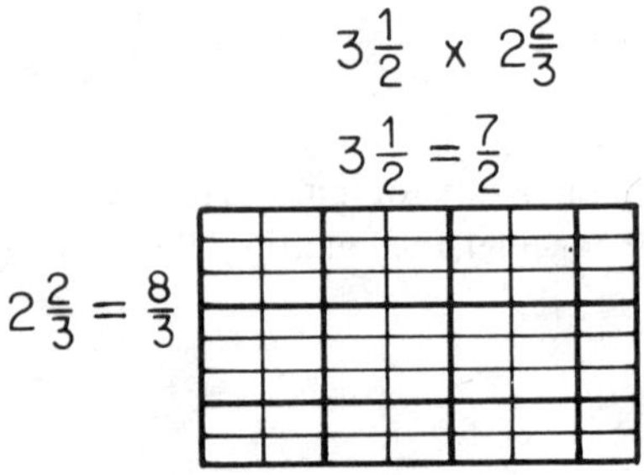

We draw the rectangle $3\frac{1}{2}$ units long and $2\frac{2}{3}$ units wide, then divide it into strips. Each vertical strip is $\frac{1}{2}$ a unit wide. Each horizontal strip is $\frac{1}{3}$ of a unit wide. So the small rectangles are each $\frac{1}{6}$ of a square unit. We have 56 of them thus $3\frac{1}{2} \times 2\frac{2}{3} = \frac{56}{6} = 9\frac{2}{6} = 9\frac{1}{3}$.

EXAMPLE 6

$$3\tfrac{1}{2} \times 2\tfrac{2}{3}$$

$$= \frac{7}{\cancel{2}_{\,1}} \times \frac{\cancel{8}^{\,4}}{3}$$

Change to improper fractions and divide out.

$$= \frac{28}{3}$$

$$= 9\tfrac{1}{3} \quad \text{Ans.}$$

EXAMPLE 7

$3\frac{1}{4} \times 1\frac{3}{11} \times \frac{2}{7}$	Change whole numbers.
$= \frac{13}{\cancel{4}_{\cancel{2}_1}} \times \frac{\cancel{14}^{\cancel{2}^1}}{11} \times \frac{\cancel{2}^1}{\cancel{7}_1}$	Divide out.
$= \frac{13}{11}$	Multiply numerators and denominators.
$= 1\frac{2}{11}$ Ans.	Change back to whole numbers.

EXERCISE 39

1. $\frac{1}{4} \times \frac{8}{9}$
2. $\frac{1}{3} \times \frac{5}{16}$
3. $\frac{1}{2} \times \frac{1}{4}$
4. $\frac{2}{3} \times \frac{15}{16}$
5. $\frac{2}{5} \times \frac{3}{10}$
6. $\frac{3}{8} \times \frac{2}{3}$
7. $\frac{3}{11} \times \frac{22}{33}$
8. $\frac{5}{6} \times \frac{3}{5}$
9. $1\frac{1}{4} \times \frac{8}{9}$
10. $1\frac{1}{3} \times \frac{15}{6}$
11. $2\frac{1}{2} \times 1\frac{1}{4}$
12. $2\frac{3}{4} \times \frac{2}{11}$
13. $3\frac{1}{8} \times 1\frac{1}{5}$
14. $2\frac{3}{8} \times \frac{3}{8}$
15. $3\frac{1}{4} \times 1\frac{3}{13}$
16. $1\frac{2}{3} \times \frac{3}{5}$
17. $3\frac{3}{8} \times 2\frac{2}{3}$
18. $\frac{1}{4} \times \frac{8}{9} \times \frac{1}{3}$
19. $\frac{1}{2} \times \frac{1}{4} \times \frac{2}{3}$
20. $\frac{3}{10} \times \frac{3}{8} \times \frac{2}{3}$
21. $\frac{15}{16} \times \frac{2}{5} \times \frac{3}{10}$
22. $\frac{2}{5} \times \frac{3}{10} \times \frac{5}{8}$
23. $\frac{5}{6} \times \frac{3}{5} \times \frac{1}{4}$
24. $\frac{8}{9} \times \frac{4}{3} \times \frac{15}{16}$
25. $1\frac{1}{4} \times \frac{8}{9} \times 3\frac{1}{3}$
26. $2\frac{1}{2} \times 3\frac{3}{4} \times \frac{2}{3}$
27. $2\frac{1}{2} \times \frac{1}{4} \times 3\frac{3}{4}$
28. $4\frac{1}{4} \times \frac{2}{17} \times 1\frac{8}{9}$
29. $2\frac{5}{8} \times 1\frac{3}{7} \times 5\frac{1}{3}$
30. $3\frac{1}{4} \times 1\frac{3}{11} \times \frac{2}{7}$

How many pieces of wood 2 m long can I cut from 6 m?

$$6 \div 2 = 3$$

Dividing 6 by 2 gives the same answer as finding $\frac{1}{2}$ of 6.

$6 \div 2 = 3$ and $6 \times \frac{1}{2} = 3$

How many pieces of string $\frac{1}{2}$ m long can be cut from 5 m?

$$5 \div \frac{1}{2} = ?$$

From 1 m we can cut 2 pieces, so from 5 m we can cut 5 x 2 pieces.
So $5 \div \frac{1}{2} = 10$, which is answer we get from 5 x 2.

$5 \div \frac{1}{2} = 10$ and $5 \times 2 = 10$

Looking at the fraction chart on page 86 we see that

$\frac{1}{4} \div 2 = \frac{1}{8}$ and $\frac{1}{4} \times \frac{1}{2} = \frac{1}{8}$

What is the connection between 2 and $\frac{1}{2}$?
Since 2 represents 2 whole ones we can write it as $\frac{2}{1}$.
Divide $\frac{1}{4}$ into 3 equal parts.

$\frac{1}{4} \div \frac{3}{1} = \frac{1}{12}$ (from the chart).

This is the same as finding $\frac{1}{3}$ of $\frac{1}{4}$ (or $\frac{1}{4}$ of $\frac{1}{3}$)

$\frac{1}{4} \times \frac{1}{3} = \frac{1}{12}$ (by multiplication).

The pairs of numbers $\frac{2}{1}$, $\frac{1}{2}$ and $\frac{3}{1}$, $\frac{1}{3}$ are called INVERSES.

What happens if we multiply them together?

$\frac{2}{1} \times \frac{1}{2} = 1$ and $\frac{3}{1} \times \frac{1}{3} = 1$

Can you find the inverses of these fractions:

$\frac{1}{5}, \frac{10}{1}, \frac{1}{7}, \frac{1}{99}, \frac{16}{1}, \frac{1}{4}, \frac{2}{3}, \frac{4}{5}, \frac{7}{9}, \frac{11}{2}$?

We can write $\frac{3}{4} \div \frac{7}{8}$ as $\dfrac{\frac{3}{4}}{\frac{7}{8}}$. We get rid of the awkward denominator by multiplying it by its inverse, $\frac{8}{7}$, and to compensate for this we must multiply the numerator by the same number.

$$\frac{\frac{3}{{}_1\not{4}} \times \frac{\not{8}^{\,2}}{7}}{\frac{7}{8} \times \frac{8}{7}} = \frac{\frac{6}{7}}{1} = \frac{6}{7}$$

Dividing by a fraction has the same result as multiplying by its inverse.

EXAMPLE 8

$\frac{2}{3} \div \frac{2}{5}$

$$\frac{\not{2}^{1}}{3} \times \frac{5}{\not{2}_{1}} = \frac{5}{3}$$

$$= \underline{1\tfrac{2}{3}} \text{ Ans}$$

EXAMPLE 9

$2\frac{2}{3} \div 3\frac{1}{5}$

$$\frac{8}{3} \div \frac{16}{5} = \frac{\not{8}^{1}}{3} \times \frac{5}{\not{16}_{2}} = \frac{5}{6}$$

$$= \underline{\tfrac{5}{6}} \text{ Ans}$$

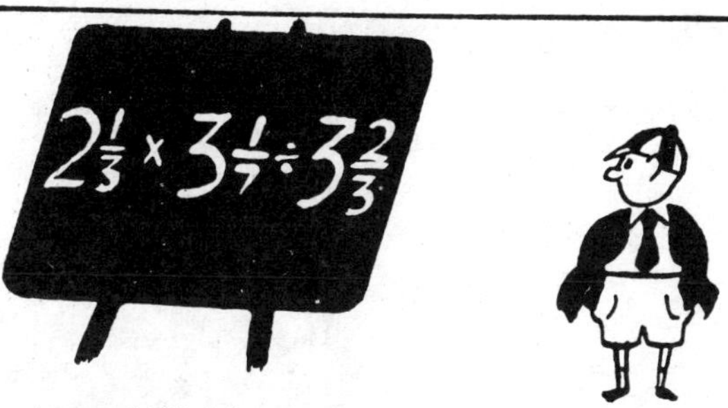

Multiplication and Division

Make " top heavy "

Turn divisor upside down and Multiply . . .

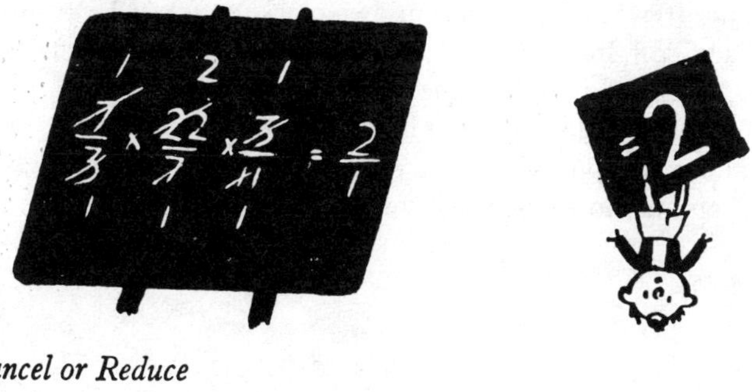

Cancel or Reduce

EXERCISE 40

1. $\frac{1}{3} \div \frac{2}{3}$
2. $\frac{1}{2} \div \frac{1}{8}$
3. $\frac{3}{4} \div \frac{4}{9}$
4. $\frac{3}{8} \div 1\frac{1}{2}$
5. $\frac{2}{5} \div 3\frac{1}{3}$
6. $1\frac{1}{3} \div 1\frac{1}{15}$
7. $2\frac{3}{4} \div 5\frac{1}{2}$
8. $1\frac{2}{3} \div 2\frac{2}{3}$
9. $2\frac{1}{2} \div 3\frac{1}{4}$
10. $1\frac{1}{9} \div 1\frac{1}{4}$
11. $3\frac{1}{8} \div 2\frac{1}{2}$
12. $10 \div 3\frac{1}{8}$
13. $\frac{1}{4} \times \frac{9}{9} \div \frac{3}{1}$
14. $\frac{1}{4} \div 1\frac{1}{8} \times \frac{1}{3}$
15. $\frac{1}{2} \times \frac{1}{4} \div 1\frac{1}{2}$
16. $\frac{15}{16} \div 2\frac{1}{2} \times \frac{3}{10}$
17. $\frac{3}{10} \div \frac{3}{8} \times \frac{2}{3}$
18. $\frac{2}{5} \div \frac{3}{10} \times \frac{3}{8}$
19. $\frac{5}{6} \times \frac{3}{5} \div \frac{1}{4}$
20. $\frac{8}{9} \times 1\frac{1}{3} \div \frac{4}{15}$

Notice the difference in these two examples.

EXAMPLE 10 $\quad 1\frac{1}{2} + \frac{1}{8} \div 2\frac{3}{4}$

$$= 1\frac{1}{2} + \left(\frac{1}{\cancel{8}_{2}} \times \frac{\cancel{4}^{1}}{11}\right)$$

$$= 1\frac{1}{2} + \frac{1}{22}$$

$$= 1\frac{12}{22}$$

$$= 1\frac{6}{11} \text{ Ans.}$$

EXAMPLE 11 $\quad (1\frac{1}{2} + \frac{1}{8}) \div 2\frac{3}{4}$

$$= (1\frac{5}{8}) \div 2\frac{3}{4}$$

$$= \frac{13}{\cancel{8}_{2}} \times \frac{\cancel{4}^{1}}{11}$$

$$= \frac{13}{22} \text{ Ans.}$$

In example 10 we had to insert our own brackets and to do the division before the addition. In example 11 the brackets are inserted for us indicating that that part must be done first.

In the following examples, insert brackets to enclose the fractions joined by multiplication and division signs, and work these parts first unless brackets are shown, in which case work the brackets first.

EXERCISE 41

1. $\frac{1}{4} + \frac{1}{2} \times \frac{1}{3}$
2. $\frac{2}{3} + \frac{1}{4} \times \frac{1}{2}$
3. $\frac{1}{2} + \frac{2}{3} \times \frac{1}{4}$
4. $\frac{1}{4} - \frac{1}{2} \times \frac{1}{3}$
5. $\frac{2}{3} - \frac{1}{4} \times \frac{1}{2}$
6. $\frac{1}{2} - \frac{2}{3} \times \frac{1}{4}$

7. $\frac{1}{4} \times \frac{1}{2} - \frac{1}{16}$
8. $\frac{2}{3} \div 1\frac{1}{3} - \frac{1}{8}$
9. $2\frac{1}{3} \times \frac{1}{4} - \frac{5}{12}$
10. $(\frac{3}{4} + \frac{1}{7}) \times \frac{2}{3}$
11. $(1\frac{3}{4} - \frac{3}{7}) \times \frac{2}{3}$
12. $(1\frac{3}{4} - \frac{3}{7}) \div \frac{1}{14}$
13. $(3\frac{1}{3} + \frac{3}{5}) \div 2$
14. $(2\frac{1}{4} + 3\frac{1}{3}) \div \frac{3}{4}$
15. $\frac{1}{4} \times \frac{2}{3} - \frac{1}{12} \div \frac{1}{2}$
16. $\frac{2}{3} \times (\frac{1}{4} - \frac{1}{12}) \div \frac{1}{2}$
17. $\frac{1}{2} \times (\frac{1}{4} - \frac{1}{6}) \div \frac{1}{3}$
18. $(\frac{1}{6} - \frac{1}{2} \times \frac{1}{4}) \div \frac{1}{3}$
19. $\frac{2}{3}$ of $\frac{1}{4} + \frac{1}{12}$ of $\frac{1}{3}$
20. $\frac{2}{3}(\frac{1}{2} + \frac{3}{8}) \times 1\frac{1}{8}$

EXERCISE 42

1. How many weeks are there in $\frac{1}{2}$ a year?
2. How many metres are there in $\frac{1}{4}$ km?
3. What fraction of a metre is 75 cm?
4. What fraction of £1 is 55p?
5. How many pence are there in $\frac{3}{25}$ of a pound?
6. The directions on a medicine bottle are 'Two 5 ml spoonfuls to be taken 4 times a day.' How much medicine would be taken in a week? What fraction of a litre is this?
7. $\frac{3}{4}$ of a sum of money is £0·33. What is the whole sum?
8. How much does Mary earn in a week if $\frac{3}{5}$ of the money she gets for her paper round is 90p?
9. John spent $\frac{2}{5}$ of his pocket money on a model aeroplane and had 45p left. How much pocket money does he get?
10. There are just 9 litres of petrol in my car and the petrol gauge reads exactly halfway between $\frac{1}{4}$ and $\frac{1}{2}$ full marks. How many litres does it hold?
11. A tank is $\frac{3}{8}$ full and needs 270 litres of water to fill it. How many litres does it hold?
12. When a tank is $\frac{3}{5}$ full there is room for another 90 litres. What is its total capacity?
13. Mary lives $1\frac{1}{2}$ km from school while her friend Betty has half as far again to walk. How far does each girl walk in a day if they go home to dinner? What is the total distance they walk?
14. I used $\frac{3}{4}$ of a tin of paint, then unfortunately spilt half of what was left. What fraction of the whole tin of paint remained?
15. In making a piece of tweed a weaver puts 14 threads to $2\frac{1}{2}$ cm. How many rows must he weave to complete $\frac{1}{4}$ of a metre? What would be the width of 56 threads?

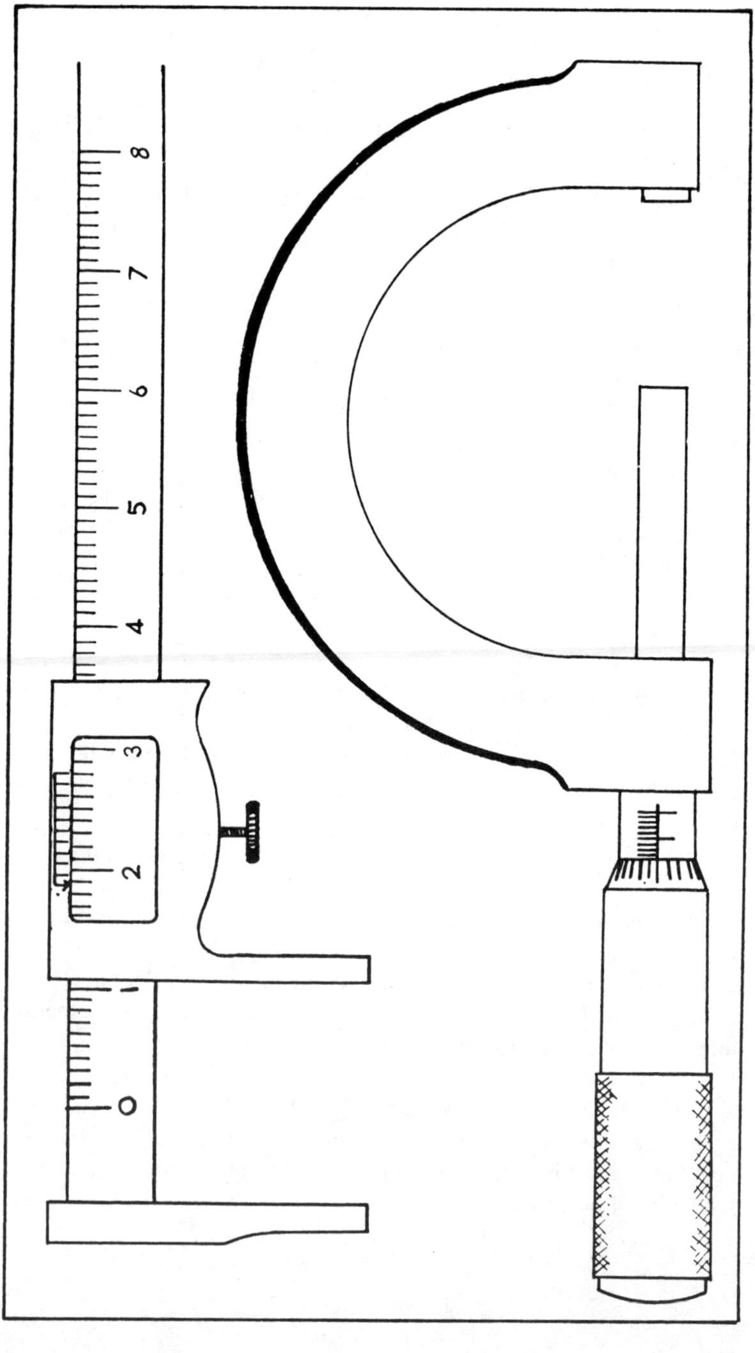
0
1
2
3
4
5
6
7
8

11 Decimals

If you were asked to count a large number of articles you would probably divide them into sets of TEN, the odd ones would be recorded as UNITS. Then the sets of TEN would be grouped into tens and the odd sets recorded as TENS, the sets of ten tens as HUNDREDS and so on. This is the DECIMAL SYSTEM.

We subdivide our units to measure very small distances. For example, we use tenths, or even hundredths, of a centimetre and probably use one of the instruments on the opposite page.

You will have seen a kilometre-counter (or mileometer) in a car. As each km is completed the number in C moves on. After 10 km, O appears in C while 1 appears in B. After 100 km, B and C have Os, while 1 appears in A. Tenths of a km are recorded, in red, in D.

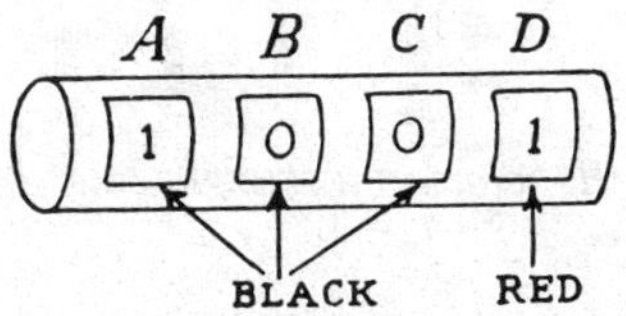

It would not be convenient to write whole numbers in black and fractions in red, so we simply separate them by inserting a decimal point. Just as the full stop shows where a sentence ends, a decimal point shows where the whole numbers end and the fractions begin. The idea of decimal fractions was first used by Simon Stevin in the sixteenth century.

When adding and subtracting decimals it is important to remember to keep the figures in their correct columns, units under units, tens under tens and tenths under tenths etc.; this will mean that the decimal points will come in a straight line under one another as well.

EXAMPLE 1

Add 85·5 + 17·61 + 11·02

```
  85·5
  17·61
  11·02
 ------
 114·13   Ans
```

EXAMPLE 2

From 13·20 subtract 5·94

```
 13·20
  5·94
 -----
  7·26   Ans
```

Keep points in a straight line

EXERCISE 43 Add these.

1. 3·5
   2·4
2. 4·7
   5·8
3. 2·4
   7·6
4. 5·45
   2·7
5. 14·75
   6·31
6. 4·091
   5·909
7. 14·082
   7·234
8. 11·002
   8·019
9. 0·012
   2·34
10. 3·013
    4·21
11. 5·002
    1·761
12. 11·01
    7·09
13. 31·86
    21·75
    5·3
    7·021
14. 14·06
    4·75
    0·007
    7·342
15. 0·0432
    2·1701
    1·002
    12·013
16. 25·16
    14·3
    4·85
    21·77
17. 13·4
    2·012
    11·001
18. 6·08
    11·172
    3·015
19. 14·01
    2·36
    91·013
20. 0·312
    2·42
    17·3
21. 4·75 + 5·909
22. 0·012 + 3·1 + 1·762
23. 31·86 + 0·0016 + 3·01 + 9·013
24. 0·31 + 0·031 + 0·3102 + 0·00032
25. 17·345 + 9·312 + 0·12 + 1·2345
26. 3·25 + 0·12 + 3·14

27. 1·7321 + 1·414 + 0·7971
28. 0·021 + 2·324 + 12·02
29. 11·12 + 10·3 + 1·01
30. 18·012 + 0·013 + 10·8

EXERCISE 44

Subtract

1. 7·5 − 3·2
2. 4·38 − 2·14
3. 8·38 − 5·16
4. 11·46 − 3·23
5. 15·38 − 8·16
6. 12·43 − 11·21
7. 6·50 − 2·38
8. 5·40 − 3·42
9. 4·30 − 1·73
10. 4·200 − 1·732
11. 0·43 − 0·12
12. 3·2 − 1·41
13. 0·815 − 0·61
14. 0·432 − 0·082
15. 9 − 0·8
16. 3 − 0·4
17. 4 − 1·7
18. 25 − 0·25
19. 5 − 0·012
20. 1 − 0·001

21. 6·4 − 2·1
22. 3·01 − 2·07
23. 1 − 0·01
24. 13·84 − 7·46
25. 7·05 − 3·14
26. 6·89 − 4·321
27. 0·02 − 0·001
28. 0·2 − 0·0708
29. 6·4 − 0·64
30. 0·1 − 0·01

EXAMPLE 3

3·1 − 4·72 + 11·64 − 0·97

Rearrange

3·1 + 11·64 − 4·72 − 0·97

(3·1 + 11·64) − (4·72 + 0·97)

Total addition − total subtraction

14·74 − 5·69

9·05 Ans

EXERCISE 45

1. 2·8 + 3·5 − 1·6
2. 4·2 − 3·1 + 2·25
3. 5·31 + 2·75 − 1·25
4. 2·675 − 1·375 + 4·5
5. 1 − 0·01 − 0·001 + 2
6. 5 − 0·3 − 0·002 + 8·4
7. 3·14 − 2·16 + 1·75 − 2·1
8. 14·01 − 2·16 + 1·75 − 2·142
9. 3 − 0·03 − 0·003
10. 10 − 0·1 − 0·01 − 0·001

1	2	3	4	5	6	7	8	9	10
10	20	30	40	50	60	70	80	90	100

In the top line are the numbers from 1 to 10 and below are the same numbers multiplied by 10. As they are all whole numbers a decimal point could be placed after each.

$$2{\cdot}0 \times 10 = 20{\cdot}$$
$$5{\cdot}0 \times 10 = 50{\cdot}$$

Can you see what has happened to the figures? They have moved one place to the left and the point appears to have moved one place to the right.

$$6{\cdot}3 \times 10 = 63{\cdot}$$
$$4{\cdot}17 \times 10 = 41{\cdot}7$$

To multiply by 100 or 1000 we use the same process

$$3{\cdot}0842 \times 100 = 308{\cdot}42$$
$$0{\cdot}4123 \times 1000 = 412{\cdot}3$$

As division is the reverse of multiplication we have the reverse process.

$$36{\cdot}5 \div 10 = 3{\cdot}65$$
$$597{\cdot}6 \div 100 = 5{\cdot}976$$
$$34267{\cdot}1 \div 1000 = 34{\cdot}2671$$

EXAMPLE 4

$2{\cdot}013 \times 300$ (multiply by 100 first)
$= 201{\cdot}3 \times 3$
$= \underline{603{\cdot}9}$ Ans.

EXAMPLE 5

$14{\cdot}016 \div 400$ (divide by 100 first)
$= 0{\cdot}14016 \div 4$
$= \underline{0{\cdot}03504}$ Ans.

EXERCISE 46

1. 3·14 x 10	**11.** 2·2 x 20	**21.** 31·12 x 400	**31.** $0{\cdot}123 \times 10^2$
2. 1·7321 x 10	**12.** 3·2 x 30	**22.** 13·001 x 500	**32.** $31{\cdot}14 \times 10^4$
3. 1·414 x 10	**13.** 1·5 x 40	**23.** 0·001 x 600	**33.** $1{\cdot}74 \times 10^2$
4. 10·3 x 10	**14.** 10·2 x 50	**24.** 700 x 2·11	**34.** $0{\cdot}11 \times 10^5$
5. 16·01 x 100	**15.** 60 x 0·12	**25.** 1·14 x 800	**35.** $63{\cdot}01 \times 10^2$
6. 2·3 x 100	**16.** 70 x 0·012	**26.** 2·041 x 900	**36.** $54{\cdot}3 \times 10^4$
7. 11·03 x 100	**17.** 80 x 2·103	**27.** 0·866 x 1000	**37.** $21{\cdot}36 \times 10^2$
8. 16·001 x 100	**18.** 8·04 x 90	**28.** 4·341 x 2000	**38.** 5·012 x 20
9. 0·002 x 1000	**19.** 4·04 x 200	**29.** 0·0012 x 3000	**39.** 4·532 x 400
10. 1·0121 x 100	**20.** 300 x 1·106	**30.** $0{\cdot}102 \times 10^3$	**40.** 31·142 x 10

EXERCISE 47

1. 5·4 ÷ 2
2. 6·8 ÷ 2
3. 11·4 ÷ 2
4. 10·4 ÷ 2
5. 15·9 ÷ 3
6. 9·9 ÷ 3
7. 12·1 ÷ 4
8. 2·79 ÷ 6
9. 16·4 ÷ 4
10. 2·04 ÷ 4
11. 4·004 ÷ 4
12. 1·44 ÷ 4
13. 6·25 ÷ 5
14. 22·5 ÷ 5
15. 0·125 ÷ 5
16. 37·5 ÷ 5
17. 0·049 ÷ 7
18. 0·63 ÷ 7
19. 0·063 ÷ 7
20. 0·0063 ÷ 7
21. 0·5 ÷ 2
22. 0·9 ÷ 3
23. 0·2 ÷ 5
24. 0·4 ÷ 5
25. 3·014 ÷ 2
26. 14·22 ÷ 3
27. 0·6024 ÷ 8
28. 2·025 ÷ 6
29. 3·0612 ÷ 6
30. 0·0054 ÷ 9

EXERCISE 48

1. 3·14 ÷ 10
2. 1·7321 ÷ 10
3. 1·414 ÷ 10
4. 100·4 ÷ 10
5. 20·01 ÷ 100
6. 0·0032 ÷ 100
7. 11·03 ÷ 100
8. 0·4 ÷ 100
9. 3·14 ÷ 200
10. 1·732 ÷ 20
11. 70·05 ÷ 30
12. 36·5 ÷ 50
13. 36·618 ÷ 60
14. 0·01428 ÷ 70
15. 10·11 ÷ 500
16. 365 ÷ 500

EXERCISE 49

1. $\frac{6{\cdot}16}{20}$
2. $\frac{1{\cdot}64}{20}$
3. $\frac{18{\cdot}6}{30}$
4. $\frac{28{\cdot}35}{70}$
5. $\frac{32{\cdot}4}{80}$
6. $\frac{2{\cdot}4}{100}$
7. $\frac{6{\cdot}18}{2}$
8. $\frac{18{\cdot}9}{90}$
9. $\frac{234{\cdot}4}{200}$
10. $\frac{6{\cdot}48}{80}$
11. $\frac{7{\cdot}15}{50}$
12. $\frac{14{\cdot}4}{20}$
13. $\frac{2{\cdot}25}{50}$
14. $\frac{18{\cdot}12}{60}$
15. $\frac{2000{\cdot}4}{400}$

When multiplying or dividing by numbers which are not multiples of 10 care must be taken over the position of the decimal point. One way of checking your answer is to make a rough estimate first.

EXAMPLE 6

4·13 x 3·05		Estimate (4 +) x (3 +) = (12 +)
4 13		
3 05		Multiply as if the numbers were 413 and 305.
2065		
1239		
125965		There are four figures after the point; count four
12·5965	Ans.	places in from the right in your answer and insert the point.

If there are not enough figures in the answer to put the decimal point in the right place, noughts must be put in to fill the places.

EXAMPLE 7

0·4 x 0·2 Estimate: $\frac{4}{10} \times \frac{2}{10} = \frac{8}{100}$

0·4
0·2
0·08 Ans.

EXERCISE 50

1. 0·2 x 1·1
2. 1·0 x 0·2
3. 0·3 x 0·2
4. 0·5 x 0·4
5. 0·6 x 0·7
6. 0·7 x 0·8
7. 0·8 x 0·9
8. 0·9 x 1·0
9. $(0{\cdot}2)^2$
10. $(0{\cdot}3)^2$
11. $(0{\cdot}4)^2$
12. $(0{\cdot}5)^2$
13. $(0{\cdot}6)^2$
14. $(0{\cdot}7)^2$
15. $(0{\cdot}8)^2$
16. $(0{\cdot}9)^2$
17. 10 x 0·2
18. 10 x 0·12
19. 0·123 x 10
20. 0·142 x 100
21. 0·321 x 100
22. 100 x 0·421
23. 0·531 x 100
24. 2·12 x 100
25. 3·1412 x 100
26. 16·231 x 1000
27. 0·0013 x 1000
28. 1·7321 x 1000
29. $(1{\cdot}1)^2 \times 0{\cdot}5$
30. 0·1 x 0·2 x 0·3

EXERCISE 51

1. 3·14 x 2·1
2. 15·4 x 3·4
3. 254 x 1·2
4. 1234 x 3·2
5. 3·42 x 1·5
6. 2·3 x 1·5
7. 302·4 x 2·2
8. 9·8 x 3·82
9. 6·04 x 4·01
10. 22 x 0·8
11. 12·04 x 3·4
12. 3·1 x 5·02
13. 13·9 x 2·2
14. 8·4 x 1·06
15. 11·2 x 7·2
16. 13·4 x 12
17. 3·04 x 11
18. 0·08 x 9
19. 320 x 0·048
20. 1200 x 0·014
21. 320 x 0·032
22. 201 x 0·201
23. 654 x 0·0032
24. 1·03 x 12·7
25. 0·02 x 0·32
26. 14·4 x 0·74
27. 0·002 x 0·031
28. 0·0024 x 0·2
29. 16·66 x 1·04
30. 0·17 x 0·12
31. 0·041 x 0·032
32. 200 x 3·012
33. 132·4 x 0·24
34. 148·4 x 0·042
35. 203 x 0·004
36. 0·013 x 0·012
37. 3000 x 0·804
38. 4200 x 0·402
39. 2002 x 0·304
40. 12·4 x 1·3
41. 21·4 x 1·05
42. 8·5 x 6·4
43. 11 x 13·02
44. 500 x 0·303
45. 3·08 x 21·04
46. 1·01 x 0·01
47. 20·02 x 3·01
48. 0·0333 x 1·7
49. 1800 x 0·402
50. 0·0011 x 302
51. 0·00101 x 20
52. 20·02 x 300
53. 400 x 30·03
54. 1234 x 0·056
55. 0·002 x 0·04
56. 0·004 x 0·14
57. 0·303 x 1600
58. $(0{\cdot}25)^2$
59. $(0{\cdot}013)^2$
60. $(2{\cdot}14)^2$

EXAMPLE 8

$$\frac{2{\cdot}4 \times 1{\cdot}8 \times 3{\cdot}3}{1{\cdot}2 \times 0{\cdot}48 \times 4{\cdot}4} \qquad \text{Estimate} \quad \frac{(2+) \times (2-) \times (3+)}{(1+) \times (\frac{1}{2}) \times (4+)} = \frac{12}{2} \quad \text{Approx.}$$

To make the denominator into a whole number we must multiply it by 10 000. If we do this we shall alter the value of the fraction, so to compensate, we multiply both numerator and denominator by the same figure.

$$\frac{2{\cdot}4 \times 1{\cdot}8 \times 3{\cdot}3 \times 10\,000}{1{\cdot}2 \times 0{\cdot}48 \times 4{\cdot}4 \times 10\,000} = \frac{\overset{\overset{1}{\cancel{2}}}{\cancel{24}} \times \overset{3}{\cancel{18}} \times \overset{3}{\cancel{33}} \times \overset{5}{\cancel{10}}}{\underset{1}{\cancel{12}} \times \underset{8}{\cancel{48}} \times \underset{\underset{\underset{1}{\cancel{2}}}{\cancel{4}}}{\cancel{44}}} = \frac{45}{8} = \underline{5{\cdot}625} \text{ Ans.}$$

EXERCISE 52

1. $\frac{0{\cdot}2 \times 0{\cdot}3}{0{\cdot}6}$
2. $\frac{0{\cdot}5 \times 0{\cdot}4}{0{\cdot}2}$
3. $\frac{0{\cdot}7 \times 0{\cdot}8}{3{\cdot}5}$
4. $\frac{0{\cdot}9 \times 1{\cdot}1}{3{\cdot}3}$
5. $\frac{1{\cdot}1 \times 2{\cdot}1}{0{\cdot}7}$
6. $\frac{3{\cdot}2 \times 0{\cdot}1}{0{\cdot}8}$
7. $\frac{1{\cdot}6 \times 3{\cdot}2}{0{\cdot}4 \times 0{\cdot}8}$
8. $\frac{8{\cdot}4 \times 1{\cdot}3}{0{\cdot}013 \times 0{\cdot}07}$
9. $\frac{16{\cdot}9 \times 0{\cdot}044}{2{\cdot}2 \times 1{\cdot}3}$
10. $\frac{5{\cdot}4 \times 4{\cdot}2 \times 0{\cdot}1}{0{\cdot}03 \times 0{\cdot}9 \times 1{\cdot}4}$
11. $\frac{0{\cdot}1 \times 2{\cdot}2}{1{\cdot}1}$
12. $\frac{3{\cdot}3 \times 2{\cdot}2}{6{\cdot}6}$
13. $\frac{15{\cdot}05}{0{\cdot}2 \times 0{\cdot}5}$
14. $\frac{4{\cdot}8}{1{\cdot}2 \times 0{\cdot}4}$
15. $\frac{5{\cdot}6}{0{\cdot}7 \times 0{\cdot}8}$
16. $\frac{6{\cdot}3}{0{\cdot}07 \times 0{\cdot}9}$
17. $\frac{0{\cdot}11 \times 1{\cdot}1}{1{\cdot}21}$
18. $\frac{3{\cdot}9 \times 2{\cdot}1}{0{\cdot}3 \times 1{\cdot}3 \times 0{\cdot}7}$
19. $\frac{0{\cdot}028 \times 4{\cdot}9 \times 2{\cdot}2}{0{\cdot}04 \times 0{\cdot}7 \times 1{\cdot}4}$
20. $\frac{2{\cdot}2 \times 1{\cdot}8}{0{\cdot}003 \times 6{\cdot}6}$
21. $\frac{0{\cdot}1 \times 0{\cdot}2}{2{\cdot}2 \times 0{\cdot}01}$
22. $\frac{2{\cdot}8 \times 0{\cdot}396}{0{\cdot}14 \times 6{\cdot}6}$
23. $\frac{6{\cdot}3 \times 3{\cdot}1}{0{\cdot}007 \times 0{\cdot}9}$
24. $\frac{0{\cdot}22 \times 0{\cdot}03}{1{\cdot}1 \times 1{\cdot}5}$
25. $\frac{0{\cdot}05 \times 0{\cdot}4}{0{\cdot}2 \times 0{\cdot}1 \times 1{\cdot}5}$
26. $\frac{2{\cdot}1 \times 0{\cdot}32 \times 6{\cdot}4}{0{\cdot}7 \times 1{\cdot}6 \times 0{\cdot}8}$
27. $\frac{3{\cdot}9 \times 1{\cdot}4}{0{\cdot}13 \times 0{\cdot}007}$
28. $\frac{3{\cdot}84 \times 0{\cdot}13 \times 0{\cdot}26}{0{\cdot}064 \times 1{\cdot}69}$
29. $\frac{(0{\cdot}2)^2 \times (0{\cdot}4)^2}{2{\cdot}0 \times 1{\cdot}6}$
30. $\frac{0{\cdot}72 \times 8{\cdot}1 \times 1{\cdot}21}{2{\cdot}2 \times 9{\cdot}9 \times 3{\cdot}6}$

EXAMPLE 9

16·75 ÷ 2·5

$= \dfrac{16{\cdot}75}{2{\cdot}5}$

= 167·5 ÷ 25

Estimate: (16+) ÷ ($2\frac{1}{2}$) = (6+)

Make the denominator a whole number. In this case it means multiplying it by 10. To compensate you must also multiply the numerator by 10.

```
        6·7   Ans.
25 )167·5
    150
     175
     175
     ...
```

EXAMPLE 10

7·368 ÷ 0·15

= 736·8 ÷ 15

Estimate: 7 ÷ $\frac{3}{20}$ = (40+)

Multiply both numerator and denominator by 100.

```
      49·12   Ans.
15)736·80
   60
   136
   135
     18
     15
      30
      30
      ..
```

EXERCISE 53

1. 16·9 ÷ 1·3
2. 20·9 ÷ 1·9
3. 13·6 ÷ 1·7
4. 20·7 ÷ 2·3
5. 12·1 ÷ 1·1
6. 1·21 ÷ 1·1
7. 0·169 ÷ 1·3
8. 0·0169 ÷ 0·13
9. 16·9 ÷ 0·13
10. 16·9 ÷ 0·013
11. 0·169 ÷ 0·13
12. 0·625 ÷ 2·5
13. 3·51 ÷ 2·7
14. 0·351 ÷ 0·27
15. 0·351 ÷ 0·027
16. 1·96 ÷ 1·4
17. 0·0169 ÷ 1·3
18. 0·064 ÷ 0·08
19. 2·31 ÷ 2·1
20. 2·25 ÷ 1·5
21. 6·25 ÷ 1·25
22. 0·09 ÷ 0·3
23. 0·04 ÷ 0·2
24. 4·5796 ÷ 2·14
25. 17·6 ÷ 0·8
26. 0·72 ÷ 9
27. 0·064 ÷ 0·32
28. 304·8 ÷ 1·2
29. 6·594 ÷ 3·14
30. 5·13 ÷ 3·42

Sometimes it is necessary to write a fraction in decimal form. It is quite easy to do this if we remember that the numerator tells us how many objects and the denominator tells into how many parts they have been divided.
So $\frac{3}{4}$ means that 3 objects have been divided into 4 parts.

EXAMPLE 11

Write $\frac{3}{4}$ in decimal form.

$3 \div 4$ we write $3 = 3{\cdot}00$

$$4\,\overline{)\,3{\cdot}00}\quad \text{quotient } 0{\cdot}75$$

$\frac{3}{4} = \underline{0{\cdot}75}$ Ans.

EXAMPLE 12

Write $3\frac{3}{5}$ in decimal form

$3\frac{3}{5} = \frac{18}{5}$ we write $18{\cdot}00 \div 5$

$$5\,\overline{)\,18{\cdot}00}\quad \text{quotient } 3{\cdot}60$$

$3\frac{3}{5} = \underline{3{\cdot}6}$ Ans.

EXERCISE 54

Change these fractions to decimals.

1. $\frac{1}{32}$	**11.** $\frac{3}{10}$	**21.** $\frac{1}{3}$
2. $\frac{1}{16}$	**12.** $\frac{7}{10}$	**22.** $\frac{2}{3}$
3. $\frac{1}{8}$	**13.** $3\frac{1}{32}$	**23.** $\frac{5}{6}$
4. $\frac{1}{4}$	**14.** $2\frac{3}{16}$	**24.** $\frac{1}{7}$
5. $\frac{3}{8}$	**15.** $1\frac{1}{8}$	**25.** $\frac{3}{7}$
6. $\frac{1}{2}$	**16.** $2\frac{5}{16}$	**26.** $\frac{4}{9}$
7. $\frac{5}{8}$	**17.** $4\frac{7}{20}$	**27.** $\frac{9}{9}$
8. $\frac{2}{5}$	**18.** $2\frac{3}{5}$	**28.** $\frac{2}{11}$
9. $\frac{7}{8}$	**19.** $3\frac{4}{5}$	**29.** $\frac{7}{11}$
10. $\frac{1}{10}$	**20.** $3\frac{5}{8}$	**30.** $\frac{4}{13}$

You will notice that the answers to the last 10 questions of this exercise do not end nicely as the others do. This is because we are working in base 10 and numbers like 3, 7 and 13 do not divide exactly into any power of 10 as numbers like 4, 5 and 20 do. However, you will have noticed that the numbers after the point do repeat themselves.

$\frac{1}{3} = 0{\cdot}33333333$ and $\frac{2}{11} = 0{\cdot}1818181818$

These are RECURRING decimals and we write

$\frac{1}{3} = 0{\cdot}\dot{3}$ and $\frac{2}{11} = 0{\cdot}\overline{18}$

EXAMPLE 13

Express 11·61503 correct to (*a*) the nearest whole number.
(*b*) two places of decimals.
(*c*) three places.
(*d*) four places.

(*a*) 11·6 is nearer to 12 than to 11.

(*b*) 11·615 is nearer to 11·62 than to 11·61

(*c*) 11·6150 correct to three places is 11·615 since the fourth figure is less than 5.

(*d*) 11·61503 correct to four places is 11·6150. If the fifth figure had been 5 or more than the answer correct to four places would have been 11·6151.

From this you will see that to write a number correct to 3 decimal places, we look at the fourth figure after the point. If this is less than 5, we write down the first three figures as they are, if it is 5 or more, then we add 1 to the third figure.

EXERCISE 55

Express the following correct to 1, 2, and 3 places of decimals.

1.	2·0645	**3.**	3·5107	**5.**	4·0555	**7.**	1·9994	**9.**	7·092
2.	12·1503	**4.**	0·0168	**6.**	5·125	8.	0·0506	**10.**	18·52862

EXAMPLE 14

Find $\frac{2}{3}$ of £3·13 correct to the nearest penny.

$$\frac{2}{3} \text{ of } £3{\cdot}13 = \frac{2}{3} \times \frac{£3{\cdot}13}{1}$$
$$= \frac{£6{\cdot}26}{3}$$
$$= £2{\cdot}0866$$
$$= \underline{£2{\cdot}09} \text{ Ans (correct to the nearest penny).}$$

EXAMPLE 15

Find the value of 1·3 x £11·43 correct to the nearest penny.

Estimate (11+) x (1+) = (11+)

```
 1143
   13
-----
 3429
11430
-----
14·859
```

1·3 x £11·43 = £14·859
= <u>£14·86</u> Ans (correct to the nearest penny).

EXERCISE 56

Write to the nearest whole number.

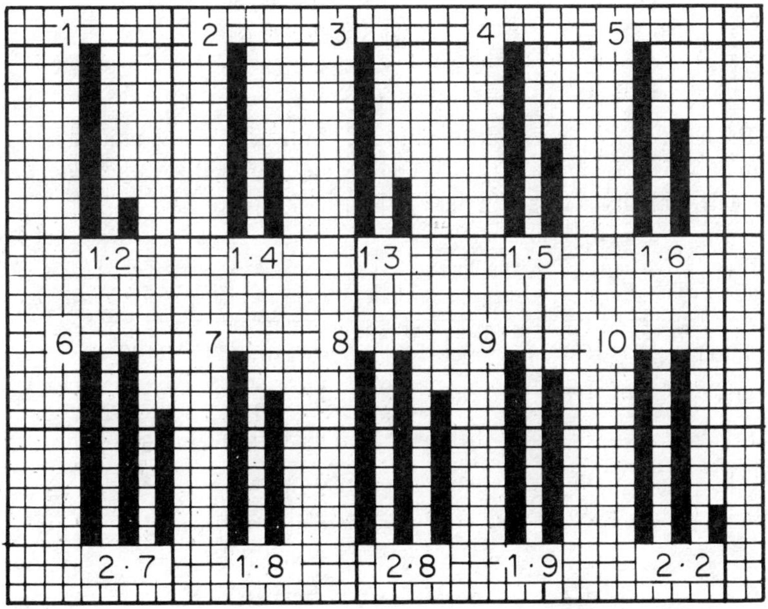

Write to the nearest pound.

11. £3·85
12. £1·45
13. £4·5
14. £2·75
15. £1·27
16. £5·49
17. £6·35
18. £2·76
19. £1·65
20. £10·99

Write to the nearest metre.

21. 2 m 30 cm
22. 1 m 95 cm
23. 4 m 61 cm
24. 5 m 65 cm
25. 1 m 15 cm
26. 4 m 75 cm
27. 2 m 33 cm
28. 3 m 50 cm
29. 2 m 82 cm
30. 6 m 11 cm

Write to the nearest kilogram.

31. 1 kg 200 g
32. 1 kg 760 g
33. 2 kg 420 g
34. 6 kg 550 g
35. 2 kg 350 g
36. 4 kg 60 g
37. 5 kg 450 g
38. 10 kg 73 g
39. 8 kg 880 g
40. 7 kg 600 g

EXERCISE 57

Give the answers correct to the nearest penny.

1. $\frac{3}{4}$ of £4·20
2. $\frac{1}{3}$ of £7·41
3. $\frac{2}{5}$ of £10·37
4. $\frac{3}{8}$ of £4·50
5. $\frac{1}{6}$ of £3·55
6. 3·5 x £1·90
7. 0·6 x £11·85
8. 1·2 x £4·63
9. 2·4 x £8·36
10. 4·7 x £5·64

The name of multiples and submultiples of the basic metric units are related to the basic units and are formed by means of prefixes which are the same for units of length, weight and capacity.

Factor by which unit is multiplied		Prefix	Example	
1 million	10^6	mega	megawatt	(MW)
1000	10^3	kilo	kilogram	(kg)
100	10^2	hecto	hectogram	(hg)
10	10^1	deca	decagram	(dag)
1	10^0		metre (m), litre (1) gram (g)	
0·1	10^{-1}	deci	decimetre	(dm)
0·01	10^{-2}	centi	centimetre	(cm)
0·001	10^{-3}	milli	millilitre	(ml)
0·000 001	10^{-6}	micro	microsecond	(μs)

You are not likely to meet the first and the last entries on this list in your mathematics lessons, but you may meet them later in science lessons or hear them mentioned on television.

Consider the length 3 km 4 hm 6 dam 2 m 1 dm 5 cm 8 mm. This can be written in seven different ways by merely moving the position of the decimal point.

(*a*) 3·462158 km
(*b*) 34·62158 hm
(*c*) 346·2158 dam
(*d*) 3462·158 m
(*e*) 34621·58 dm
(*f*) 346215·8 cm
(*g*) 3462158 mm

EXAMPLE 16

Change to cm 3 dam. 1 m. 5 mm.

3 dam. 1 m. 0 cm. 5 mm.

310·5 cm. Ans.

EXERCISE 58

Change to grams.

1. 14 gm. 7 cg.
2. 22 mg.
3. 4 hg. 3 dag. 2 mg.
4. 1 kg. 4 hg.
5. 5 cg. 8 mg.
6. 6 dag. 2 gm.
7. 2 gm. 1 dg. 5 cg.
8. 46 hg.
9. 6 dag. 1 dg.
10. 2 dg. 1 cg.

Change to metres.

11. 2 dam. 4 m.
12. 3 hm. 5 dam.
13. 1 km. 6 hm.
14. 2 m. 1 cm.
15. 7 m. 4 dm.
16. 5 dm. 1 cm.
17. 3 cm. 7 mm.
18. 9 dm. 8 mm.
19. 5 km. 2 cm.
20. 1 hm. 4 mm.

Change to litres.

21. 4 dal. 21.
22. 1 hl. 31.
23. 7 kl. 51.
24. 91. 9 dl.
25. 2 dl. 3 cl.
26. 4 cl. 5 ml.
27. 2 dal. 7 cl.
28. 8 hl. 9 dl.
29. 7 kl. 6 cl.
30. 5 hl. 4 cl.

In the next exercise we reverse the process.

EXAMPLE 17

Change 4·201 kg to metric units.

4 kg. 2 hg. 0 dag. 1 g. or 4 kg. 2 hg. 1 g. Ans.

EXERCISE 59

Change to metric units.

1. 12·6 hg.
2. 5·006 g.
3. 16·01 dg.
4. 16 hg.
5. 1·11 dg.
6. 3·142 hg.
7. 1·7321 dag.
8. 200·3 kg.
9. 704·021 g.
10. 0·123 kg.
11. 0·0023 hm.
12. 14·01 km.
13. 16·2 m.
14. 12·3 dam.
15. 80·02 dm.
16. 5·06 m.
17. 10·2 km.
18. 753·1 mm.
19. 2·002 m.
20. 10·001 m.
21. 72·4 ml.
22. 3·2611.
23. 732·1 dl.
24. 141·4 ml.
25. 0·1628 dal.
26. 1·728 hl.
27. 14·1 kl.
28. 32·06 dal.
29. 100·4 dl.
30. 27·51 dal.

Averages

1m 10cm 25kg — 1m 20cm 30kg — 1m 45cm 45kg

What is the total height of the children?

How many children are there?

Divide the total height by the number of children.

The answer should be 1 m. 25 cm. (1·25 m.). This is known as their average height.

Find their average weight from the information you are given.

Of what use are averages? If you look at some of the weighing machines to be found in chemists' shops or at railway stations you will see a table you can tell whether or not you are overweight. A tall person's weight can be greater than a short person's weight. Averages are used for comparison, for example.

(*a*) the average speeds of drivers in a race.
(*b*) the average of a cricketer's scores.
(*c*) the average attendance of children.
(*d*) the average number of goals scored by a team.
(*e*) the average amount of rain or hours of sunshine in different places.

The find any of these you divide the total quantity by the number of quantities added:

$$\text{Average} = \frac{\text{Total of quantities}}{\text{Number of quantities}}$$

Measure the height of some of your friends and find the average. Find the average age of your form. Add your last ten examination marks and find the average of them. If you turn to the chapter on graphs, you will see how to illustrate these results graphically.

EXERCISE 60

1. This table shows marks gained by a group of second year children.
   (*a*) Add up the marks for each subject and find the average mark.
   (*b*) How many children gained an above average mark?
   (*c*) Find the average mark gained by each child and put the children in order of merit.

	MATHS	ENGLISH	HISTORY	GEOGRAPHY	SCIENCE
Jennifer Jacobs	80	70	20	70	50
Mary Jones	40	50	50	30	40
Joan Forsyth	35	40	40	50	60
Rhoda Benton	46	20	30	40	28
Greta May	20	40	16	30	30
John Billings	65	55	50	45	40
Peter Coleman	54	60	65	50	50
Barry Stroud	42	20	40	60	60
Jim Scallop	30	30	70	25	30
Philip Palmer	20	40	10	70	20

2. In five boxes we found a total of 225 matches; what was the average number per box?

The following were the top of the bowling averages after a tour of the West Indies.

3.

	OVERS	RUNS	WICKETS
Greig	277·1	766	30
Hendrick	108·2	320	12
Birkenshaw	165·5	467	15
Willis	140	526	15
Underwood	263·5	573	13
Pocock	327·3	844	19

Find the average runs per wicket for each.

4. The Smiths left home for their holiday at 8 a.m. and arrived at their destination 468 km. away at 5 p.m.

   Their neighbours, the Joneses, left an hour later and were 432 km. away by 5 p.m.

   Which family travelled at the greater average speed?

5. If the rainfall figures for London in centimetres were

Jan.	Feb.	Mar.	Ap.	May	Jun.
4·75	4·0	4·75	4·0	4·5	5·0

Jul.	Aug.	Sep.	Oct.	Nov.	Dec.
5·75	5·25	4·5	6·25	5·5	5·75

(*a*) What is the total rainfall for the year?

(*b*) What is the average monthly rainfall?

6. The rainfall in Manchester in centimetres was

Jan.	Feb.	Mar.	Ap.	May	Jun.
7	5·75	6·25	5·75	6	7

Jul.	Aug.	Sep.	Oct.	Nov.	Dec.
8	9·5	6·25	8·75	8	8·75

(*a*) Find the total rainfall for the year.

(*b*) Find the monthly average.

7. Rainfall in West Africa.

Jan.	Feb.	Mar.	Ap.	May	Jun.
1·25	1·25	2·5	7·5	27·5	50

Jul.	Aug.	Sep.	Oct.	Nov.	Dec.
90	92·5	65	32·5	12·5	7·5

(*a*) Find the average monthly rainfall.

(*b*) Compare with figure of previous two questions.

8. Draw graphs to illustrate the figures given in the last three questions. Put in the average with a dotted line.

Which months have more than average rainfall, which have less?

12 Area

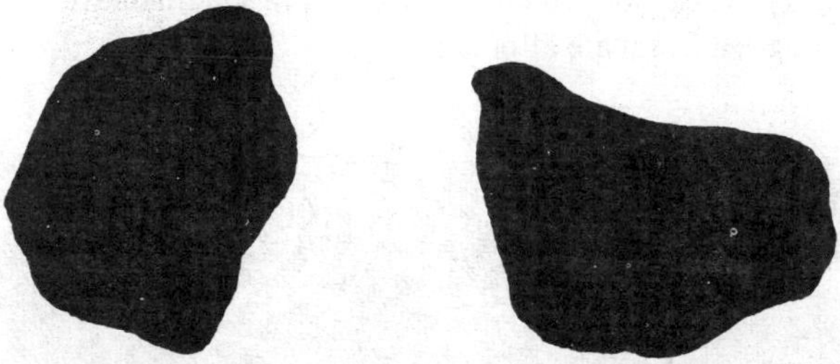

Most of you will have had an accident at some time or other and spilt ink, but how big was the blot that you made? Which of the two blots above is the bigger? Which covers more paper?

To answer that question accurately you will need to find the area of both blots. *AREA is the amount of surface covered.* It is measured in squares. Draw some squares on tracing paper, place it over the blots and count the number of squares. If a square is half covered then count it as a whole square and omit any that are less than half covered.

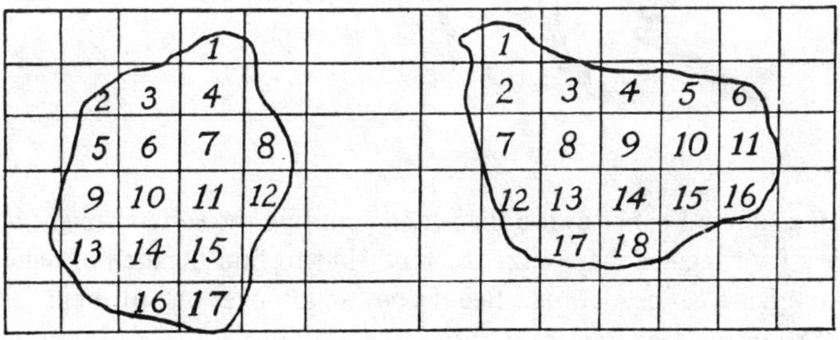

Your answer will depend on how big you have made your squares.

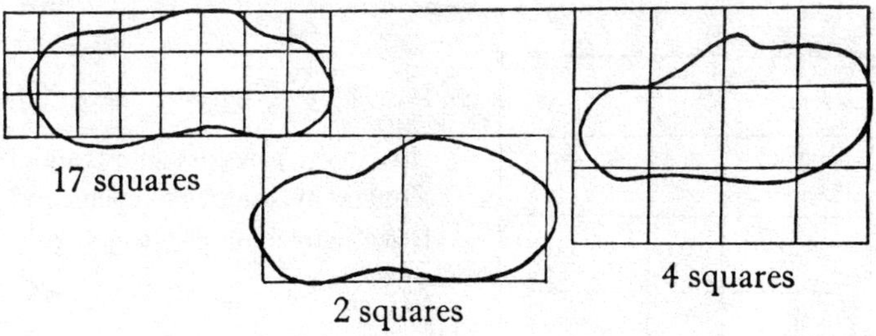

From this you will see that it is essential that our squares are of a standard size and for convenience we use the square centimetre, square metre and square kilometre, written cm^2, m^2 and km^2.

Small areas like the page of this book would be measured in square centimetres. Your classroom could be measured in square metres. A country is measured in square kilometres.

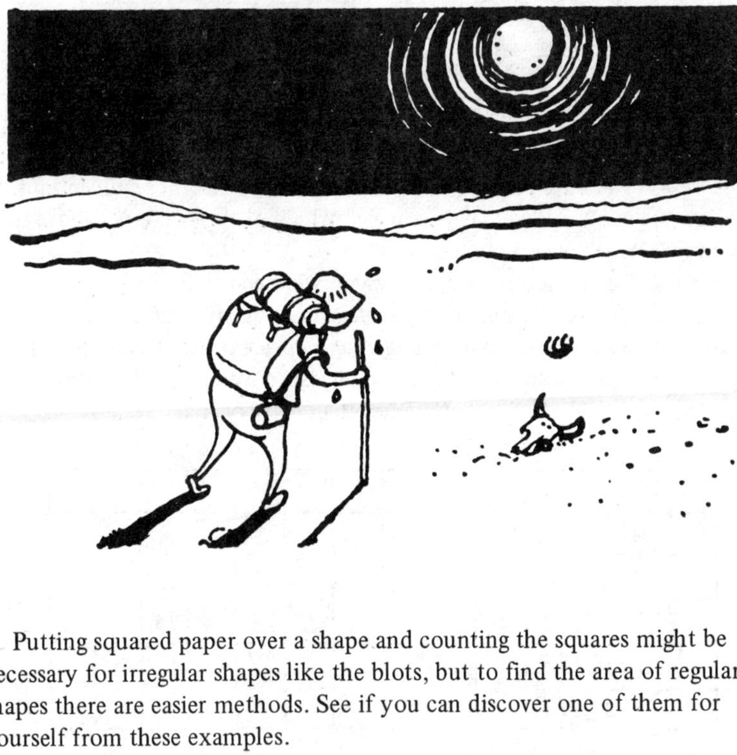

Putting squared paper over a shape and counting the squares might be necessary for irregular shapes like the blots, but to find the area of regular shapes there are easier methods. See if you can discover one of them for yourself from these examples.

Draw a rectangle 6 cm. by 4 cm. and divide it up into 1 cm. squares.

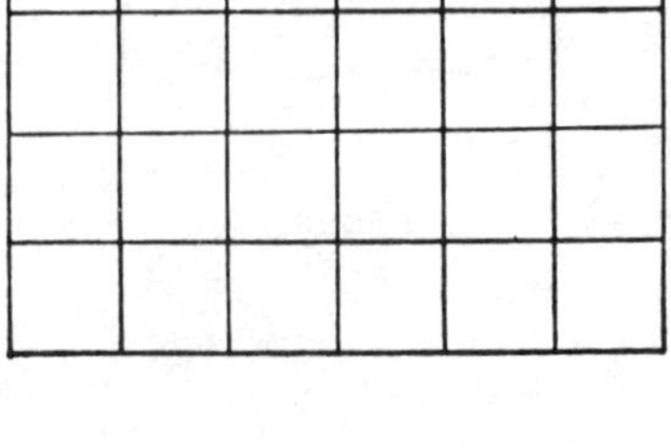

How many rows of squares are there?
How many squares in each row?
How many squares altogether?

This rectangle is 5 cm. by 4 cm.

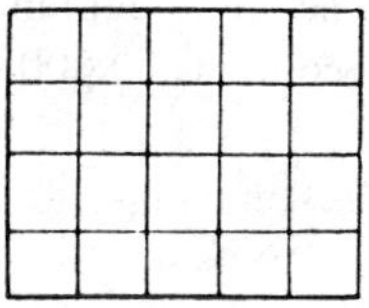

There are 4 rows of squares.
There are 5 squares in each row.
There are 20 squares altogether.

As each square is 1 square cm. the area of this rectangle is 20 square centimetres.

EXERCISE 61

Draw these figures, divide them into squares and work out their area.

1. 5 cm. x 3 cm.
2. 3 cm.x 3 cm.
3. 6 cm.x 2 cm.
4. 8 cm. x 9 cm.
5. 12 cm. x 4 cm.
6. 4 cm. x 2 cm.
7. 2 cm x 2 cm.
8. 5 cm. x 4 cm.
9. 12 cm. x 9 cm.
10. 7 cm. x 8 cm.

The total square units contained in these rectangles can be found by counting the squares but have you found an easier way? I expect that most of you have discovered that to multiply the number of squares along the longer side by the number of squares along the shorter side is a quicker way of counting them.

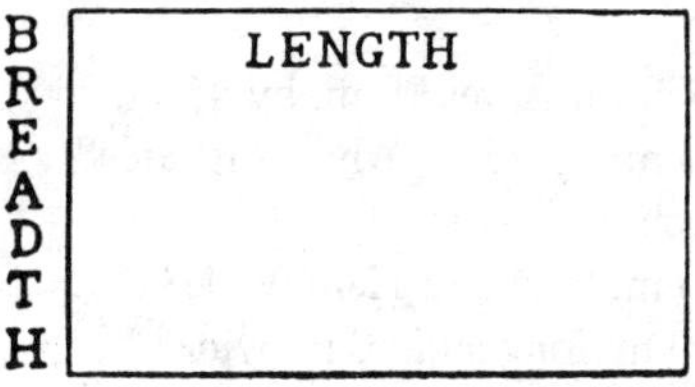

For convenience we call the longer side of the rectangle the length and the shorter the breadth. From this we have the rule:

To find the area of a rectangle multiply the length by the breadth.

In shorter form A = lb.

You will see that in the above exercise we did not need to draw the rectangles. We could have set our problems out like this.

EXAMPLE 1 Length = 5 cm. breadth = 3 cm.

$$A = lb.$$
$$= 5 \times 3 \text{ cm}^2$$
$$= \underline{15 \text{ cm}^2}. \quad \text{Ans.}$$

EXERCISE 62

Find the areas of these rectangles.

1. 6 cm. x 4 cm.
2. 7 cm. x 3 cm.
3. 9 cm. x 4 cm.
4. 6·5 cm. x 4 cm.
5. 12 cm. x 10 cm.
6. 12 cm. x 12 cm.
7. 6 cm.x 4·5 cm.
8. 6 cm. x 3 cm.
9. 10 cm. x 10 cm.
10. 8 cm. x 6 cm.

We have solved problems containing the square centimetre but what if we have bigger surfaces to measure? The square decimetre contains $10 \times 10 = 100$ square centimetres. A square metre contains 10 000 square centimetres.
A square metre contains 1 million square millimetres.

EXERCISE 63

Find the area of these rectangles.

	LENGTH	BREADTH	ANSWER IN
1.	6 cm.	4 cm.	cm^2
2.	8 m.	6 m.	m^2
3.	5 m.	3 m.	m^2
4.	6 dm.	2 dm.	dm^2
5.	10 m.	3 m.	m^2
6.	2 cm.	1 m.	cm^2
7.	3 cm.	2 mm.	cm^2
8.	9 m.	4 m.	m^2
9.	10 dm.	3 dm.	dm^2
10.	12 cm.	2 m.	m^2

11. Find the area of a football pitch which measures 90 m. by 45 m.
12. An Association Football pitch is 110 m. by 75 m. What is its area? How much bigger is this than the pitch in question 11?
13. What is the area of a netball pitch 30 m. long and 15 m. wide?
14. What is the area of a hockey pitch 90 m. long and 50 m. wide?
15. A tennis court is 24 m. by 11 m. What is its area?

While we have used the square centimetre and square metre for finding the area of various spaces, it might be necessary to find the area of a larger space and a larger unit would be needed. A farmer might say that the area of a field is 5 hectares.

1 hectare = 10 000 square metres.

We do not refer to square hectares, since it is not a measure of length like the metre, but of area.

$$1 \text{ are} = 100 \text{ m}^2$$
$$1 \text{ hectare} = 100 \text{ ares} = 100 \times 100 \text{ m}^2.$$

EXAMPLE 2

Find the area of the border.

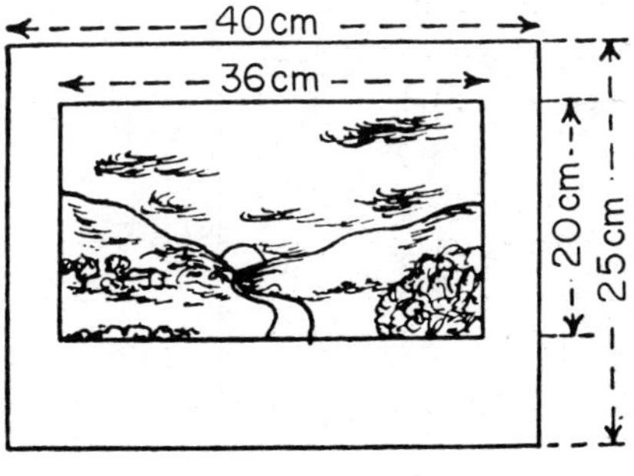

Area of border $= (LB) - (lb)$
$= (40 \times 25) - (36 \times 20)\text{cm}^2$
$= (1000 - 720)\text{cm}^2$
$= \underline{280 \text{ cm}^2}$ Ans.

EXAMPLE 3

Find area of figure.

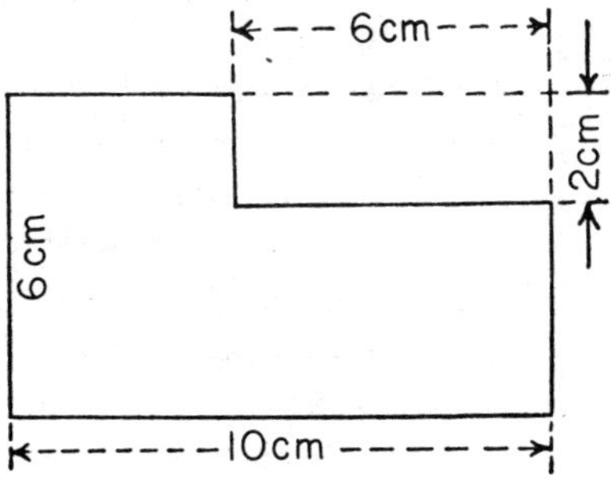

Area of figure $= (LB) - (lb)$
$= (10 \times 6) - (6 \times 2)\text{cm}^2$
$= 60 - 12 \text{ cm}^2$
$= \underline{48 \text{ cm}^2}$ Ans.

EXERCISE 64

Numbers 1 to 6 find the area of the border.

1.

2.

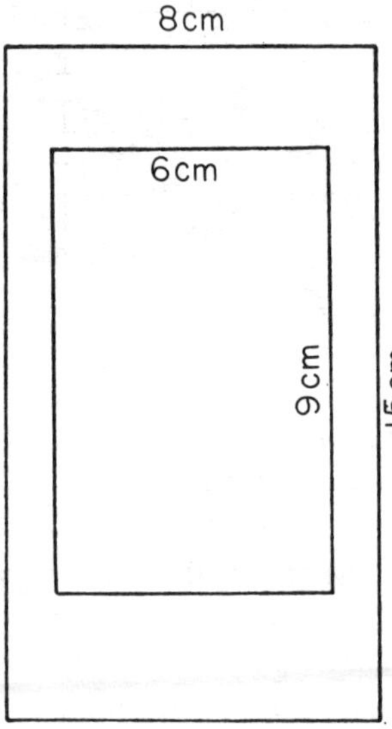

3.

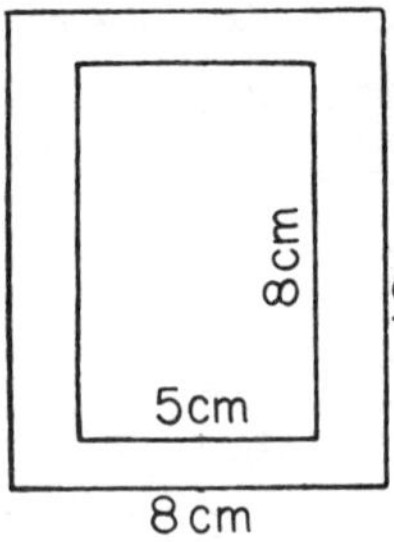

4.

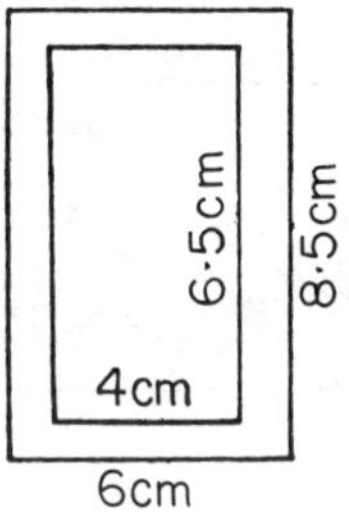

5.

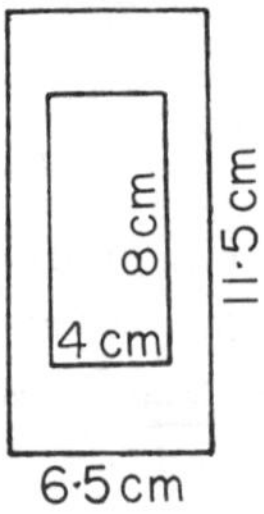

6.

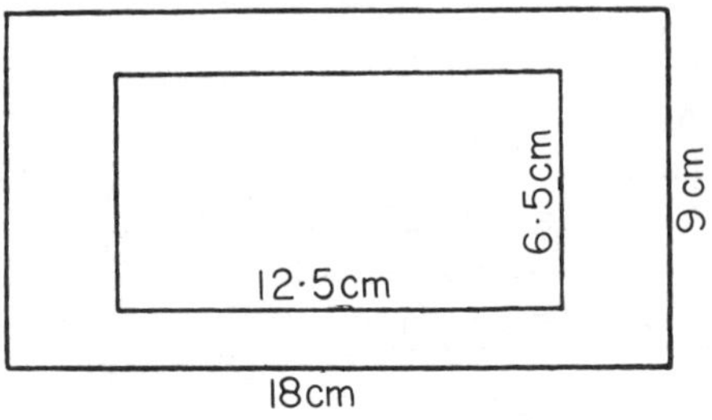

7. Find area of bricks.

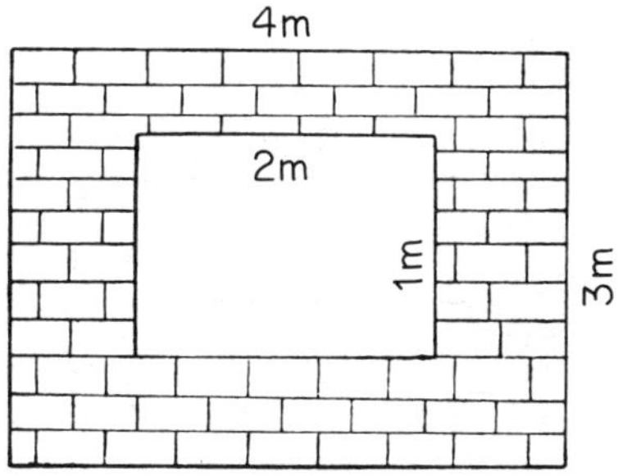

8. Find area of boards.

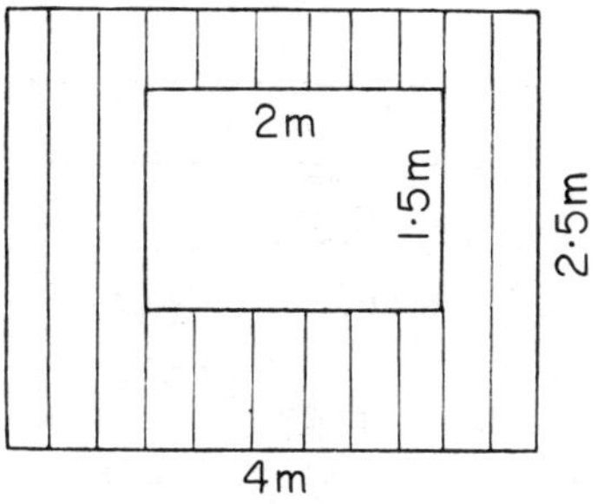

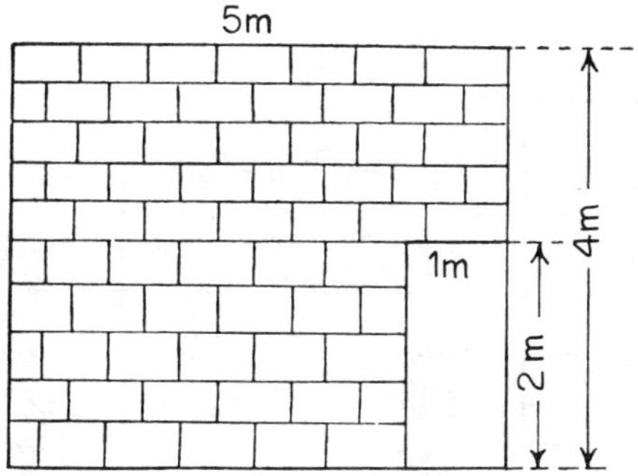

9. Find area of stone.

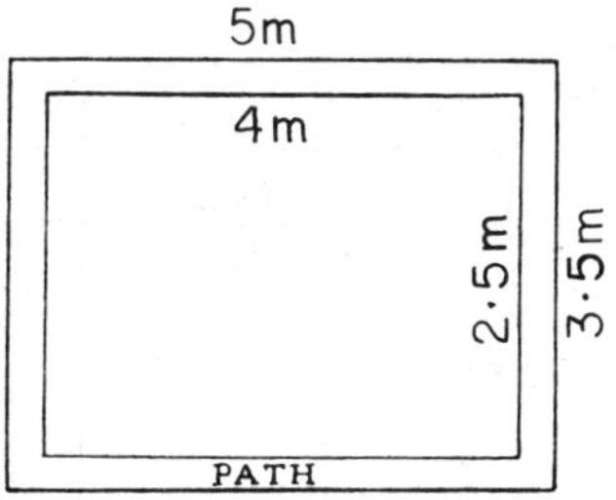

10. Find area of path.

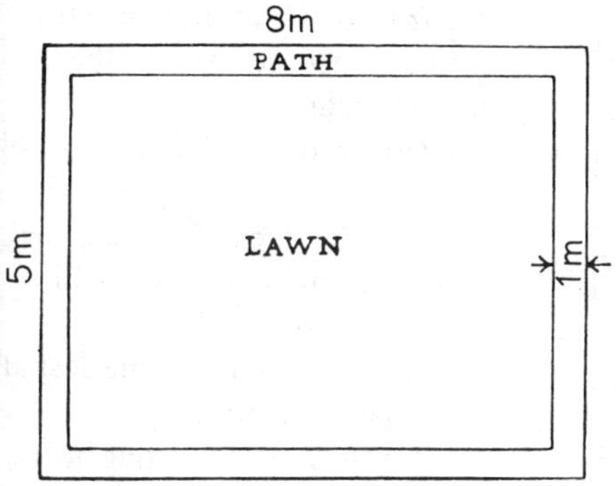

11. Find area of path if it is the same width all round.

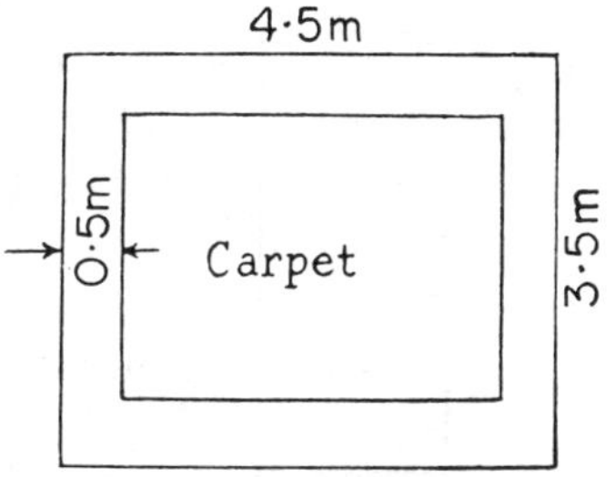

12. (*a*) Find area of carpet.
(*b*) Find area of border.

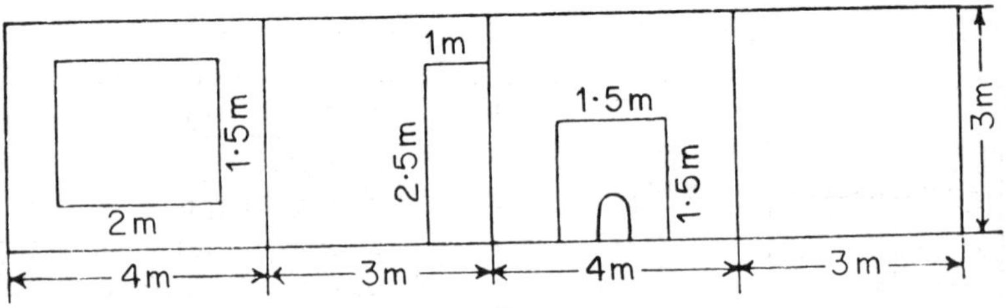

13. These are four walls of a room flattened out.

(*a*) What is the total area of the walls including the fireplace, window and door?

(*b*) What is the area of the brickwork?

14. Find area of glass.

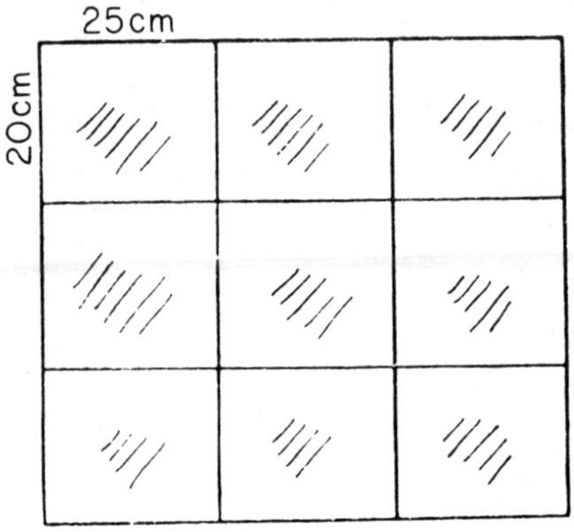

15. Find shaded area.

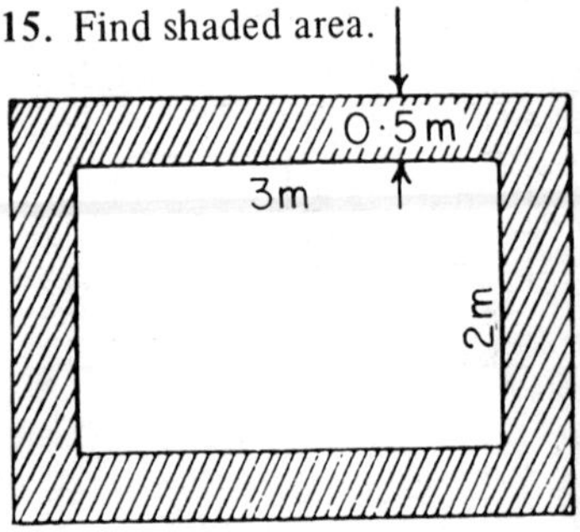

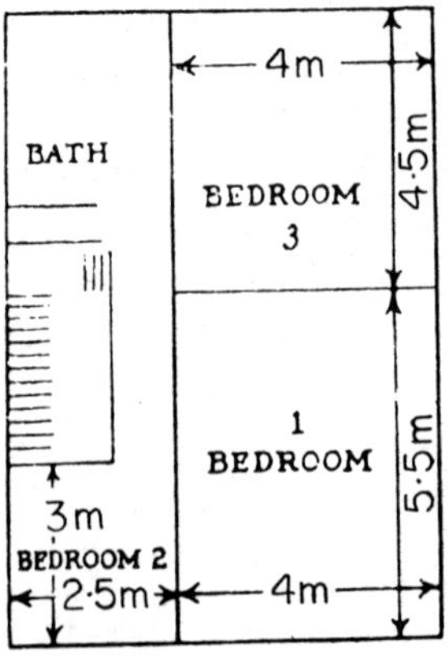

16. (*a*) What area of linoleum would be needed to cover the floor of bedroom 3?

(*b*) What would be the cost of material at $42\frac{1}{2}$p per sq. metre?

(*c*) What area of fitted carpet would be required to cover the floor of bedroom 1?

(*d*) What would be the cost at £1·62 per sq. metre?

(*e*) How much skirting board is required to go round bedroom 3 if the doorway is 1 m. wide?

17.

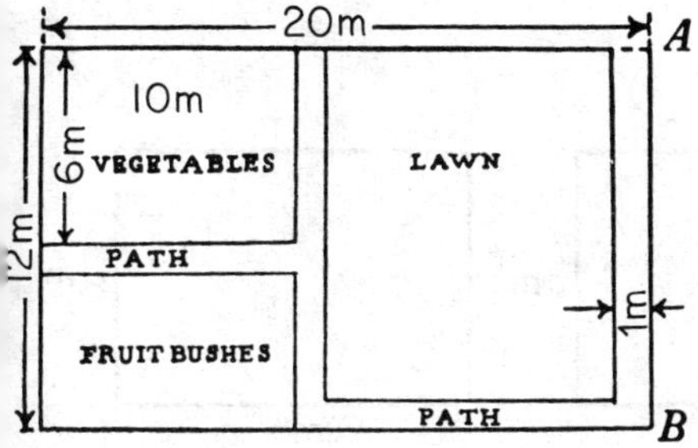

Find the area of

(*a*) The vegetable patch.

(*b*) The fruit garden.

(*c*) The lawn.

(*d*) The gravel path.

(*e*) How much fencing would be needed to fence from *A* to *B* round three sides?

ALL PATHS 1 M WIDE

18.

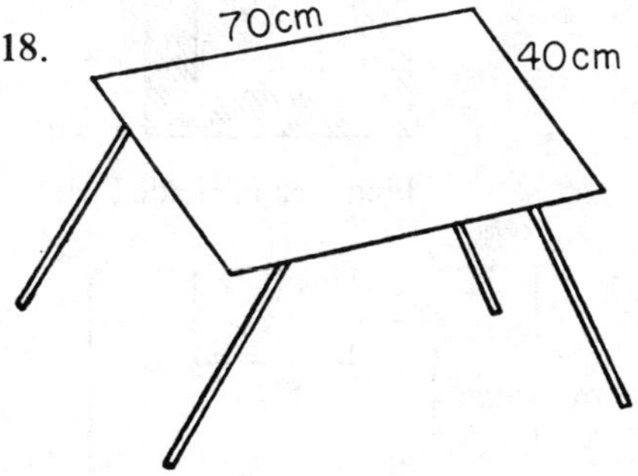

Find the area of timber required for the top of this table.

19.

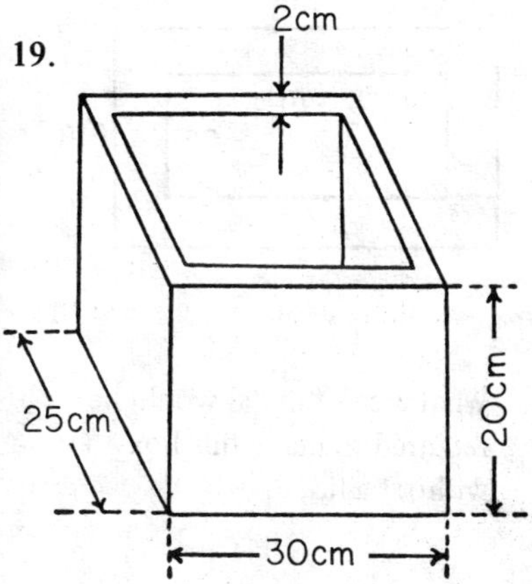

This figure shows an open box.

(*a*) What is its outside surface area?

(*b*) What is its inside surface area?

(*c*) What is the difference between the outside and inside areas?

EXERCISE 65

Find the area of the following.

1.

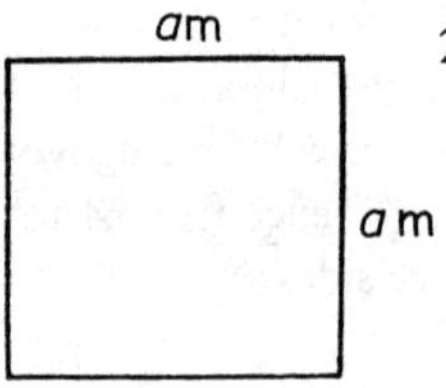

2.

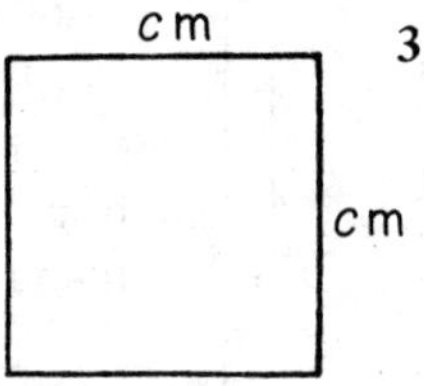

3.

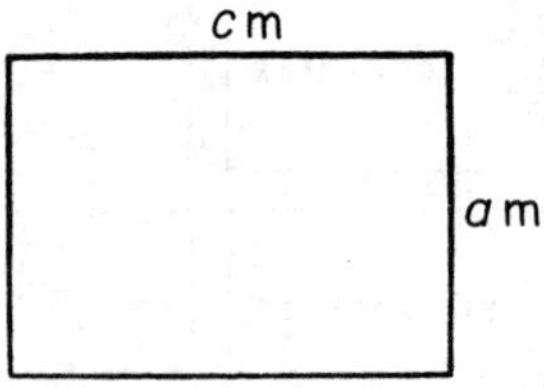

4.

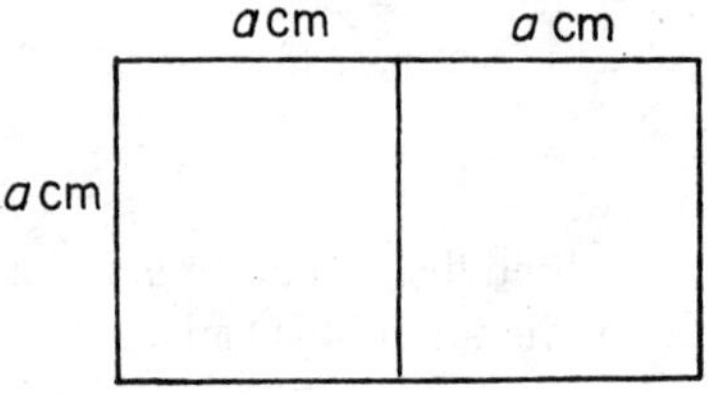

5.

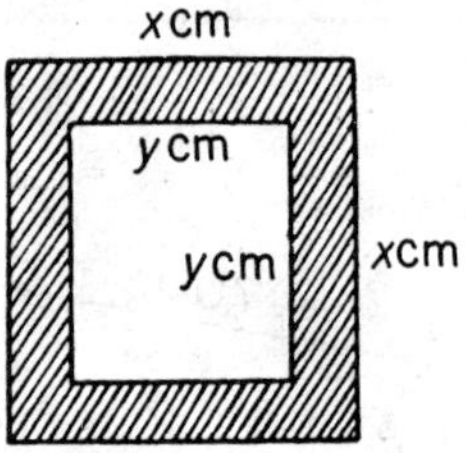

Find area of shaded part.

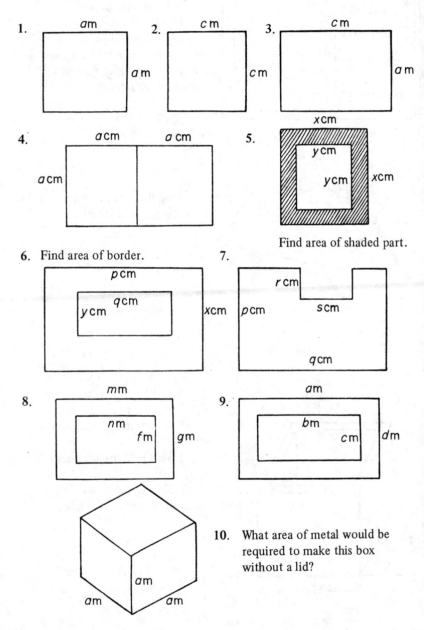

6. Find area of border.

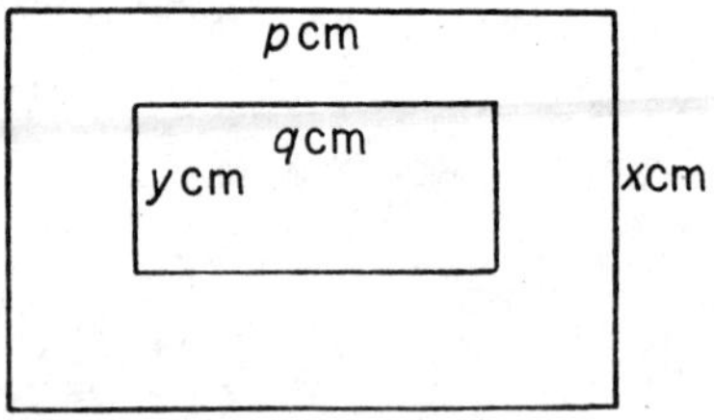

7.

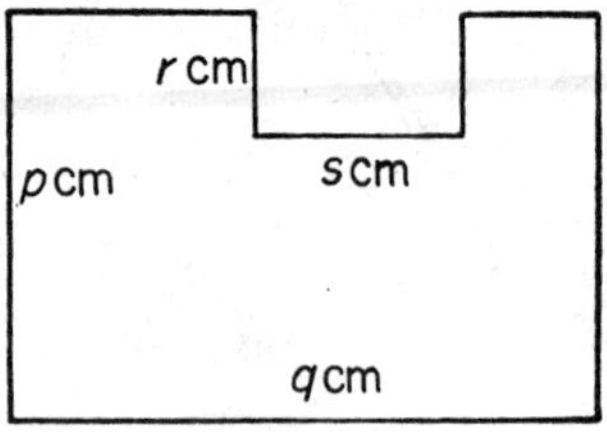

8.

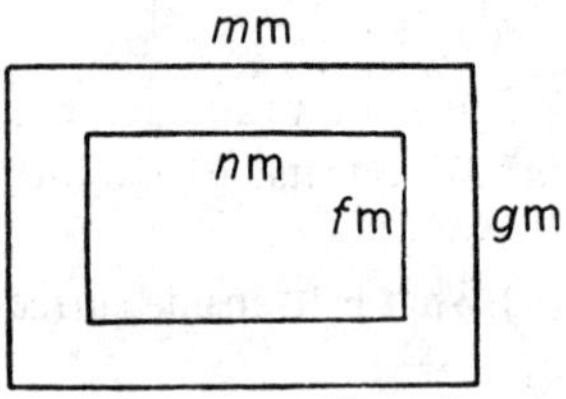

9.

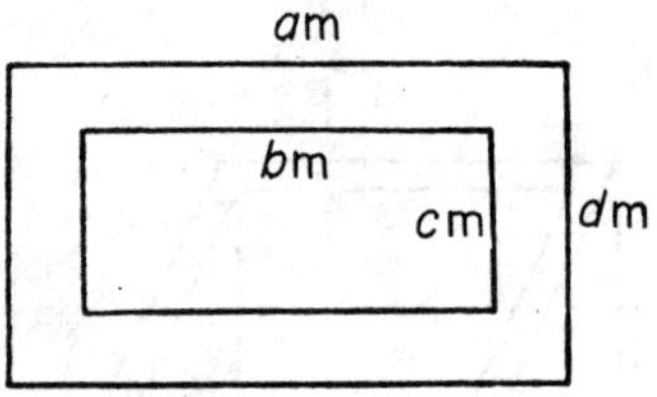

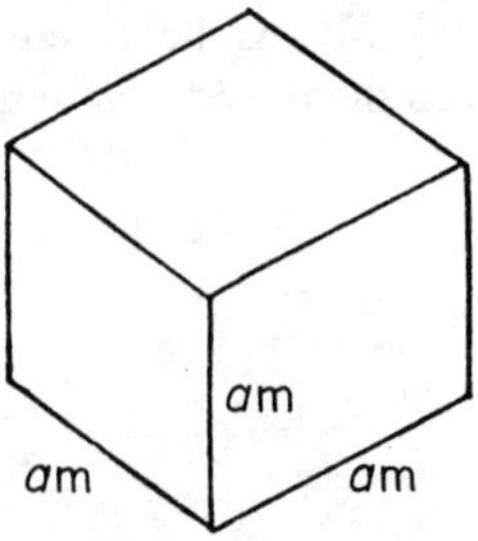

10. What area of metal would be required to make this box without a lid?

Triangle

On a piece of paper draw a rectangle 10 cm. by 8 cm.

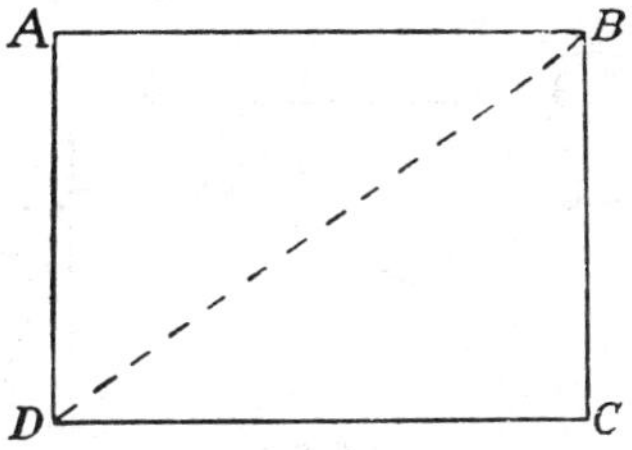

Cut along *BD*. What shapes have you made from the rectangle?

You will find that the triangles are equal in area so we can say that the area of one triangle is equal to half the area of the rectangle.

On another piece of paper draw another rectangle and mark as shown in this diagram.

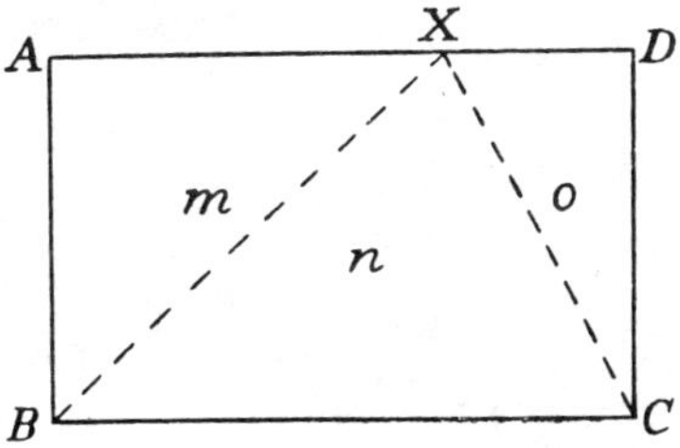

Cut along *XB* and *XC*. Have you halved the rectangle this time? You can check this by covering *n* with the pieces *o* and *m*.

The two triangles look like this.

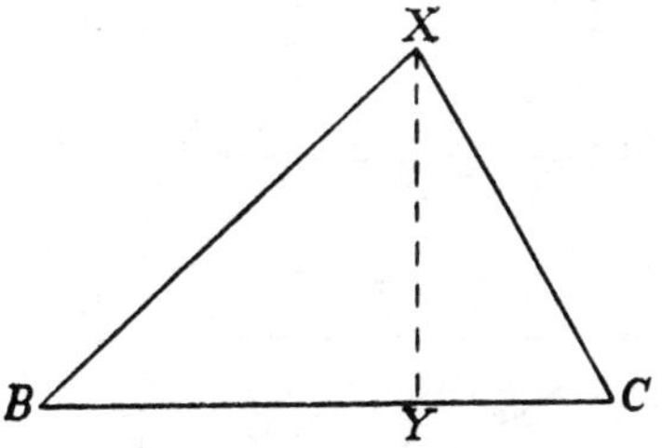

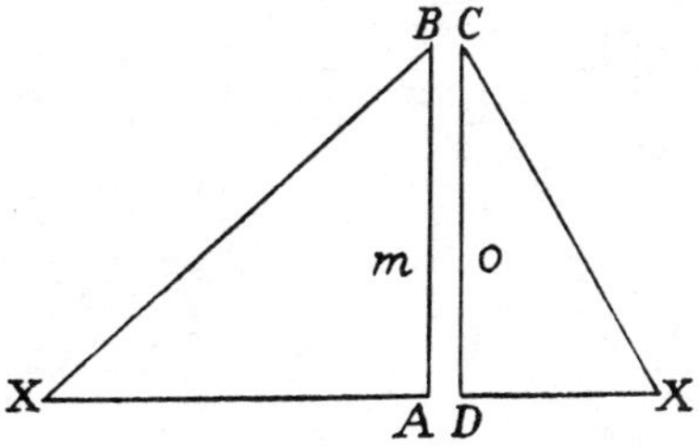

The area of the rectangle is *lb.* i.e. *BC* x *AB*

So the area of the triangle will be half of *BC* x *AB.*

However we do not refer to a triangle as having length and breadth but as having base and a height which is perpendicular to the base. In this particular example *BC* is the base and *AB* the height.

We now have the rule: Area of Triangle = $\frac{1}{2}$ base x height.

$= \frac{1}{2} bh.$

EXAMPLE 4

Base = 8 cm. Height = 4 cm.

$$= \frac{1}{\not{2}_{\,1}} \times \frac{8}{1} \times \frac{\not{4}^{\,2}}{1}$$

$= \underline{16 \text{ cm}^2}$ Ans.

EXAMPLE 5

Base $7\frac{1}{2}$ m. Height $1\frac{2}{3}$ m.
Area of triangle = $\frac{1}{2}bh$.

$$= \frac{1}{2} \times \frac{\overset{5}{\cancel{15}}}{2} \times \frac{5}{\underset{1}{\cancel{3}}}\ \text{m}^2$$

$$= \frac{25}{4}\ \text{m}^2$$

$$= \underline{6\tfrac{1}{4}\ \text{m}^2} \qquad \text{Ans.}$$

EXERCISE 66

Find the area of these triangles.

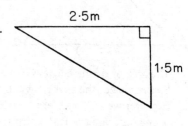

1.
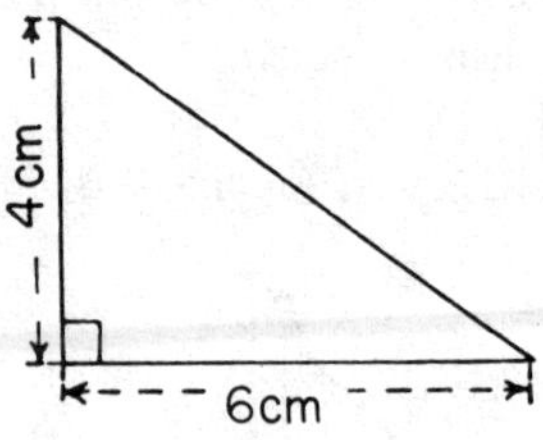

2.
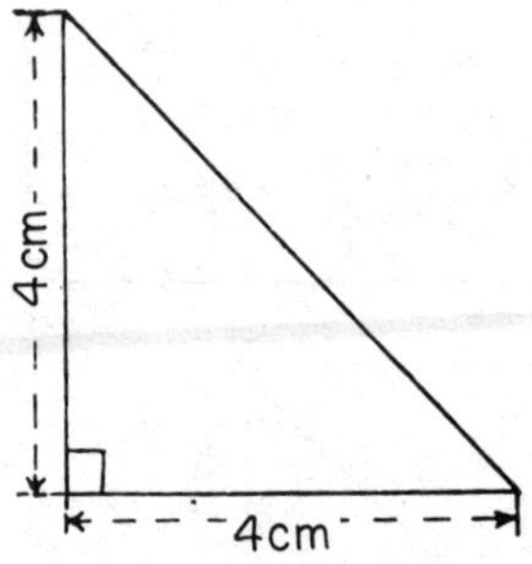

3.
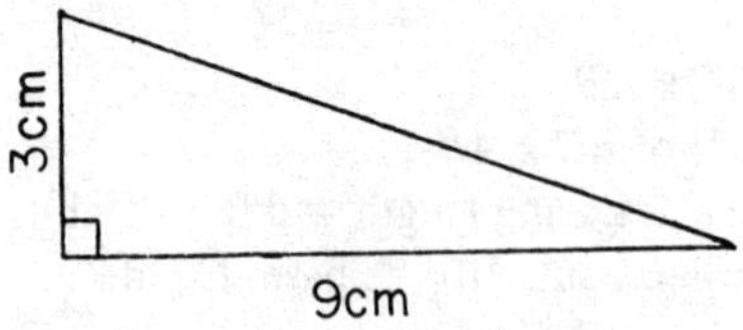

4.
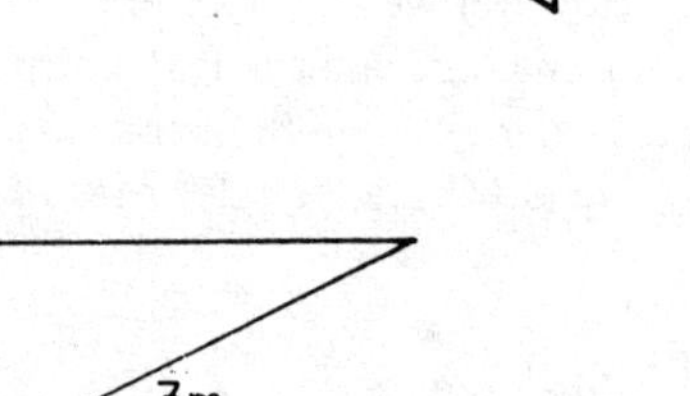

5.
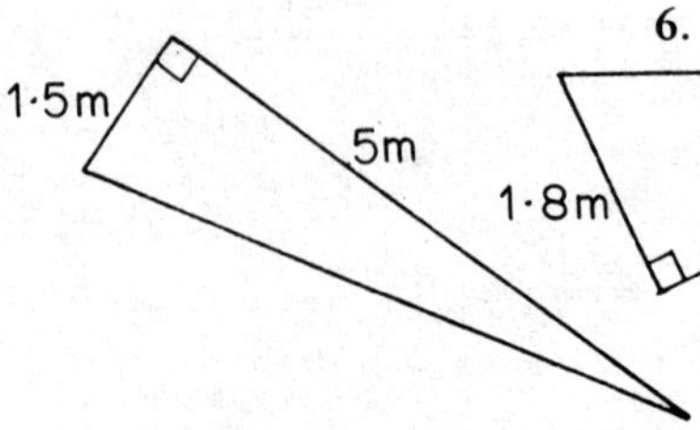

6.
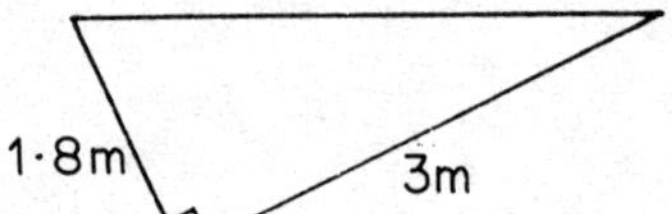

7.

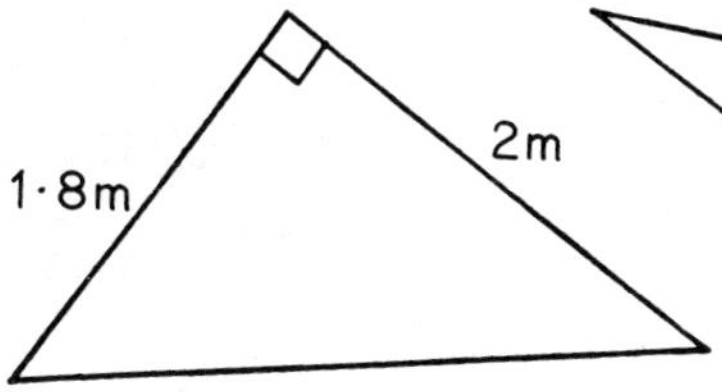

8.

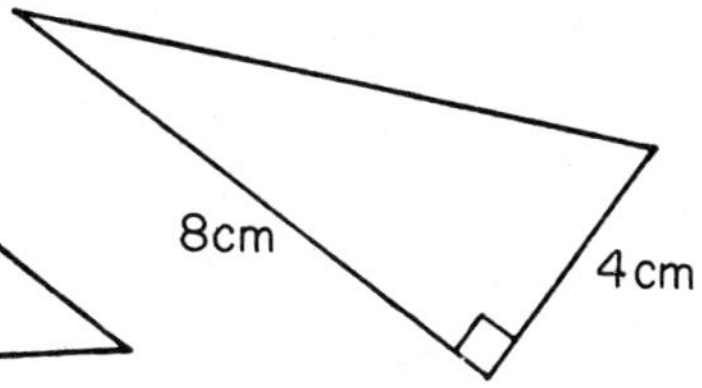

9.

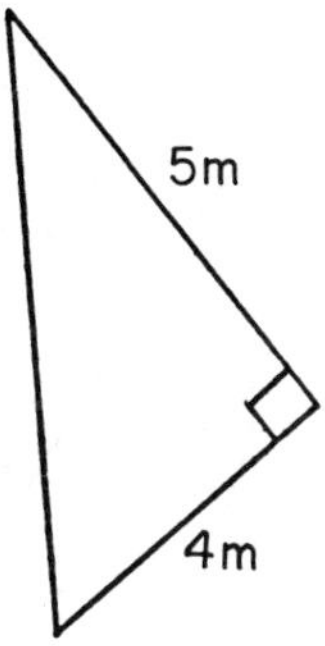

10.

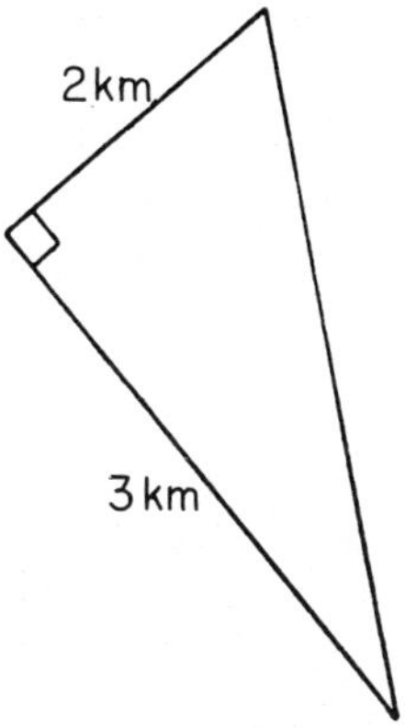

11.

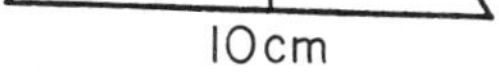

12.

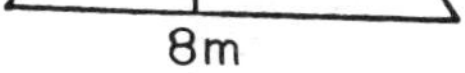

13.

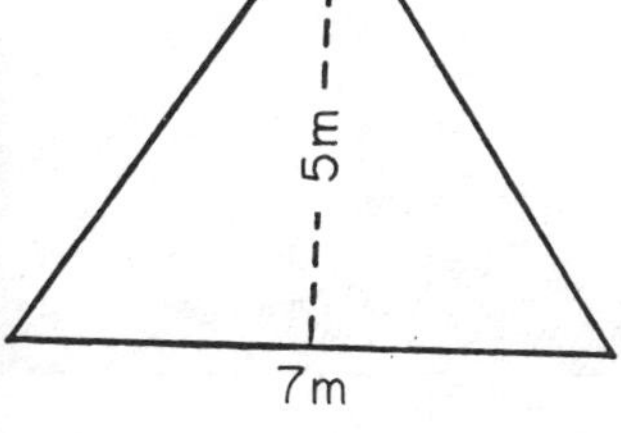

14.

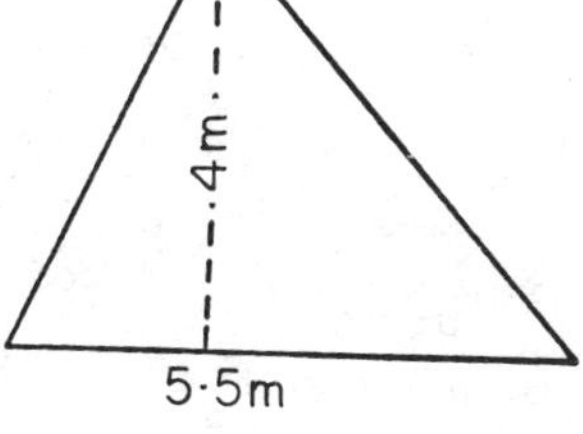

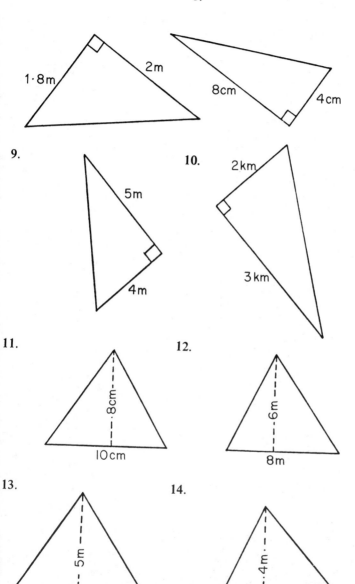

15.

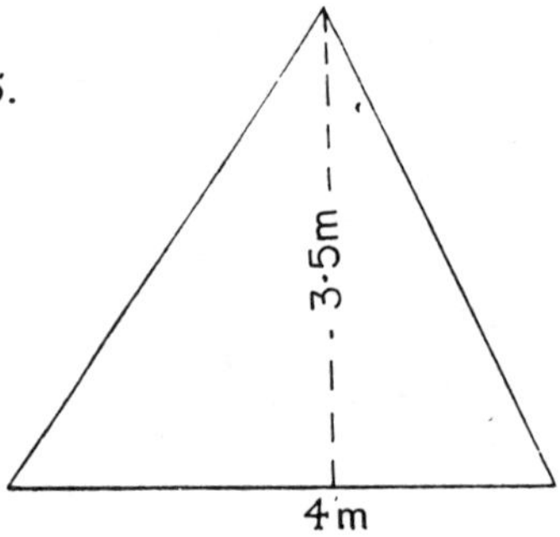

16.

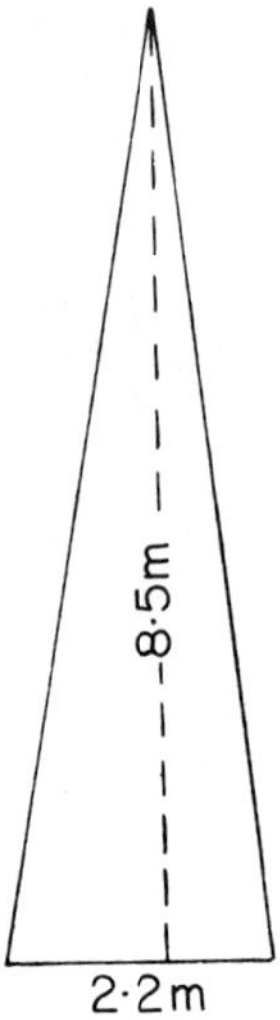

17.

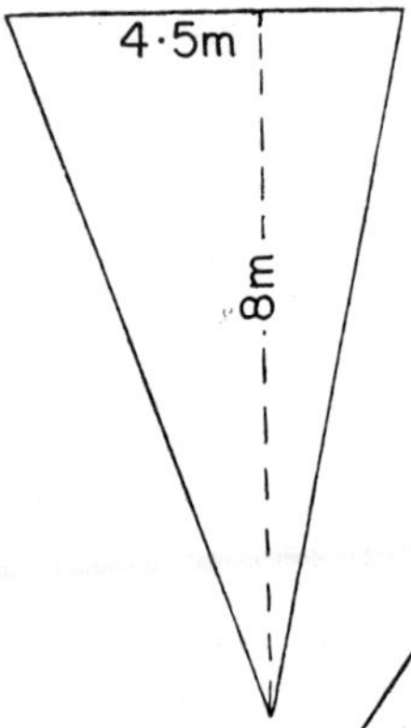

18.

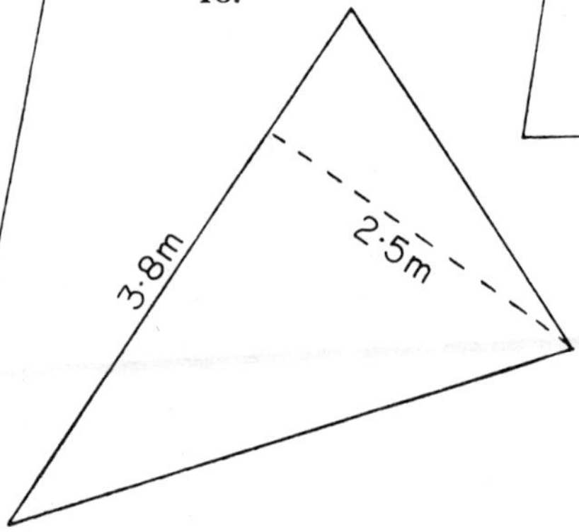

19.

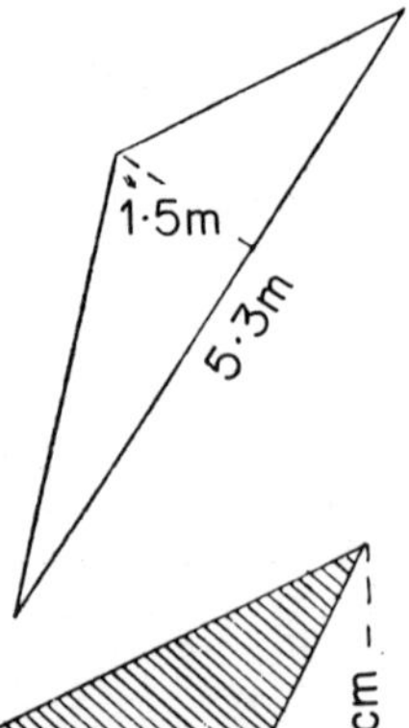

20.

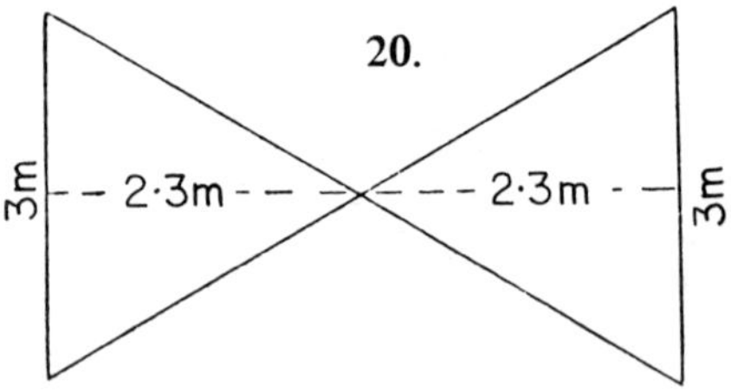

21.

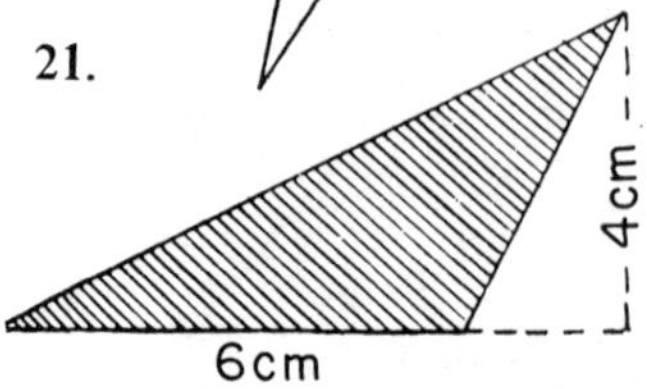

22.

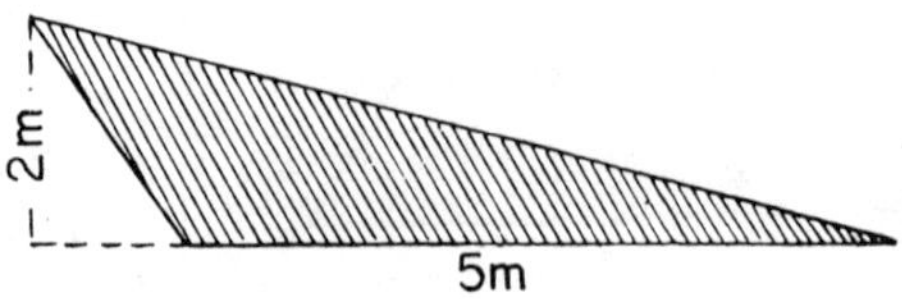

23.

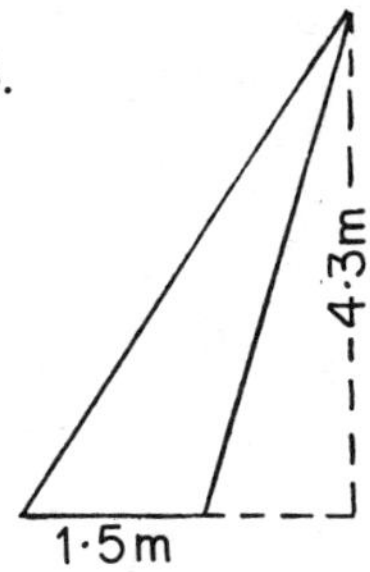

24.

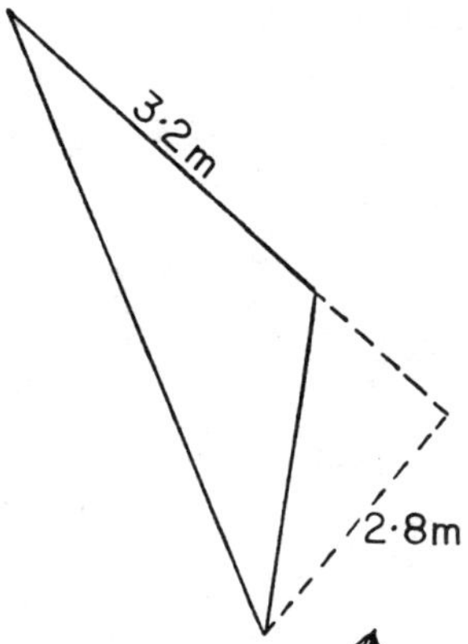

25.

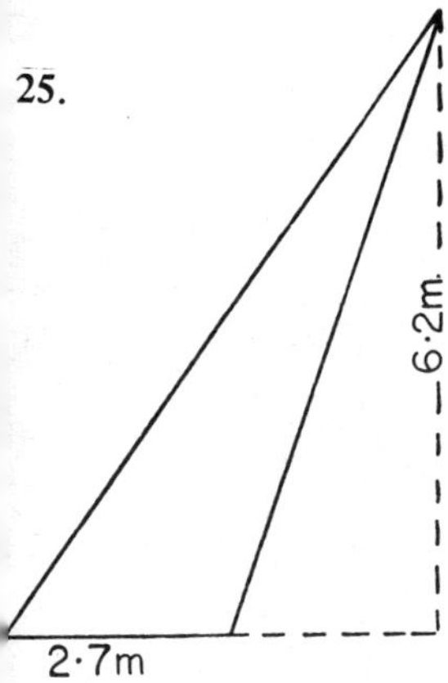

26.

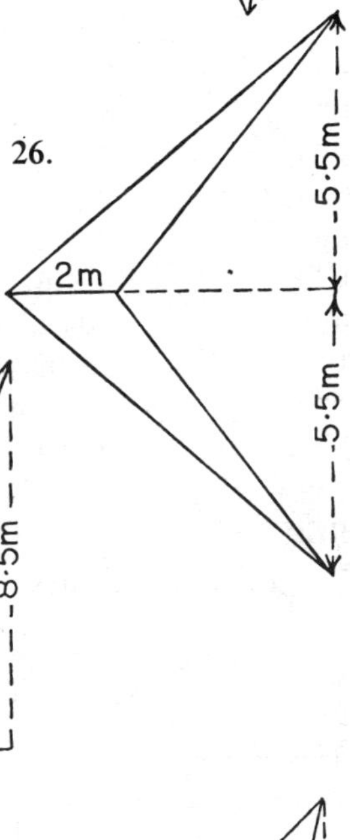

27.

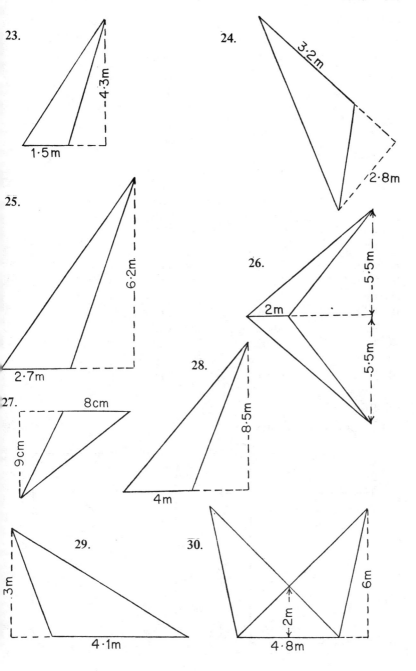

28.

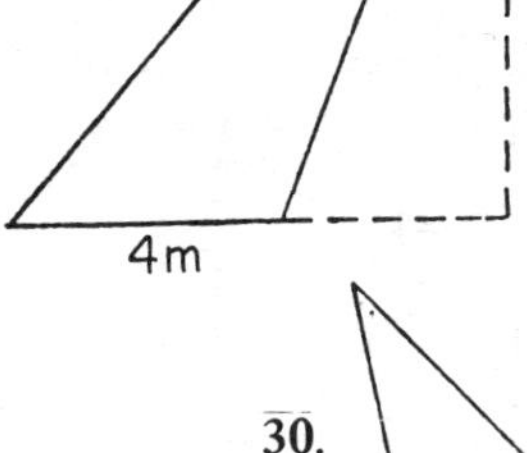

29.

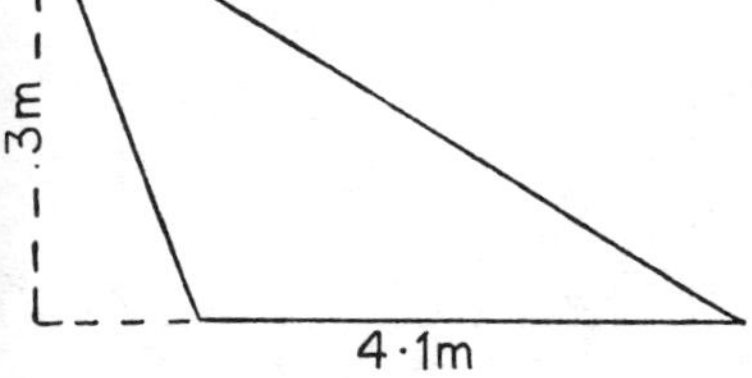

30.

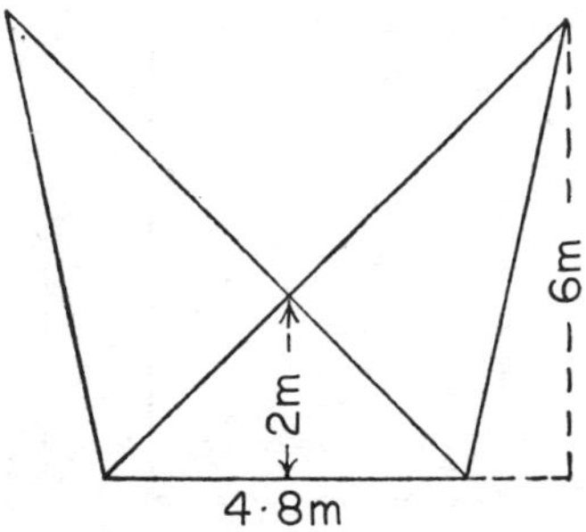

EXERCISE 67

Find the area of these triangles.

	BASE	HEIGHT	ANSWER IN
1.	3 cm.	15 cm.	cm^2.
2.	6 cm.	9 cm.	cm^2.
3.	5 cm.	8 cm.	cm^2.
4.	2 km.	1·42 km	km^2.
5.	30 m.	8 m.	m^2.
6.	15 m.	9 m.	m^2.
7.	9 m.	6·3 m.	m^2.
8.	4 dm.	8 cm.	cm^2.
9.	8 mm.	5 cm.	mm^2.
10.	4·5 cm.	1·8 cm.	cm^2.
11.	6·8 m.	2·1 m.	m^2.
12.	2·3 cm.	4·7 cm.	cm^2.
13.	3·5 m.	7 m.	m^2.
14.	8·3 m.	2·8 m.	m^2.
15.	1·6 m.	2·5 m.	m^2.
16.	4·3 m.	3·5 m.	m^2.
17.	50 cm.	70 cm.	cm^2.
18.	35 cm.	75 cm.	cm^2.
19.	1·1 km.	1·14 km.	km^2.
20.	2 km.	34 km.	km^2.

EXERCISE 68

Find area of these figures.

1.

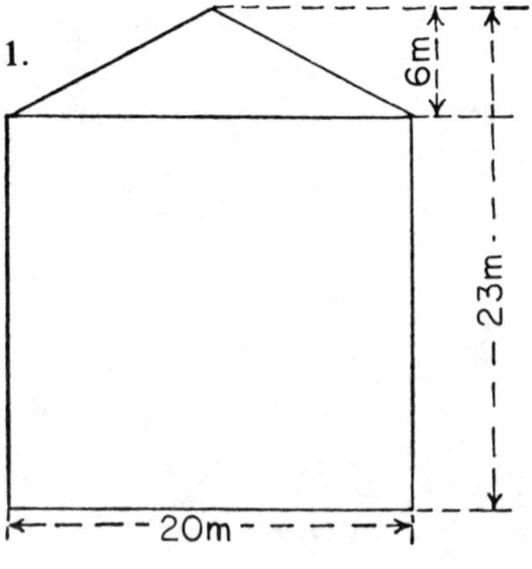

2.

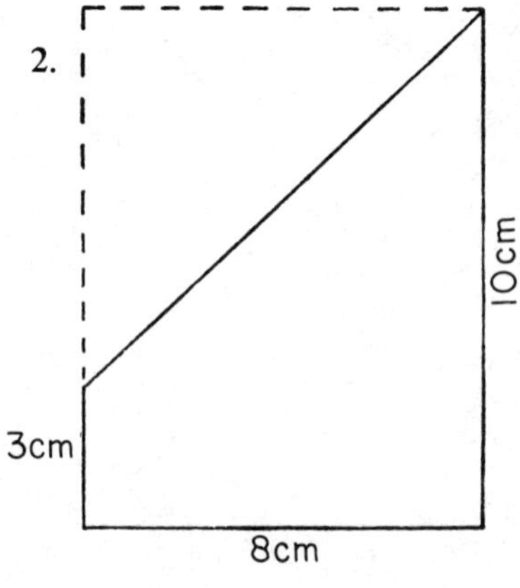

3. Area of shaded part.

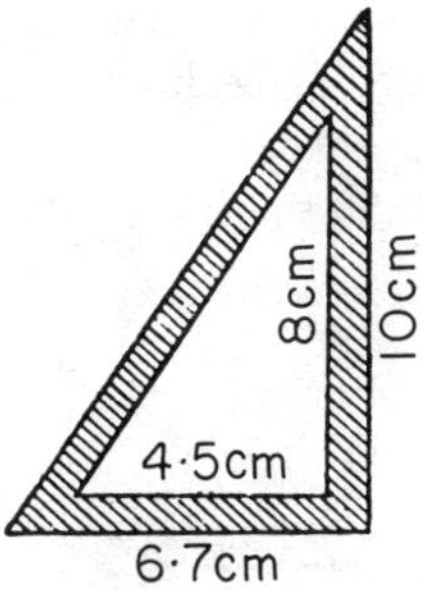

4. Shaded part.

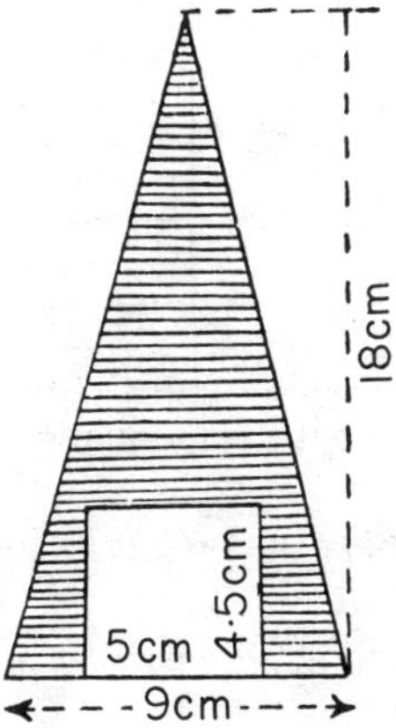

5.

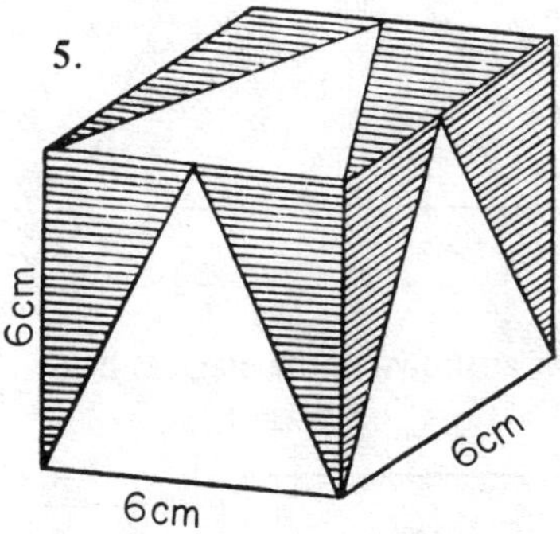

Shaded part.

6. Find area of shadow.

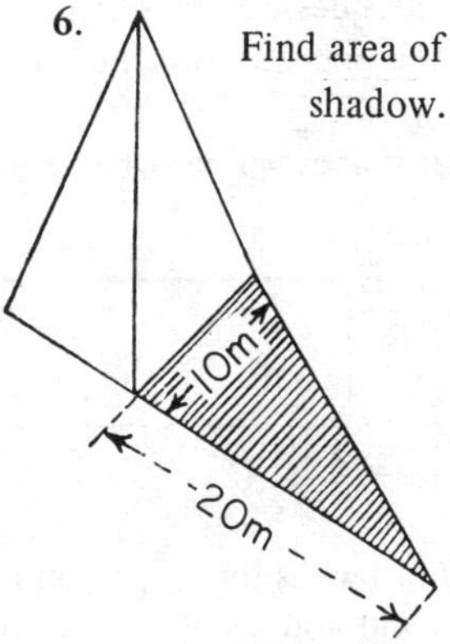

7.

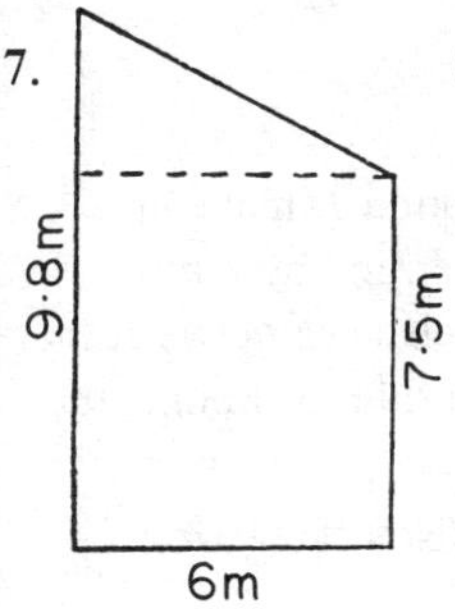

8.

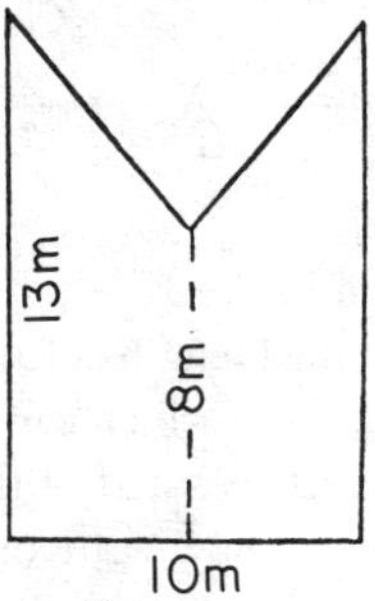

9.

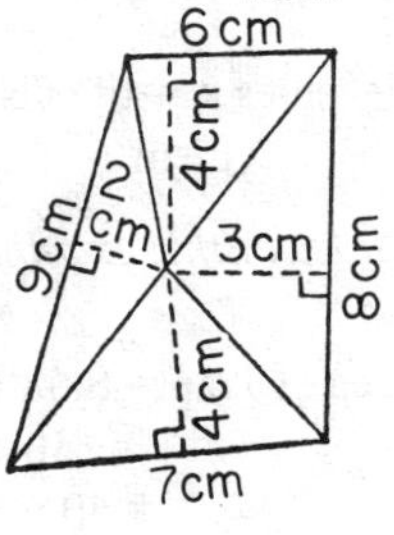

10.

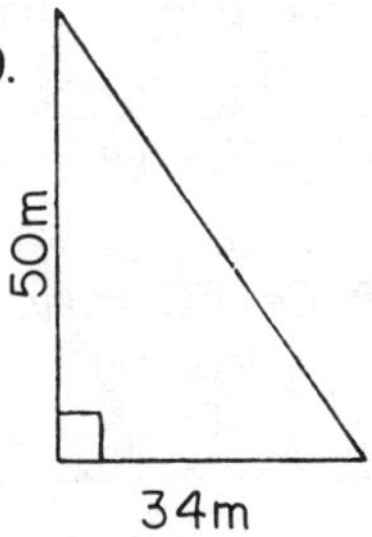

Trapezium

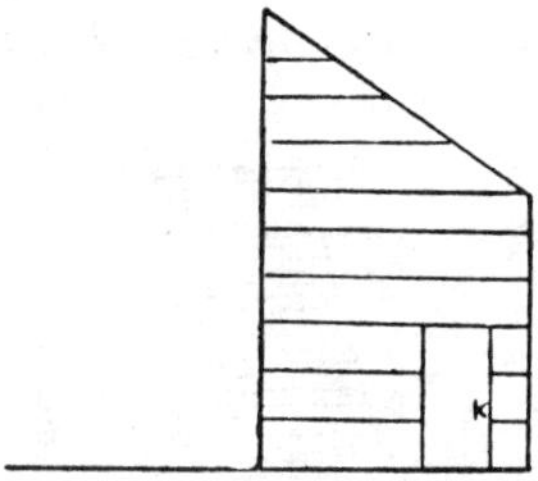

How many of you have seen a shed, garage or greenhouse with ends of this shape?

We call this shape a TRAPEZIUM.

Cut out a trapezium such as this.

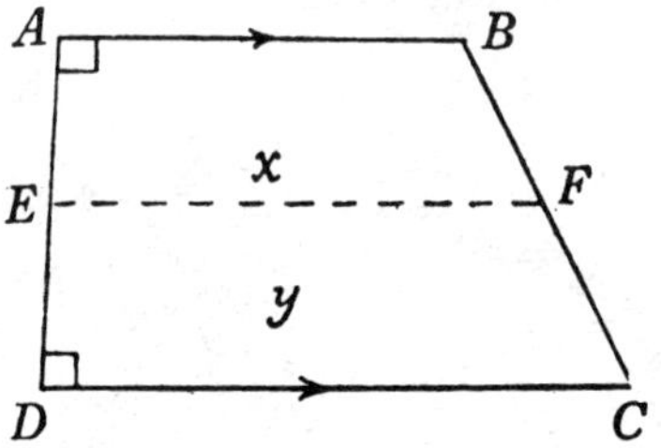

What can you say about the sides *AB* and *CD*?

Yes, they are parallel to each other.

Draw a dotted line halfway between the sides *AB* and *CD* and rotate the upper section *x* around the point *F*.

Your shapes will now be fitted together as shown in the diagram below.

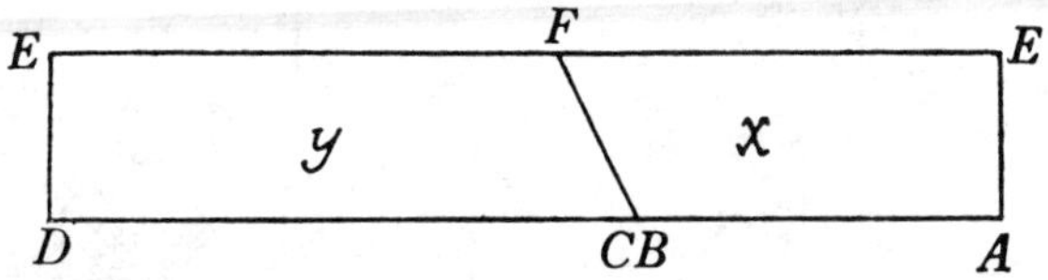

What shape is this? Can you find its area?

You will now see that the combined lengths of *AB* and *CD* make up the side *AD* in our rectangle. The area of which can be found by multiplying *ED* by *AD*. But the distance *ED* is a half of the distance between the parallel sides of our trapezium. We can now write out a rule for finding the area of a trapezium, i.e.

Area of Trapezium = Sum of parallel sides x $\frac{1}{2}$ the distance between them.

EXAMPLE 6 (Note that none of the sides in this example are at right angles to each other)

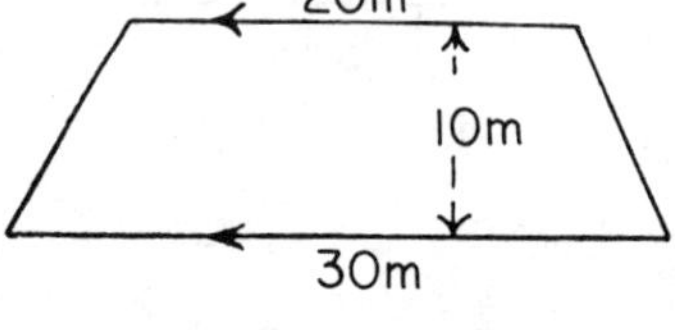

Area of Trap. = Sum of || sides x $\frac{1}{2}$ D

$= (30 + 20) \times \frac{1}{2} \times 10$

$= 50 \times 5\ \text{m}^2$

$= 250\ \text{m}^2$ Ans.

EXERCISE 69

1.

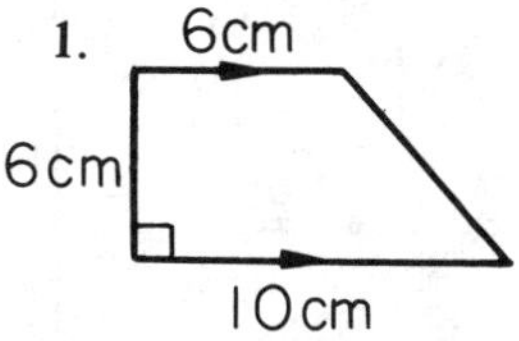

2.

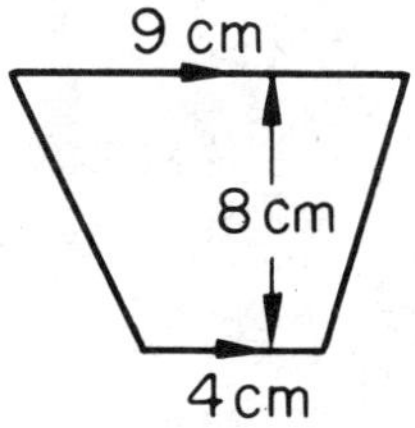

3.

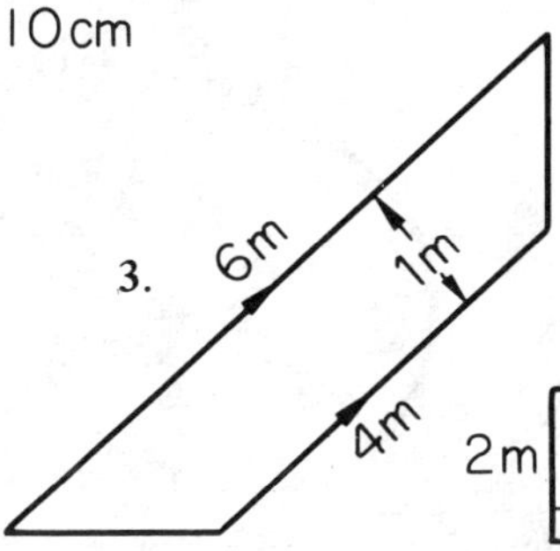

4.

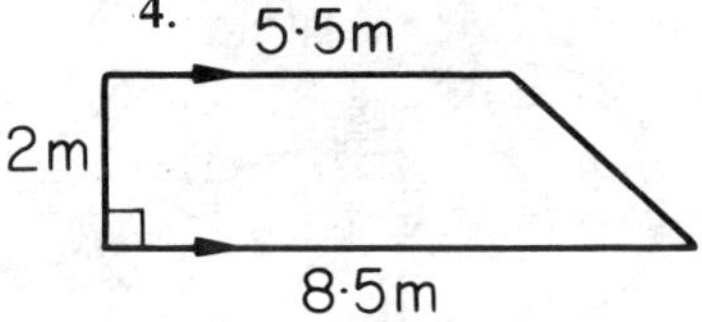

5.

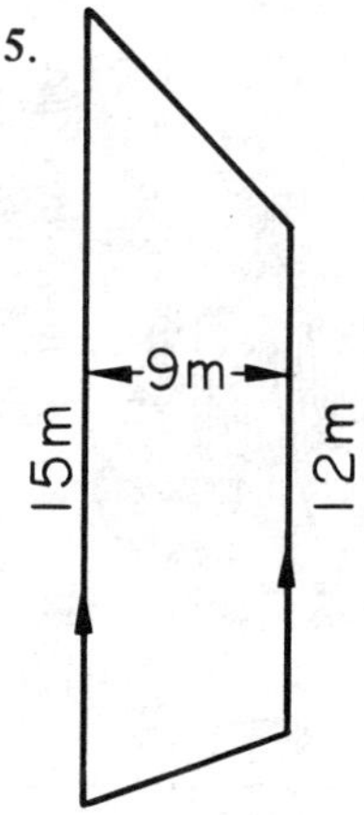

6.

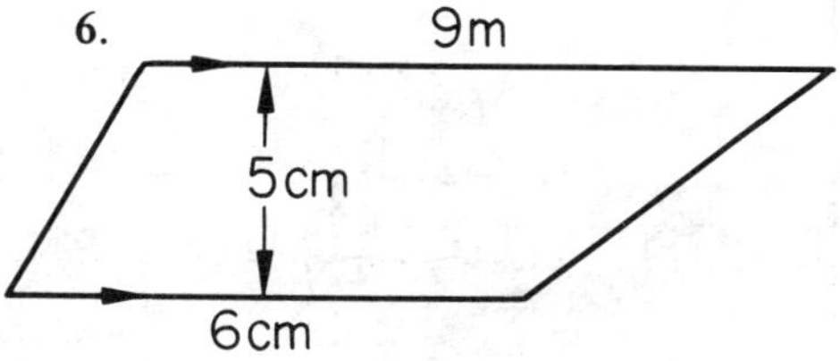

7.

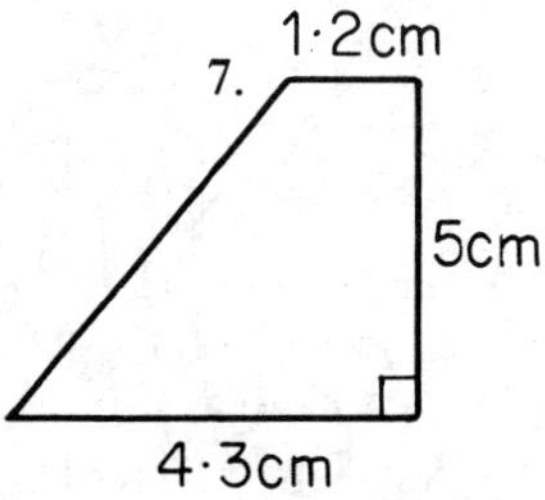

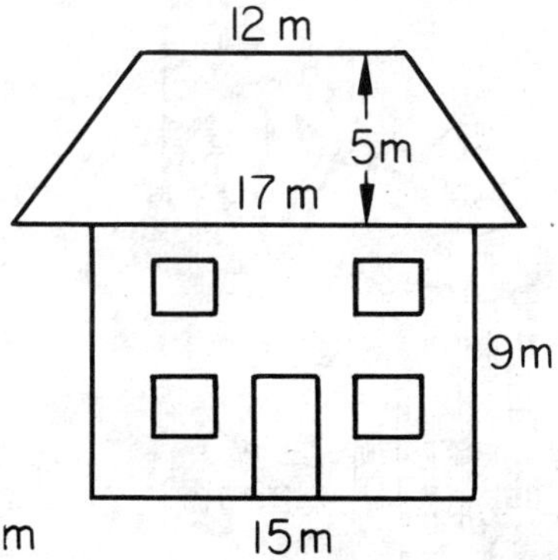

8. Find area of this picture of a house.

13 The Circle

On the opposite page are some examples of the circle. How can we measure the circle? The longest distance across it? The distance round it? The area of it?

Put a round tin on a flat surface and place rectangular blocks on each side of it as in this diagram. If we now measure the distance *AB* it will represent the greatest distance across the circle. This is known as the DIAMETER.

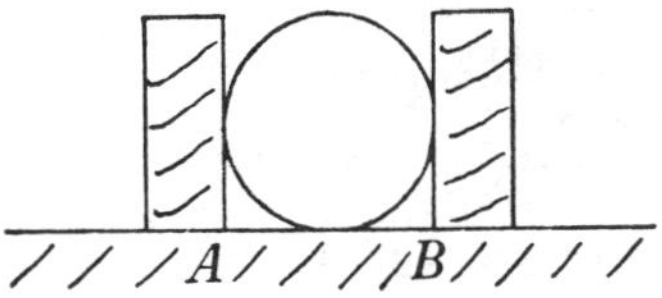

You can make another instrument to measure the diameter like this.

Fix two laths at an angle. Bisect the angle with another lath. The bisector may be graduated to take measurements across the centre of the circle.

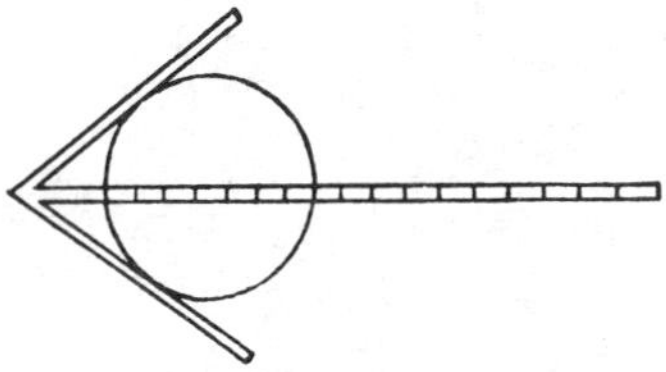

Make a classroom collection of 20 circular objects and measure their diameters by one or both of these methods.

How can we find the distance round the circles? Here is one method.

Place a bicycle wheel so that the tyre valve is nearest to the ground? mark this point.

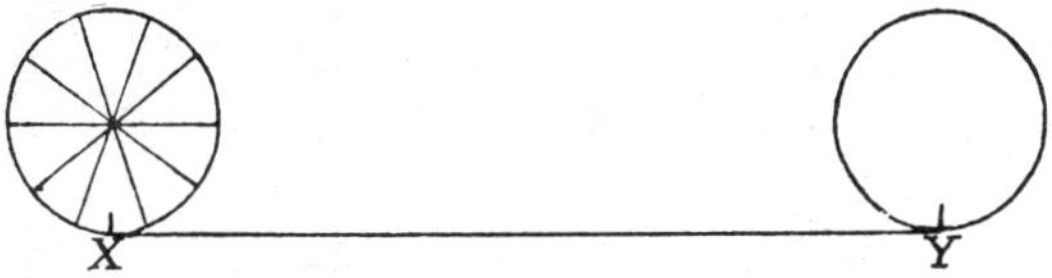

Roll the wheel until the valve is again next to the ground. Make another mark and measure the distance between the two marks. This is the distance round the wheel. The distance round the wheel is called the CIRCUMFERENCE. Distance round figures with straight sides is called the PERIMETER.

Now measure the circumference of the objects in your collection. Choose three of them and draw lines to represent the diameter and circumference of each.

________ Diameter of 1st circle.

____________________________________ Circumference.

________ Diameter of 2nd circle.
________________________ Circumference.
________ Diameter of 3rd circle.
________________________ Circumference.

In each case divide the circumference by the diameter and tabulate your answers. You will see that each time the diameter will go into the circumference just over three times. It goes approximately $3\frac{1}{7}$ times (3·14). This is an important ratio which we frequently use in problems involving the circle.

We call this ratio Pi and give it the Greek symbol π.

$$\pi = \frac{C}{d}$$

$$\therefore\ C = \pi d$$

But the diameter is equal to 2 radii

$$\therefore\ C = 2\pi r$$

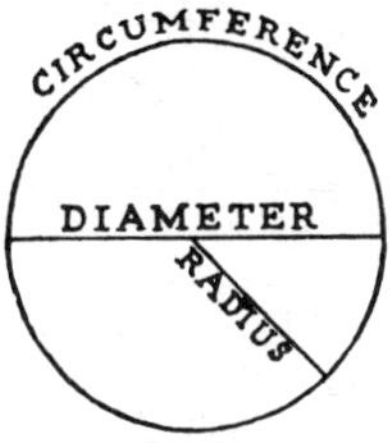

EXAMPLE 1

The diameter of a circle is 14 cm. What is its circumference?

$d = 14$ cm.

$C = \pi d$

$$= \frac{22}{\not{7}_{1}} \times \frac{\not{14}^{2}}{1} \text{ cm.}$$

$= \underline{44 \text{ cm.}}$ Ans.

$r = 7$ cm.

$C = 2\pi r$

$$= \frac{2}{1} \times \frac{22}{\not{7}_{1}} \times \frac{\not{7}^{1}}{1} \text{ cm.}$$

$= \underline{44 \text{ cm.}}$ Ans.

EXERCISE 70

Find the circumference of these circles.

1.	d = 28 cm.	11.	r = 35 m.
2.	14 cm.	12.	28 m.
3.	7 cm.	13.	14 m.
4.	3·5 cm.	14.	7 m.
5.	1·75 cm.	15.	3·5 m.
6.	0·875 m.	16.	2·1 m.
7.	1·4 m.	17.	10·5 cm.
8.	70 cm.	18.	4·2 m.
9.	49 cm.	19.	84 cm.
10.	56 cm.	20.	2·8 m.

EXERCISE 71

($\pi = 3\frac{1}{7}$ unless told otherwise)

1. The diameter of a roller is $2\frac{5}{8}$ cm. What distance does a point on the circumference move in three revolutions?
2. The diameter of the roller of the lawn mower is 14 cm. How many turns does it make in mowing a strip of lawn 6·60 m. long?
3. The diameter of a bicycle wheel is 56 cm. How far does it travel in 10 revolutions?
4. The diameter of the roller of a steam roller is 1·4 m. How much road does it roll in 8 revolutions?
5. A motor-cyclist rides 20 laps on Wall of Death, the diameter of which is 28 m. What distance does he travel?
6. The radius of the Earth is (approx.) 6364 km. How far is the North Pole from the Equator?
7. The diameter of the axle in the winch to this well is 35 cm. How many times must the handle be turned to draw up the bucket if the depth to water is 5 m?
8. A cricket pitch is 20 m. long. How many revolutions would be made by a roller, with a radius of 50 cm., in rolling the pitch?
9. The diameter of a bicycle wheel is 70 cm. How many revolutions will it make in travelling a distance of 1 km?

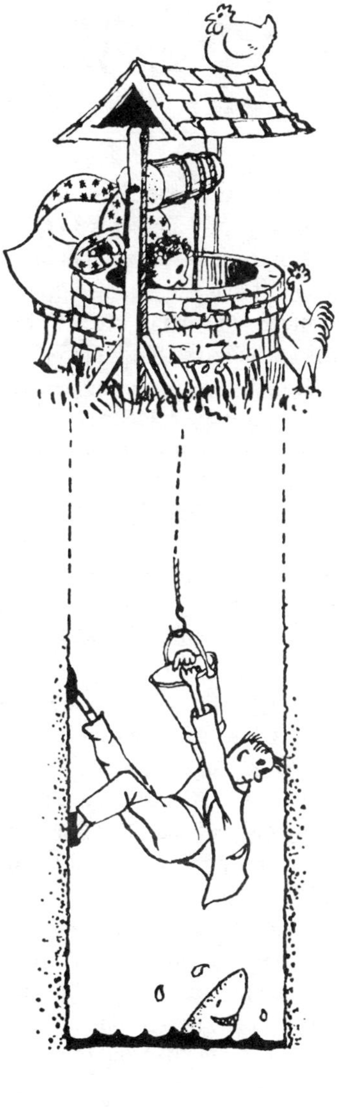

10. An old penny rolls a distance of 1 m. How many revolutions would it make if its diameter is 3 cm.
11. The diameter of an old halfpenny is 2·5 cm. How many revolutions would it make in rolling 6·28 m. ($\pi = 3{\cdot}14$)
12. If the Earth's radius is 6364 km., how long would it take to fly an aeroplane round the Equator at an average speed of 1600 km/h? (Give your answer to the nearest hour.)
13. A skirt is to be made of a circular piece of material with a radius of 77 cm. How much binding would be required to go round the bottom of the skirt? What would this cost at 9p per metre?
14. How much trimming should I need to put a double row round a Christmas cake, the diameter of which was 28 cm.

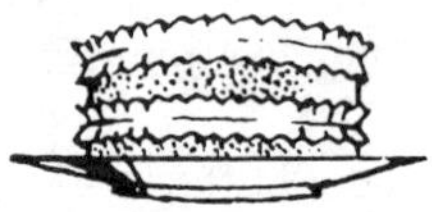

15. The minute hand of a clock is 7 cm. long. How far does the tip travel in 20 minutes?
16. Two boys wish to get to the opposite side of a circular pool, radius 112 m. One of them decided to swim across, while the other ran round the edge. How far did each have to go?
17. Cotton is wound round a reel of 3 cm. diameter 1000 times. Is this more or less than the 100 m. which the reel is supposed to hold?
18. Chairs are arranged around a circular stage 14 m. in diameter. How many people can sit on the front row if each chair takes up 50 cm. of the circumference?
19. How much trimming would be needed to finish off a lampshade whose upper radius is 10 cm. and lower radius is 15 cm? (To the nearest cm.) ($\pi = 3{\cdot}14$)
20. How many plants would be needed to set them at 8 cm. intervals round a circular garden plot with a radius of 2 m.

Draw a circle of 5 cm. radius, and cut it into four equal parts as shown in this diagram.

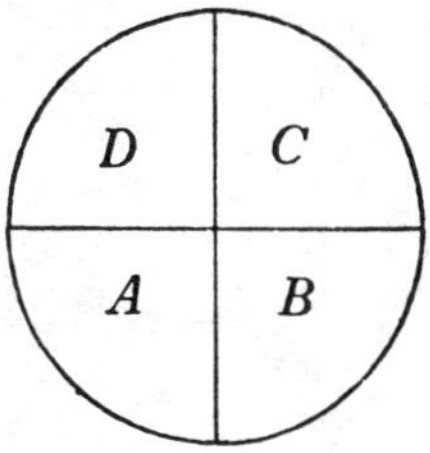

Place these parts together, i.e.

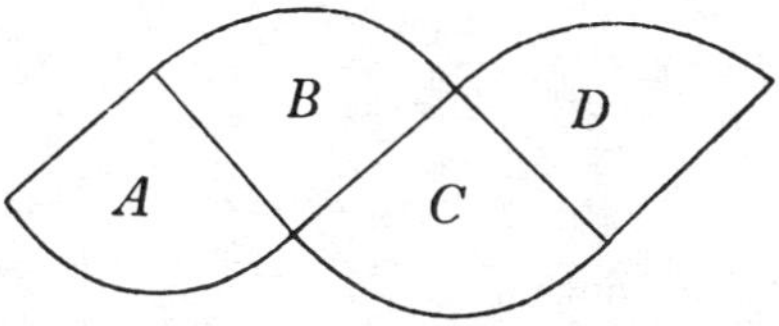

Cut a similar circle into 8 parts and arrange as follows.

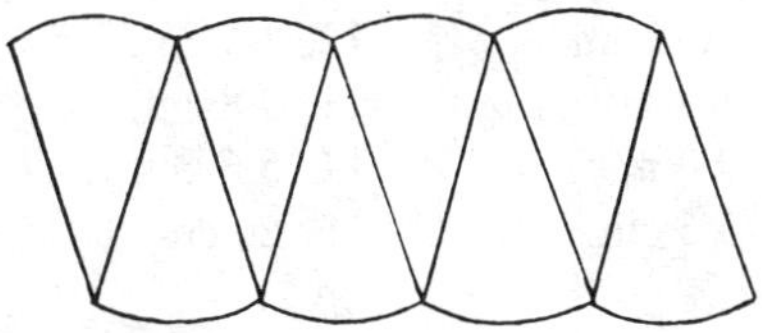

Cut a third circle into 16 equal parts, arrange as before.

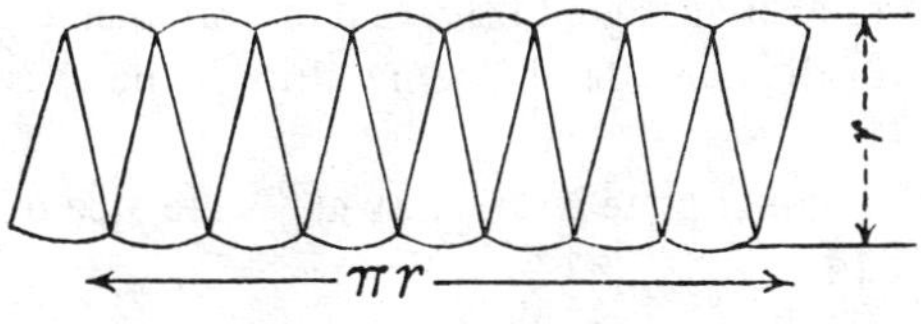

What general shape is slowly being formed by the parts of the circle which you have placed together?

What do you notice happening to the curved parts of the sections?

Yes, our curved parts are becoming less pronounced and are in fact approaching a straight line. The general shape of our mosaic is becoming a rectangle.

You will see that the length of this rectangle is equal to half the circumference of the circle, i.e. πr. The width of the rectangle is equal to the radius of the circle.

We can now say that the area of the rectangle which we have formed from the sections of the circle is $\pi r \times r$ or πr^2 units. Therefore the area of a circle is found by using this new rule:

$$\text{Area of circle} = \pi r^2$$

EXAMPLE 2

To find the area of a circle with a radius of 7 cm. ($\pi = 3\frac{1}{7}$)

$$\text{Area of circle} = \pi r^2$$

$$= \frac{22}{\not{7}_1} \times \frac{\not{7}^1}{1} \times \frac{7}{1} \text{ cm}^2$$

$$= \underline{154 \text{ cm}^2} \quad \text{Ans.}$$

EXERCISE 72

Find the areas of the following circles.

RADIUS ($\pi = 3\frac{1}{7}$)		DIAMETER ($\pi = 3{\cdot}14$)	
1. 35 m.	**6.** 1·75 cm.	**11.** 0·6 cm.	**16.** 5 cm.
2. 28 cm.	**7.** 10·5 cm.	**12.** 1·4 cm.	**17.** 3 cm.
3. 14 m.	**8.** 1·4 m.	**13.** 2·8 cm.	**18.** 24 mm.
4. 7 cm.	**9.** 2·8 m.	**14.** 5·6 cm.	**19.** 16 dm.
5. 3·5 m.	**10.** 3·5 cm.	**15.** 26 cm.	**20.** 18 m.

EXERCISE 73

($\pi = 3\frac{1}{7}$ unless told otherwise)

1. What is the area of a circular gold-fish pond of 7 m. diameter?
2. What is the area of the one side of a penny if its diameter is 2 cm? ($\pi = 3{\cdot}14$)
3. The diameter of a dinner plate is 25 cm. What is the area of the one side of it? ($\pi = 3{\cdot}14$)
4. The radius of a cake tin is 8 cm. What area of greaseproof paper would be required to line the bottom of it? ($\pi = 3{\cdot}14$)
5. Find the area of the shaded part.

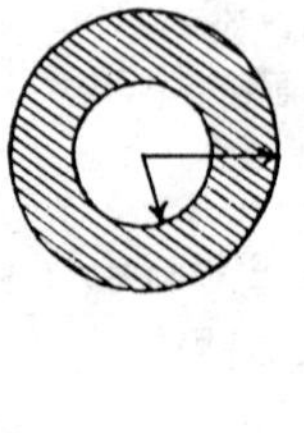

R = 7m

r = 3·5m

6. A circular pond is 14 m. in diameter. Round it is a path 1 m. wide. What area of gravel would be required to cover the path?
7. This shaded area is a diagram of the run up for the high jump.

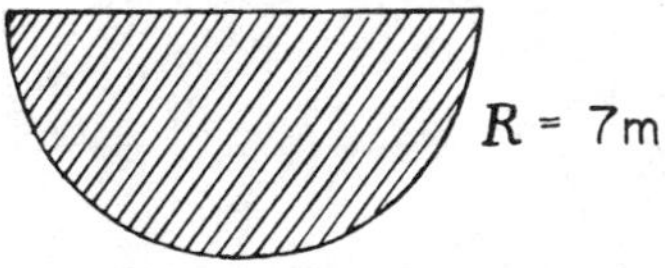

What area of ash is required to cover it?

8. A flower pot is 10 cm. diameter at the top and 6 cm. at its base. What is the difference in the area of these two circles?
9. 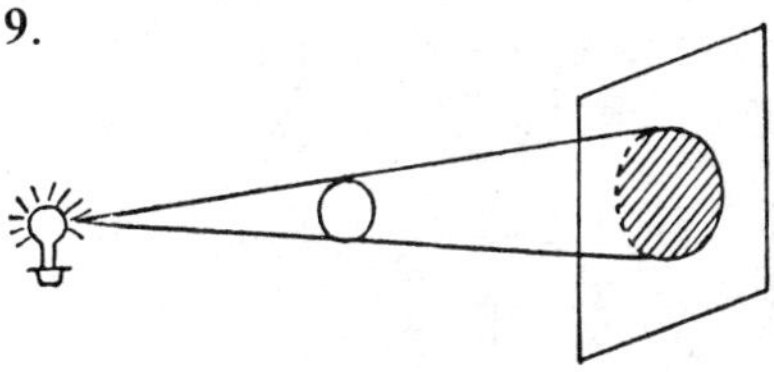

A $3\frac{1}{2}$ m. diameter disc casts a shadow of 7 m. diameter. How much bigger in area is the shadow?

10. A spot light illuminates a circle with diameter of 35 m. What is the area illuminated?

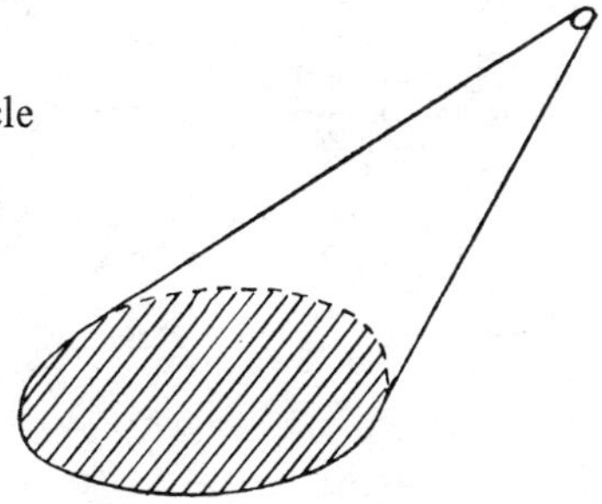

EXERCISE 74

Find the area of the shaded part where indicated.

1. 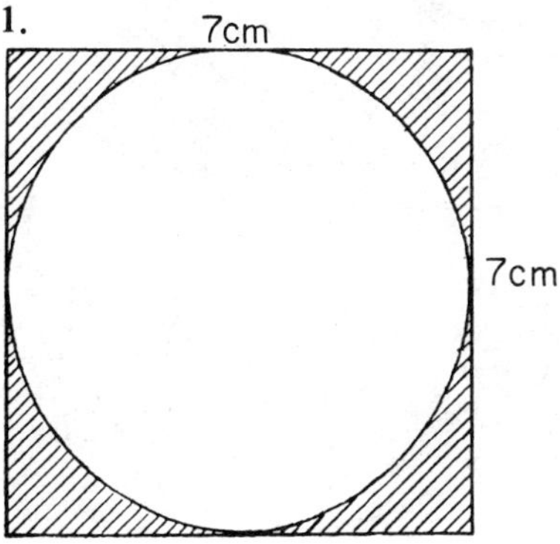

2. 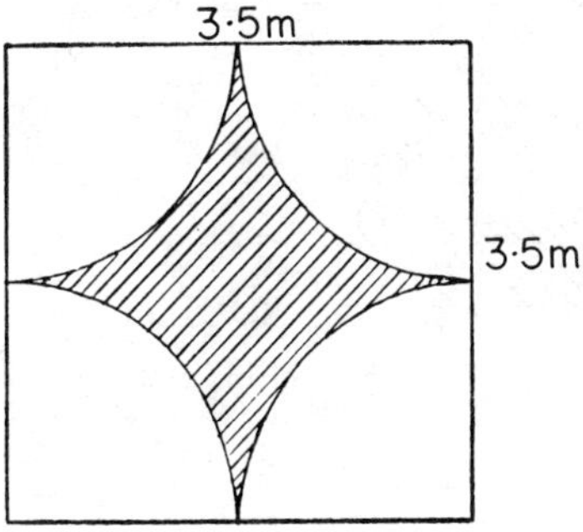

3. ($\pi = 3{\cdot}14$)

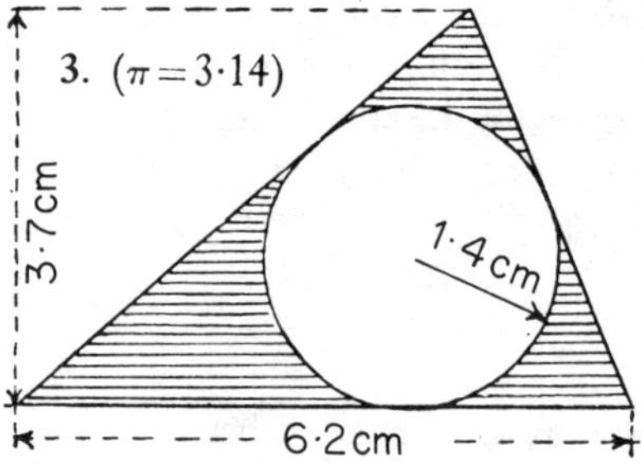

4.

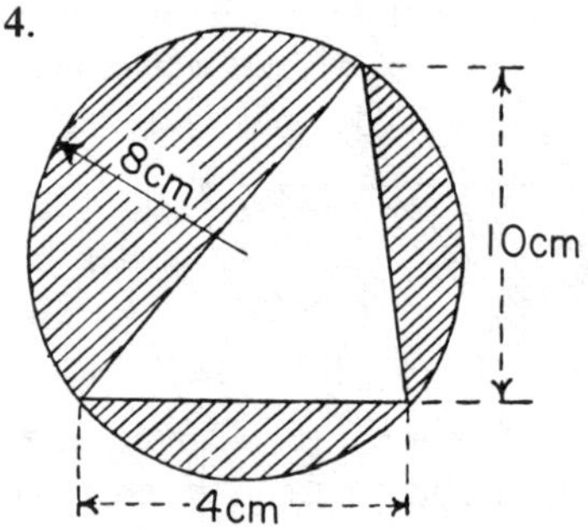

5. ($\pi = 3{\cdot}14$)

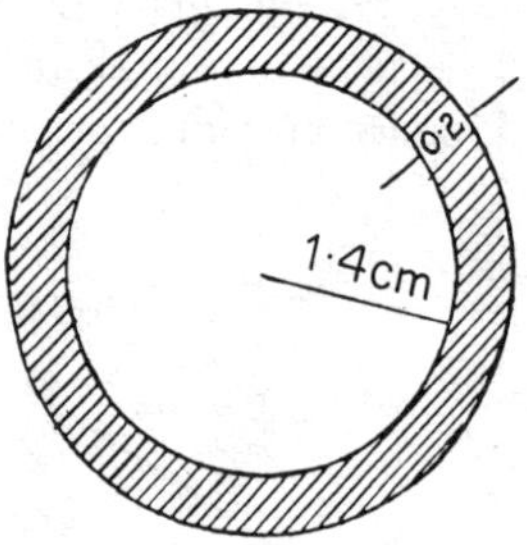

6. Find total area.

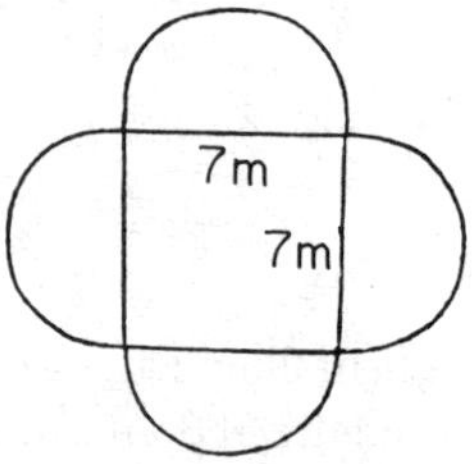

7. Total area.

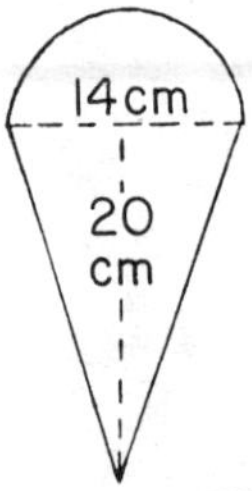

8.

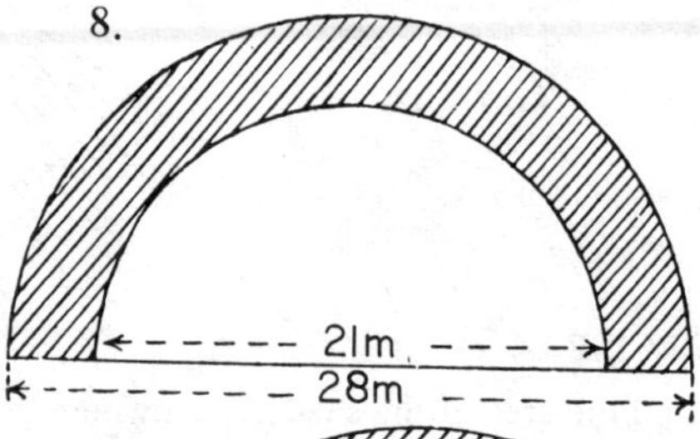

9.

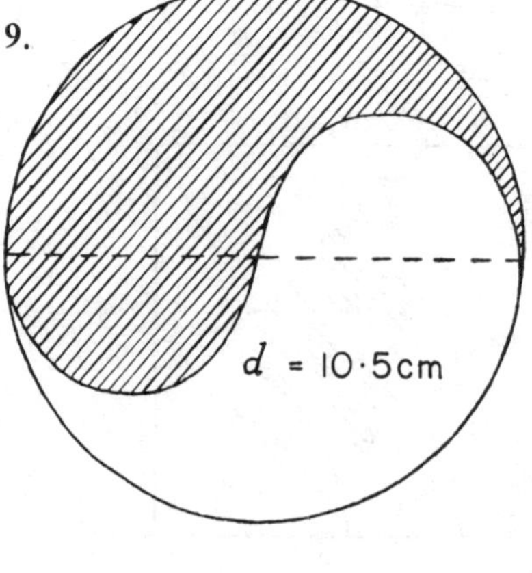

10.

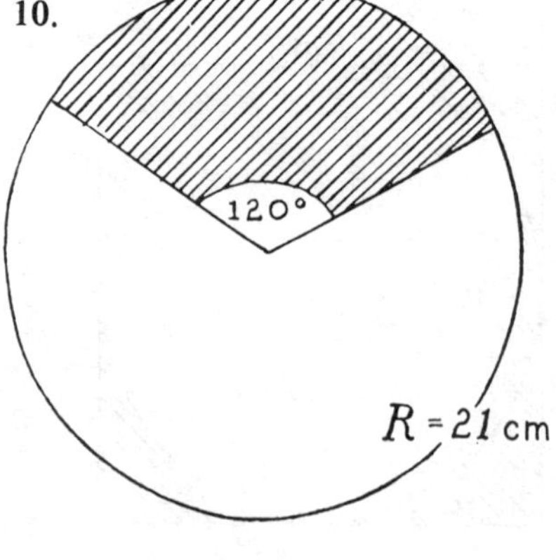

To change the subject of a formula:

EXAMPLE 3

If $12 = x \times 3$

$\underline{x = 4.}$ Ans.

EXAMPLE 4

If $x^2 = 9$

$x = \sqrt{9}$

$\underline{x = 3.}$ Ans.

EXAMPLE 5

If $50 = \frac{1}{3} \times \frac{25}{1} \times b$

$$b = \frac{50}{\frac{1}{3} \times 25}$$

$$= \frac{\overset{2}{\cancel{50}}}{1} \times \frac{3}{1} \times \frac{1}{\underset{1}{\cancel{25}}}$$

$\underline{b = 6.}$ Ans.

EXERCISE 75

Change the subjects in the following formulae.

1. $A = \pi r^2$, find r.
2. $A = \frac{1}{2}bh$, find h.
3. $A = lb$, find l and then b.
4. $C = 2\pi r$, find r.
5. $C = \pi d$, find d.
6. Write an expression for the perimeter of a rectangle with length l and breadth 'b'. Find b in terms of P and l.
7. Using the formula you have made in question 6, find the breadth of a rectangle the length of which is 14 cm. if its perimeter is 52 cm. long.
8. What would be the area of a square if its sides are each 'a' cm. long?
9. The sides of a triangle are x cm., $(x + 2)$ cm., and $(x + 3)$ cm. long. Find the lengths of each side if the perimeter is 26 cm. long.
10. The area of a triangle is 25 cm^2. What is its height if the base is 10 cm. long?

EXERCISE 76

1. The area of a rectangle is 12 m^2. Its length is 4 m., what is the width?
2. The length of a rectangle is 7 dm. Find the width if the area is 56 dm^2.
3. What would be the length of a room 8 m. wide, if 72 m^2 of carpet was needed to cover the floor?
4. What would be the length of a corridor 2 m. wide, if 70 m^2 of fitted carpet is required to fit it?
5. The area of my garden is 800 m^2 and its width is 20 m.
   (*a*) What is its length?
   (*b*) How much fence would I require to surround it?
6. The area of a triangle is 40 m^2. Its height is 10 m. What is the length of its base?
7. Find the height of a triangle if its base is 8 cm. The area of this triangle is 32 cm^2.
8. A 6 cm. high triangle has an area of 48 cm^2. What is the length of the base?
9. What would be the height of a triangle if the area is 40 sq. cm. and its base 5 cm?
10. The area of a triangle is 39 cm^2. What is the height of this triangle if the base is 13 cm?
11. The distance round the inner lane of a circular running track is 200 m. What radius was used when constructing the track?
12. On another track an athlete has to complete five circuits of the track in order to cover 1500 m. What is the radius of this track?
13. The circumference of an old halfpenny is 7·85 cm. What is its radius?
14. Each time a bicycle wheel revolves it covers a distance of 2·2 m. What is the diameter of the wheel?
15. 4·4 m. of fence is required to surround a circular pond. What is the diameter of the pond?
16. A goat is tethered in the middle of a field. If it can graze over an area of 154 m^2, how long is the chain?
17. The area of a traffic island is 1386 m^2. (*a*) What is the distance across it? (*b*) What is the distance round it?
18. A girl has a cake frill 82·5 cm. long, and wants to make the biggest cake possible to fit it. (*a*) What would be the radius if she made a round cake? (*b*) How long would each side be if she made a square one? (*c*) Which cake would have the bigger area?
19. The pendulum of a clock is 70 cm. long and it swings through an angle of 20°. If it swings 5 times in 10 seconds, how far does the tip travel in 1 hour?

20. The area of a circular protractor is 314 sq. cm. ($\pi = 3{\cdot}14$)

(*a*) What is its radius?

(*b*) What would be the distance round the edge between 140° and 230°, measured in a clockwise direction?

21.

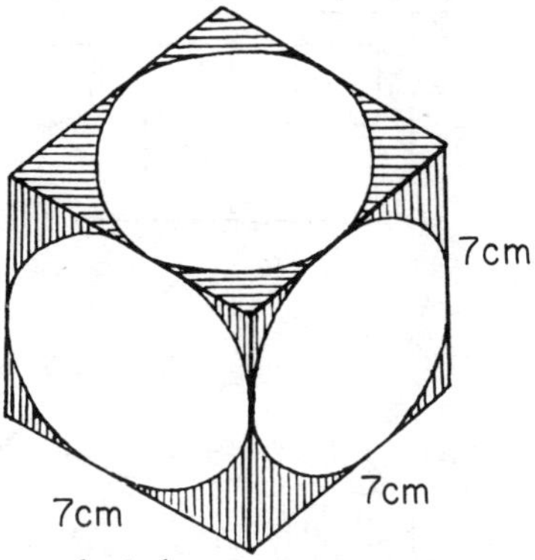

Find area shaded.

22.

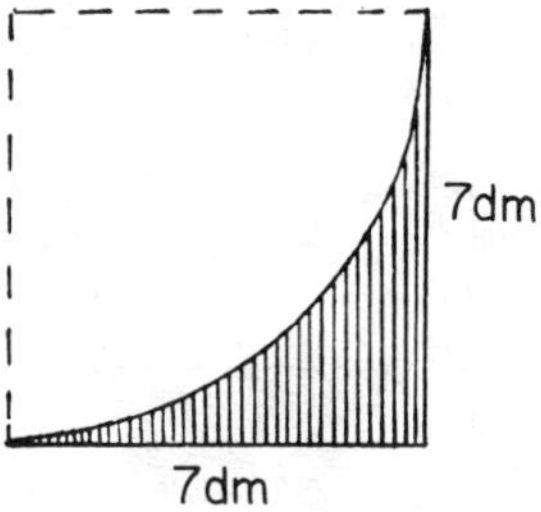

23.

$R = 10\frac{1}{2}$ m²

Find shaded area.

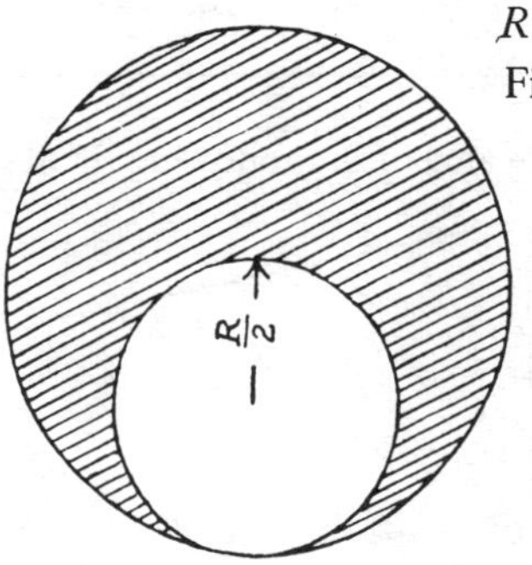

24.

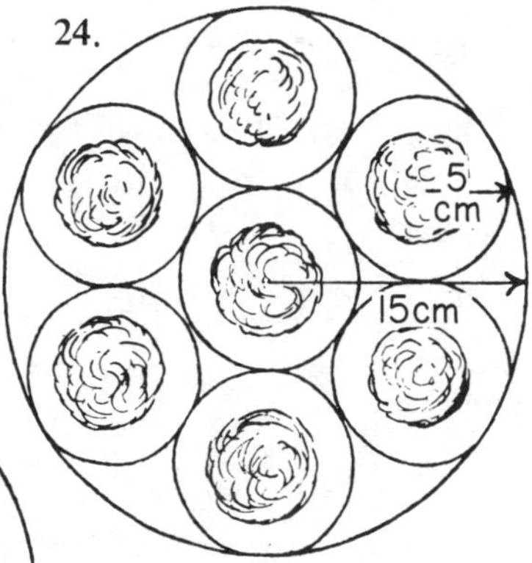

Find area of plate not covered by tarts.

25.

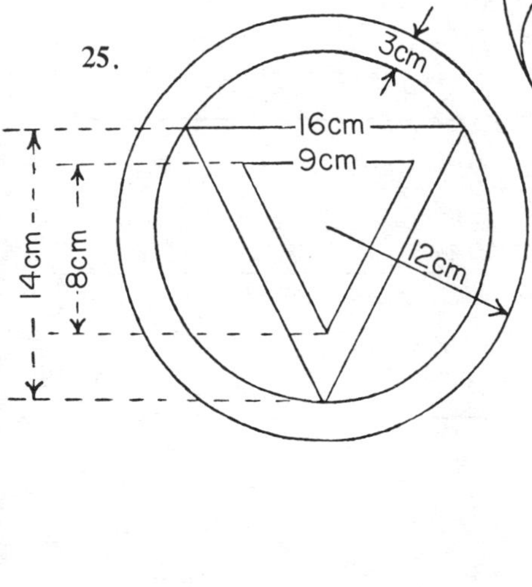

Find area shaded.

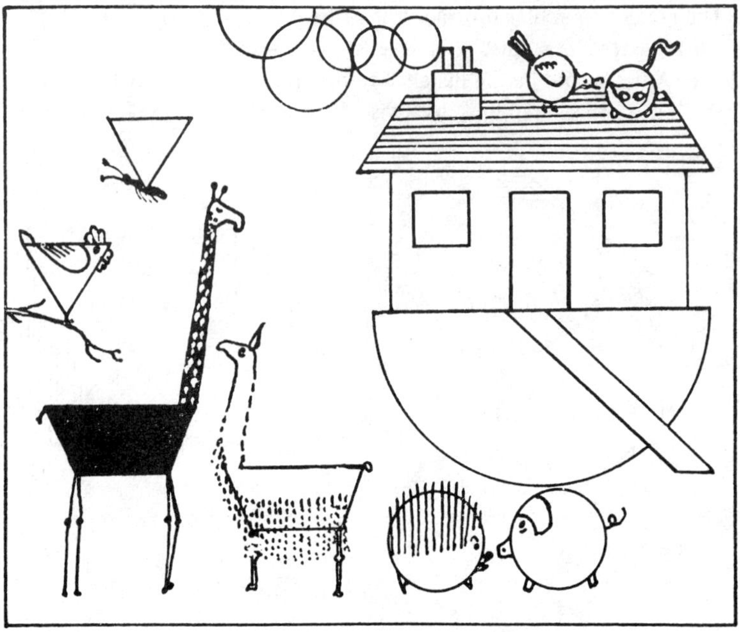

EXERCISE 77 Find the area of the following figures

1.

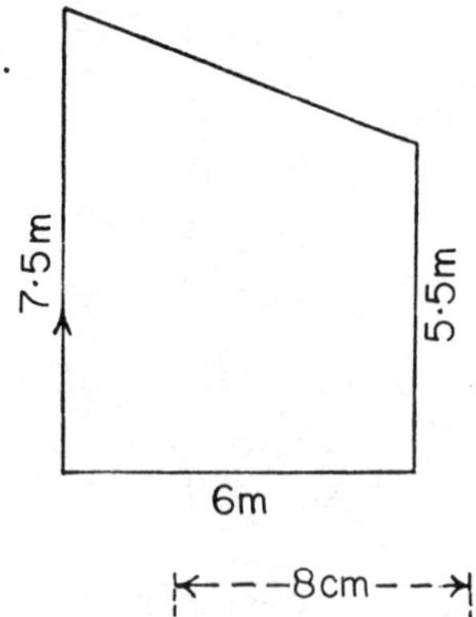

2.

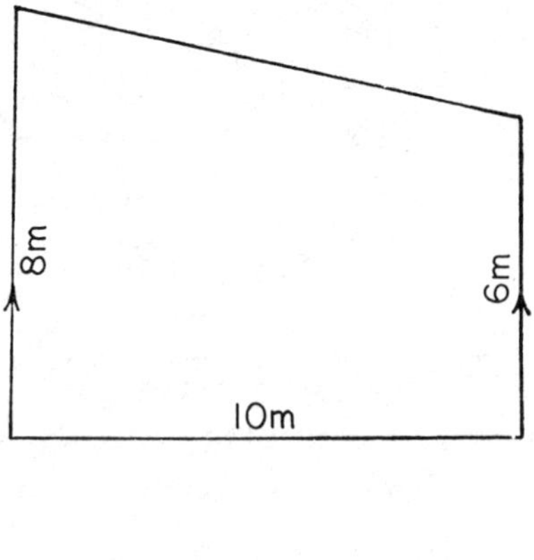

3.

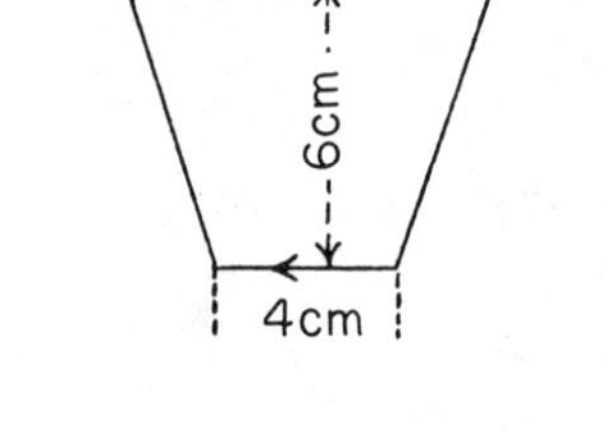

4.

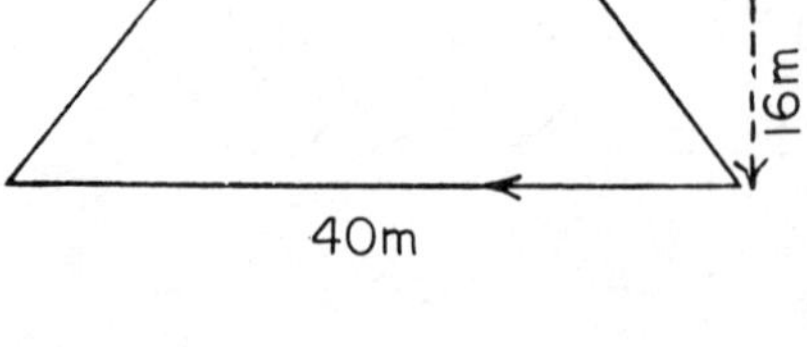

5. Find area of the 4 sloping sides.

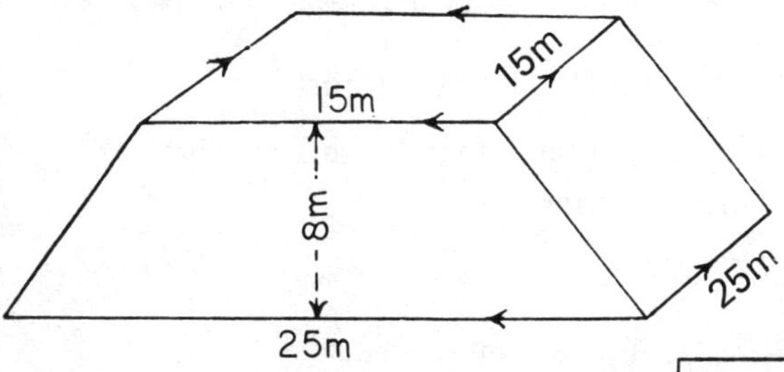

6.

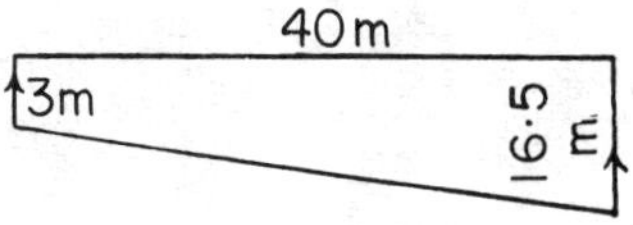

7.

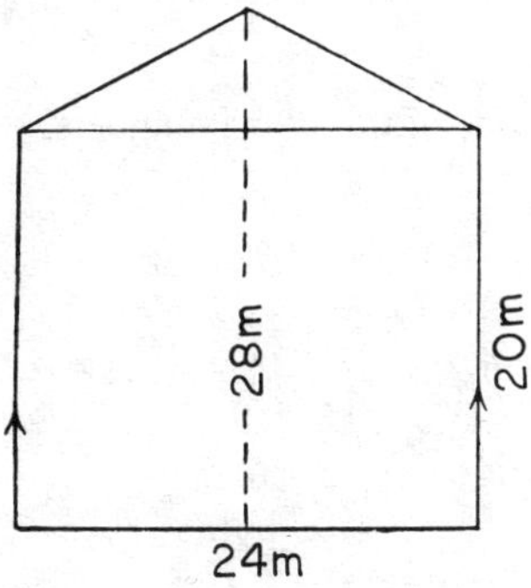

8.

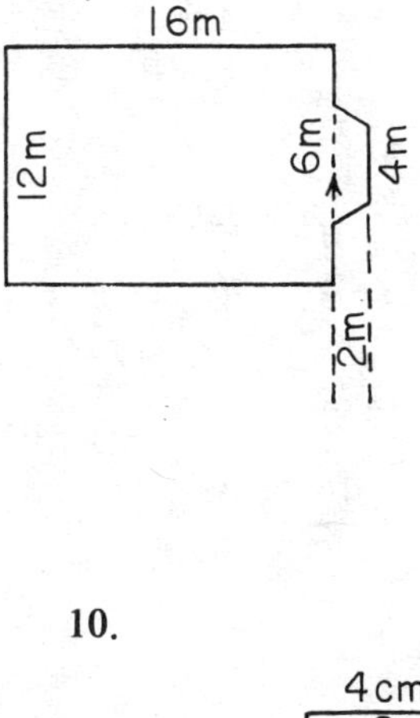

9.

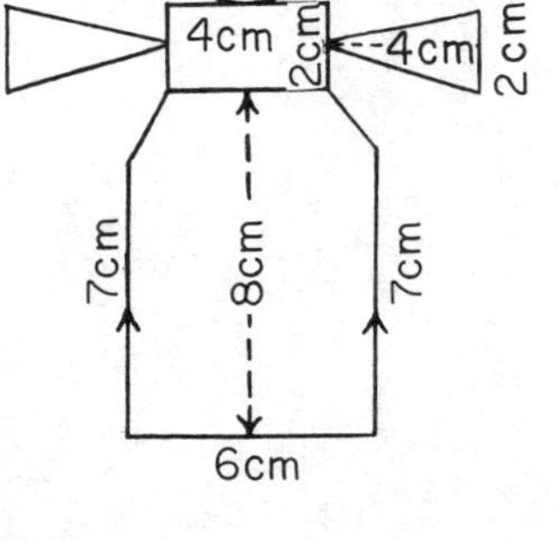

10.

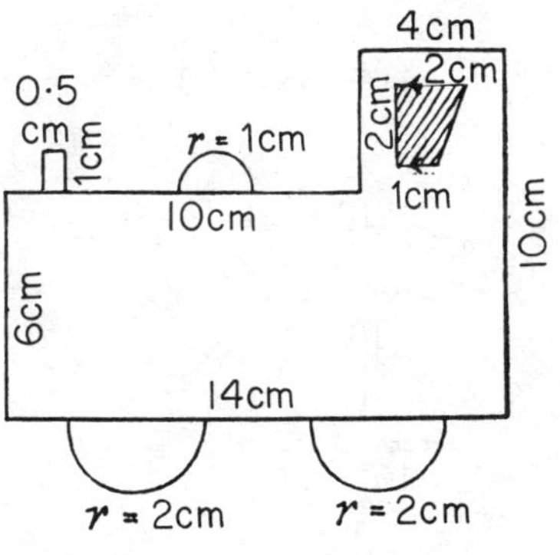

Find total area.

Find area. Subtract window area.

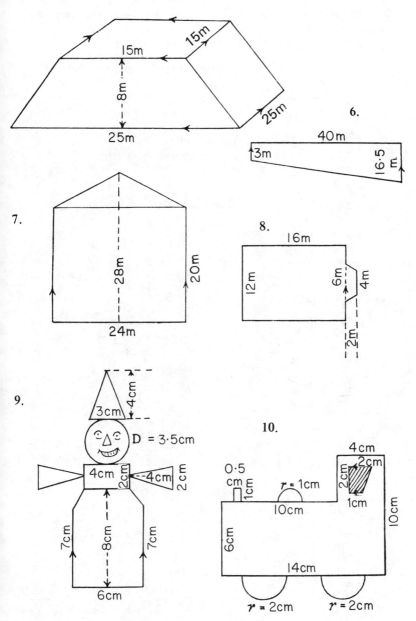

Things you have discovered.

1. Before fractions can be arranged in order they must have the same denominator.

Arrange $\frac{3}{4}, \frac{1}{3}, \frac{5}{6}, \frac{4}{9}, \frac{1}{2}$ in order greatest first.

$\frac{27}{36}, \frac{12}{36}, \frac{30}{36}, \frac{16}{36}, \frac{18}{36}$, change to the same denominator.

$\frac{5}{6}, \frac{3}{4}, \frac{1}{2}, \frac{4}{9}\ \frac{1}{3}$ Ans.

2. $3\frac{1}{4} + 2\frac{1}{10} - 1\frac{4}{5}$

$= 4\frac{5}{20} + \frac{2}{20} - \frac{16}{20}$ Whole numbers first. Find common denominator.

$= 4\frac{7}{20} - \frac{16}{20}$ Add first.

$= 3\frac{27}{20} - \frac{16}{20}$ Change one of the whole numbers.

$= 3\frac{11}{20}$ Ans.

3. $5\frac{1}{4} \times 3\frac{1}{7}$

$= \frac{\overset{3}{\cancel{21}}}{\underset{2}{\cancel{4}}} \times \frac{\overset{11}{\cancel{22}}}{\underset{1}{\cancel{7}}}$

$= \frac{33}{2}$

$= 16\frac{1}{2}$ Ans.

4. $7\frac{1}{2} \div 2\frac{1}{4}$

$= \frac{15}{2} \div \frac{9}{4}$

$\frac{\overset{5}{\cancel{15}}}{\underset{1}{\cancel{2}}} \times \frac{\overset{2}{\cancel{4}}}{\underset{3}{\cancel{9}}}$

$= \frac{10}{3}$

$= 3\frac{1}{3}$ Ans.

5. $(1\frac{1}{5} + 1\frac{1}{2}) \div 5\frac{2}{5}$

$= (2\frac{2}{10} + \frac{5}{10}) \div 5\frac{2}{5}$

$= 2\frac{7}{10} \div 5\frac{2}{5}$

$= \frac{\overset{1}{\cancel{27}}}{\underset{2}{\cancel{10}}} \times \frac{\overset{1}{\cancel{5}}}{\underset{1}{\cancel{27}}}$

$= \frac{1}{2}$ Ans.

6. To multiply decimals by 10 or 100 move the figures 1 or 2 places to the left.

To divide decimals by 10 or 100 move the figures 1 or 2 places to the right.

7. 3·25 x 11·2

```
   3·25
   11·2
 ------
 325
   325
    650
 ------
 36·400  Ans.
```

8. Divide 36·4 by 3·25

3640 ÷ 325

```
       11·2   Ans.
     -----
325)3640
    325
     390
     325
      650
      650
      ...
```

9. Change $\frac{7}{16}$ to a decimal

```
      0·4375   Ans.
    --------
16) 7·000
     6 4
      60
      48
      120
      112
       80
       80
       --
       ...
```

10. Find $\frac{3}{5}$ of £7·92 to the nearest penny.

$\frac{3}{5}$ of £7·92 = $\frac{£23·76}{5}$

= £4·752

= £4·75 to the nearest penny.

11. Find 2·13 x £4·17 to the nearest penny.

```
  £4·17
   2·13
 ------
 834
   417
   1251
 ------
 8·8821
```

2·13 x £4·17 = £8·88 to the nearest penny.

12. Average = $\dfrac{\text{Total of quantities}}{\text{Number of quantities}}$

13. Area of rectangle = Length x breadth.
 or $A = lb$.

14. Perimeter of rectangle = $2(l + b)$
 or $P = 2(l + b)$

15. Area of triangle = $\frac{1}{2}$ x base x perpendicular height
 or $A = \frac{1}{2} bh$.

16. Area of Trapezium = sum of parallel sides x $\frac{1}{2}$ perpendicular distance between them.
 or $A = \dfrac{d}{2}$ (sum of || sides)

17. Pi or $\pi = \dfrac{\text{circumference}}{\text{diameter}}$

18. Circumference of circle = 2 x π x radius
 or $C = 2\pi r$
 Circumference of circle = π x diameter
 or $C = \pi d$.

19. Area of circle = $\pi \times r \times r$
 or $A = \pi r^2$

Algebra

14 Use of Letters

This basket of fruit can be described in more detail by saying:

(*a*) The basket contains 3 kg. of fruit.

(*b*) The basket is full of apples, pears and oranges.

(*c*) The basket contains 3 kg. of apples, pears and oranges.

To give an exact description we must empty the basket,

sort out the different kinds and count them.

Five apples + six oranges + four pears.

This can be written $5a + 6o + 4p$.

12 fifty-penny pieces + 6 ten-penny pieces x 8 five-penny pieces + 7 two-penny pieces + 10 pennies, or $12F + 6T + 8f + 7t + 10p$,

Only like things can be added together. If John has 2 apples and Joan has 3 apples, together they have 5 apples.

$$2a + 3a = 5a$$

Sitting on the low-level station at Tamworth I saw three passenger trains and two goods go through. At the same time on the high level line, one passenger train, two goods trains and two diesel trains went through. We can record these facts like this.

Low-level	$3p + 2g$
High-level	$1p + 2g + 2d$
Total number of trains	$4p + 4g + 2d$

This could have been written $3p + 2g + p + 2g + 2d = 4p + 4g + 2d$.

This is called collecting terms. Just as apples and pears cannot be added, different letters cannot be added except by putting a plus sign between them. To add a to b we write $a + b$.

$a + a + a + a$ is the same as $4 \times a$ which is written as $4a$.

EXERCISE 78

Write more shortly and add the terms.

1. In a field are 10 cows, 25 sheep and 4 horses. In the next field are 15 cows, 12 sheep and 1 horse.
2. A postman delivered 40 letters, 16 postcards and 4 parcels in one street and 26 letters, 9 cards and 8 parcels in the next street.
3. Mr Brown set 3 rows of beans, 8 rows of potatoes and 1 row of lettuce in his garden while Mr Smith set 3 rows of beans, 12 rows of potatoes and one one row of lettuce.
4. One week we had 2 sunny days, 4 wet days and one foggy day and the next week we had 3 foggy days, 3 wet days and one sunny one.
5. In one group, 4 children had fair hair, 5 were dark and one had red hair, in another group 7 were dark and 2 were fair.
6. At an athletics match, 8 Englishmen, 5 Scots and 4 Welshmen took part. The officials were 4 Englishmen, 4 Scots and 4 Welshmen.
7. Three people bought cakes. The first had 4 cream-slices, 3 doughnuts and 2 iced buns; the second had 3 iced buns, 4 doughnuts and 2

3 foggy days.

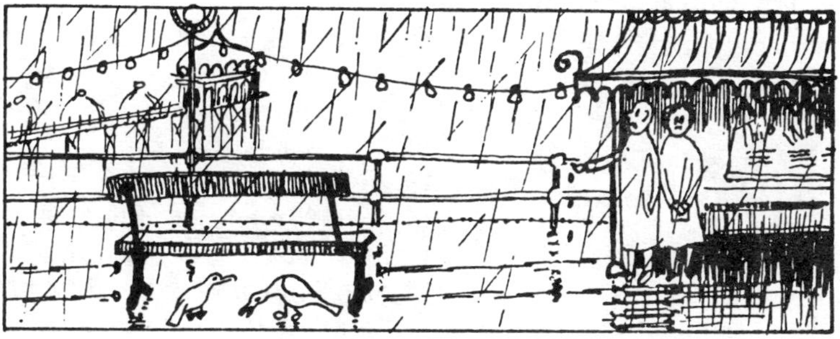

3 wet days.

1 sunny day.

chocolate éclairs; the third bought 4 éclairs, 3 doughnuts and 2 cream-slices.

8. Two girls and a boy selected the following books from the library;
   First girl: 2 adventure and 1 romance.
   Second girl: 1 cookery book and 2 adventure.
   Boy: 1 adventure and 1 western.
9. On a nature ramble a boy found 3 thrushes' nests, 2 sparrows' nests and 2 blackbirds' nests, his friend found 1 yellowhammer's nest and 2 thrushes' nests.
10. On a plate were 6 raspberry, 3 lemon and 4 strawberry tarts and on another plate were 4 strawberry, 5 raspberry and 3 lemon tarts.

EXERCISE 79

Collect terms in the following expressions.

1. $a + a + a \div a$
2. $2b + b$
3. $3d + 4d$
4. $5c + 2c + c$
5. $5f + 6f + 2f$
6. $5p + 6p + 2p \div 9p$
7. $4s + 6s + s + 4s \div 5s$
8. $s + s + 5s + 7s + s$
9. $8f + 9f + 6f + f + 2f$
10. $e + 5e + 9e + 3e + 6e$
11. $a + b + a + b + a + b + a$
12. $2a + 3b + 2c + 4a$
13. $5g + 6d + g + 2d + f$
14. $6y + 3x + 4y + y + 2x$
15. $4m + 3n + 5m + 2n + 3m$
16. $q + p + 5p + 4q + 3r + 6p$
17. $r + s + 6t + 4r + 4t + 3s + 5t$
18. $5n + 5n + 6m + 3n + 2m + 9n + m$
19. $3b + 4h + 3b + 3h + 6h + 7b$
20. $3z + 5x + 7y + x + 4z + y + 4x$

So far we have only considered Positive Numbers. We HAD 5 apples and we HAD 8 marbles which we wrote as $5a$ and $8m$. Consider this problem:

John and Peter played marbles on their way home from school. At the end of the game John had gained 4 marbles while Peter had lost 4.

We can write John's gain as +4 and Peter's loss as –4, a negative quantity.

If we lost 2 of our 8 marbles we would write $8m - 2m$ leaving $6m$.

EXAMPLE 1

I had 40p left in my purse when the milkman called and told me that my week's bill was 50 pence. How can this be written?

40p – 50p = –10p (I was short of 10 pence)

EXAMPLE 2

If we had £8 and received bills for £5 and £9 then we would write £8–£5–£9 = £8–£14 (we add up the debt before subtracting.) = –£6.

Another method of working out problems with Positive and Negative numbers is to plot the points on a ladder or staircase. Positive numbers go up and Negative numbers go down.

EXAMPLE 3

$$3 - 5 + 4 - 6 = 7 - 11$$
$$= \underline{-4} \quad \text{Ans.}$$

The sign is usually missed out from the first term when it is positive.

3 – 5 is the same as +3 – 5

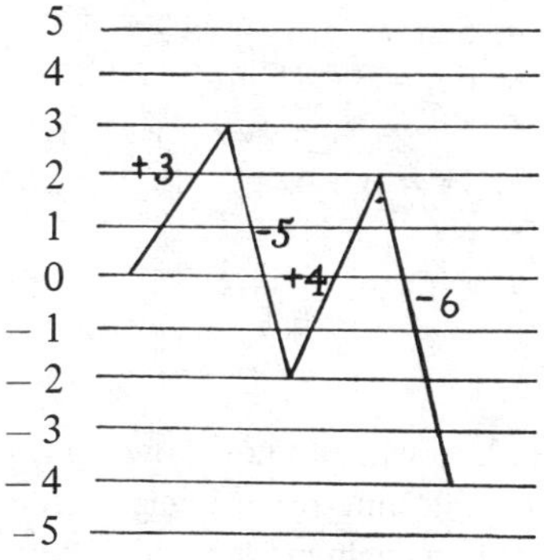

EXERCISE 80

Draw stairs in your exercise book and work out these.

1. 5 – 2 + 6 + 4
2. – 3 + 7 – 2 – 3
3. – 5 + 8 + 2 – 6
4. 4 – 6 + 3 – 6
5. – 4 + 2 – 3 + 10
6. 4 – 6 + 2 – 4 + 5
7. 3 – 4 + 5 – 6 – 3
8. – 2 + 1 – 4 + 9 – 4
9. 1 – 6 + 3 – 1 + 7 – 5
10. 0 – 3 – 2 + 10 – 4 + 1 – 5

Negative numbers do not only refer to shillings or to the rungs on a ladder. Anything that is lost or short needs a minus sign, e.g.

(*a*) Peter lost 4 marbles; he was $-4m$.
(*b*) Thieves stole 25 rings; the jeweller was $-25r$.
(*c*) A train running from London to York at -50 km/h. was in fact running at 50 km/h. from York to London.

EXERCISE 81

Collect the terms in the following, remember that only terms of the same kind can be added together.

1. $4a + 6a$
2. $7b - 5b$
3. $x + x + x - x$
4. $5r + 2r - 3r$
5. $6m - 3m + 2m$
6. $11y - 3y + 4y - 8y$
7. $7s - 3s + 4s$
8. $6n - 2n - 5n$
9. $-5t - 2t - 4t$
10. $-8z + 2z + 3z$
11. $5a + 3b + 2a$
12. $6c - 3d + 4d$
13. $7f + 2g - 3f + g$
14. $4m - 3n + 2m + 6n$
15. $2h - 4h - 3j + 5j$
16. $-6x - 3y + 10x + 5y$
17. $4h - 3h - 5k - 2k$
18. $-3r + 2s - 5r - 10s$
19. $-2y - 3z - 4z - 5y$
20. $5b + 3c - 4c - 4b + c$

You will all know how many centimetres there are in a decimetre or in three decimetres or in eight decimetres, but do you know how many there are in *p* decimetres, *r* decimetres or *y* decimetres?

In 1 dm. there are 10 cm.			= 10 cm.
In 3 dm.	" "	10×3	= 30 cm.
In 8 dm.	" "	10×8	= 80 cm.
In *p* dm.	" "	$10 \times p$	= $10p$ cm.
In *r* dm.	" "	$10 \times r$	= $10r$ cm.
In *y* dm.	" "	$10 \times y$	= $10y$ cm.

Notice that the last three answers can only be written as $10p$, $10r$ and $12y$ cm. since we do not know the value of *p*, *r* and *y*.

Write down the number of grams in three kg; five kg; *a* kg. and *c* kg. This time you must multiply by 1000 as there are 1000 g. in 1 kg.

When changing minutes to seconds multiply by 60.

EXERCISE 82

How many . . . .

1. cm. in z dm?
2. cm. in x dm?
3. g. in m kg;
4. pence in £m?
5. seconds in r minutes?
6. ten-penny pieces in £y?
7. hours in k days?
8. cm. in p metres?
9. cl. in m litres.
10. g. in r kg.
11. m. in d km.
12. kg. in e metric tons.
13. ml. in r l.
14. weeks in t years.
15. fifty-penny pieces in £m?
16. minutes in z hours?
17. dm. in y m?
18. g in f kg.
19. days in p weeks?
20. cl. in f litres?

These examples all had to be multiplied as large units were being changed into smaller ones. To change small units into large ones (pence to pounds) you must do the opposite of multiply, that is, divide.

In 48 cm. there are $\frac{48}{100}$ m.
In x cm. there are $\frac{x}{100}$ m.
In 250 pence there are £$\frac{250}{100}$.
In x pence there are £$\frac{x}{100}$.

$\frac{48}{100}$ m. cancels and becomes 0·48 m, £$\frac{250}{100}$ becomes £2·5 but $\frac{x}{100}$m. and £$\frac{x}{100}$ cannot be written in any simplier form.

EXERCISE 83

How many . . . .

1. metres in b cm?
2. litres in n cl?
3. kg. in k g?
4. kg. in f g?
5. minutes in p seconds?
6. pounds in y ten-penny pieces?
7. kg. in p g?
8. litres in k cl.
9. weeks in m days?
10. km. in n metres?
11. km. in d metres.
12. hours in s minutes?
13. pounds in n two-penny pieces?
14. years in y months?
15. metres in c dm?
16. kg. in m grams.
17. l. in n cl?
18. metres in n cm?
19. days in t hours?
20. litres in j cl?

Multiply 3 by 4 is written 3 x 4, but multiply a by b is written ab without the multiplication sign. Just as 3 x 4 is equal to 4 x 3, ab is equal to ba.

$p \times q \times r = pqr$ $\qquad$ $3a \times 5b = 15ab$ $\qquad$ $6mn \times 3r = 18mnr$

15 Indices

EXERCISE 84

Find the area of these rectangles.

1.
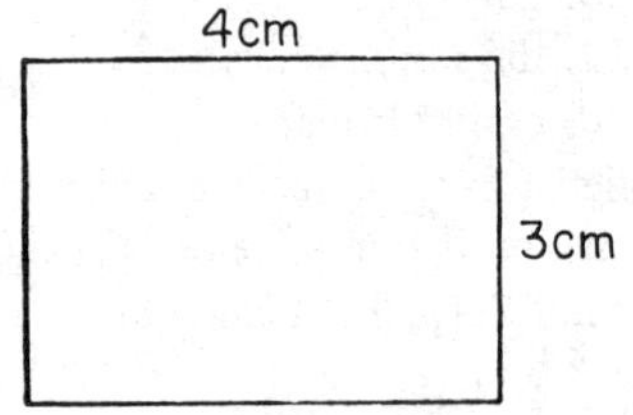

2.
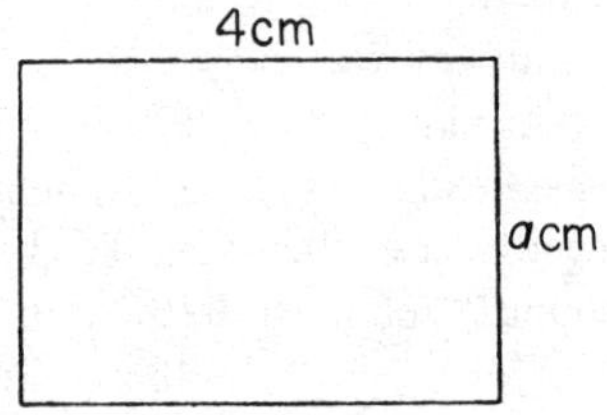

3.
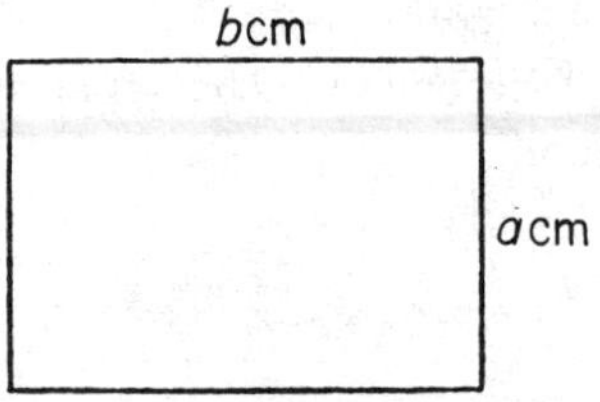

4.
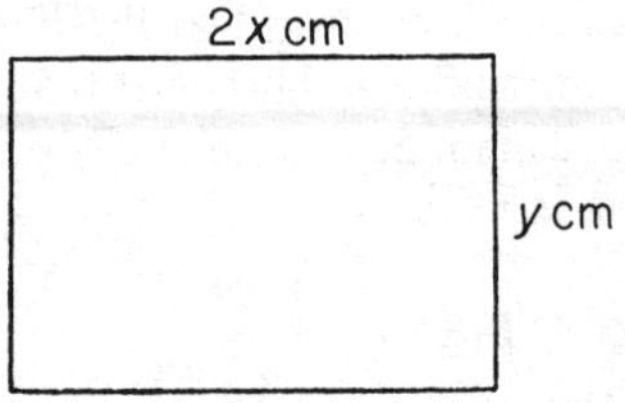

5.
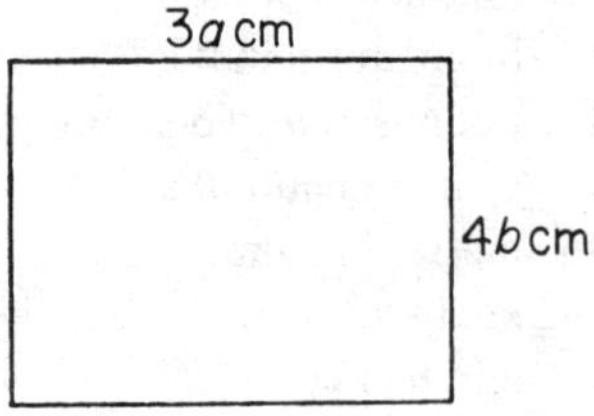

6.
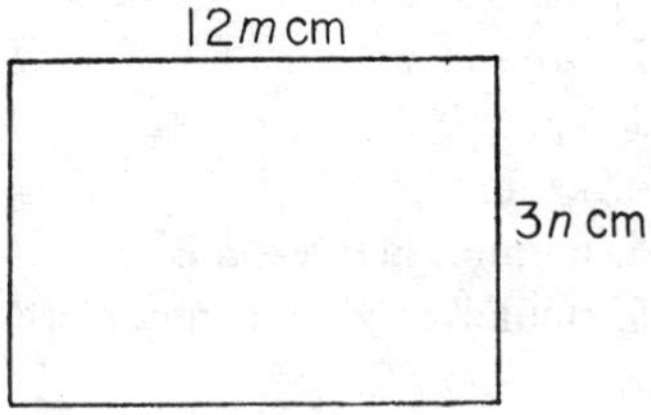

7.
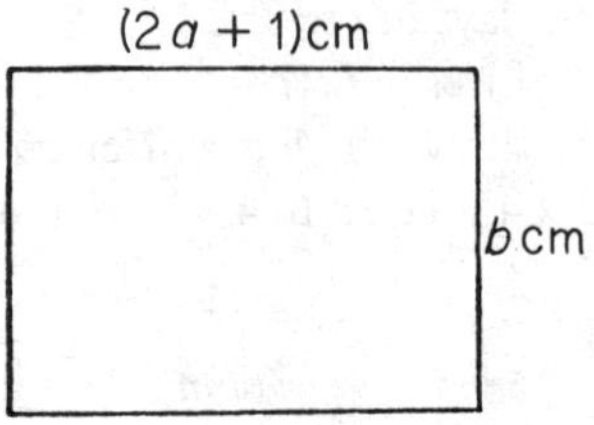

8.
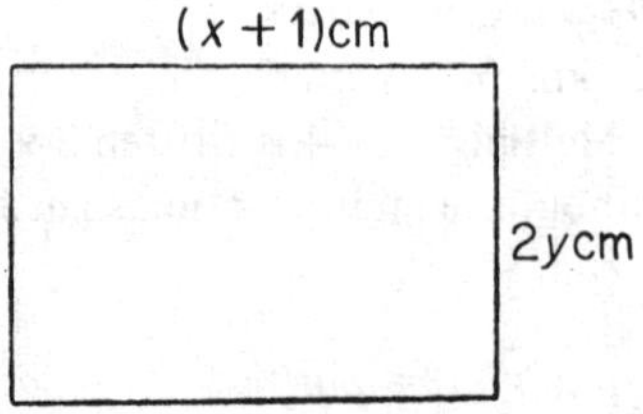

EXERCISE 85

Multiply these terms together.

1. $m \times n$
2. $p \times q$
3. $y \times 2$
4. $e \times r$
5. $g \times f$
6. $h \times b \times d$
7. $c \times a \times f$
8. $d \times r \times n$
9. $h \times k \times m$
10. $z \times w \times y$
11. $3a \times 2b$
12. $4r \times 2p$
13. $3z \times y$
14. $4m \times 4n$
15. $3s \times 2t$
16. $5y \times 3z$
17. $p \times 4r \times 2s$
18. $3f \times 2g \times 4h$
19. $3c \times 4d \times a$
20. $5b \times 2a \times 3c$

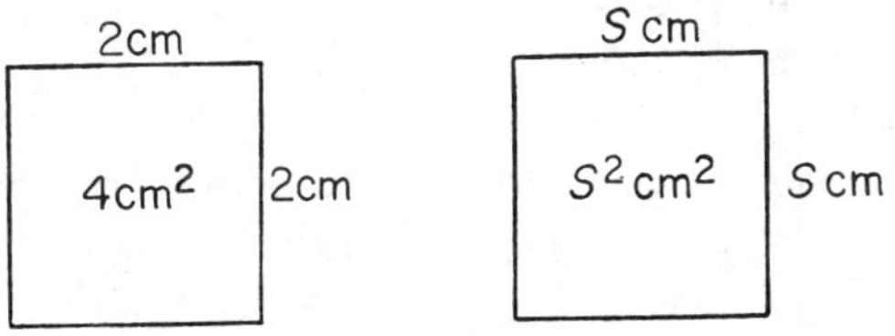

What is the area of each of these two squares?

The first one is 2cm. x 2cm. which is 4 square centimetres, but the other is *s*cm. x *s*cm.

This is written s^2 and called *s* squared. So $a \times a$ is a^2, $y \times y$ is y^2, etc.

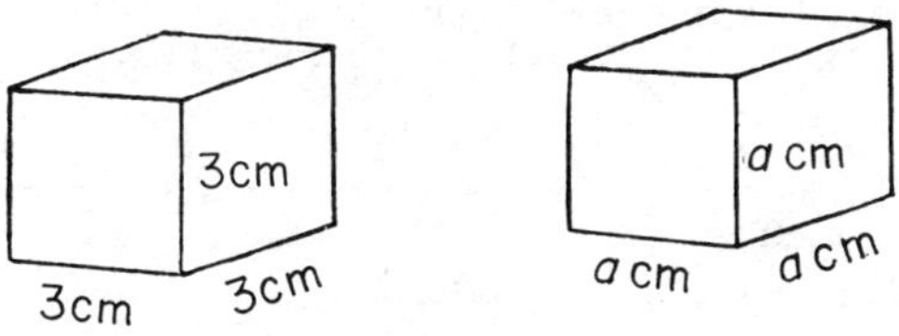

The volume of the first cube is 3 cm. x 3 cm. x 3 cm. which is 27 cm^3, but the volume of the other is $a \times a \times a$ which is written a^3, (*a* cubed.)

$b \times b \times b \times b$ is b^4 (*b* to the fourth)

$c \times c \times c \times c \times c$ is c^5 (*c* to the fifth) and so on.

The little figure at the top is called the index and indicates how many times the term has been multiplied by itself.

What is the result when s^3 is multiplied by s^2.

$$s^3 \times s^2 = s \times s \times s \quad \times \quad s \times s$$
$$= s^5$$
$$a^3 \times a^4 = a \times a \times a \quad \times \quad a \times a \times a \times a$$
$$= a^7$$

$$b \times b^3 = b \times b \times b \times b$$
$$= b^4$$
$$d^2 \times d^4 = d \times d \times d \times d \times d \times d$$
$$= d^6$$

Can you find a rule for multiplying these without writing out all the letters separately? Look at the indices.

$$a^3 \times a^4 = a^7$$
$$b \times b^3 = b^4 \quad (b \text{ is the same as } b^1.)$$

From this you will see that to multiply we *add* the indices.

EXAMPLE 86

Multiply these together.

1. $a \times a^3$
2. $m^4 \times m^2$
3. $n^5 \times n^3$
4. $p \times p^2$
5. $r^4 \times r$
6. $t^2 \times t^2$
7. $s^4 \times s^5$
8. $d^8 \times d^2$
9. $e^5 \times e^5$
10. $q^7 \times q^3$
11. $b^2 \times b^3 \times b^4$
12. $g^7 \times g \times g^2$
13. $k^2 \times k^9 \times k^7$
14. $u^4 \times u^6 \times u^3$
15. $c^3 \times c \times c^2$
16. $y^2 \times y^7 \times y^4 \times y$
17. $h \times h^5 \times h^3 \times h^6$
18. $z^4 \times z \times z^3 \times z^5$
19. $x^4 \times x^3 \times x^6 \times x^2$
20. $f^3 \times f^2 \times f^5 \times f^2 \times f^4$

What will be the result if $3c^2$ is multiplied by $4c^3$ or ab^2 by a^3b?

$$3c^2 \times 4c^3 = 3 \times 4 \times c^2 \times c^3 \text{ or } 3\ c \times c \times 4 \times c \times c \times c$$
$$= \underline{12c^5} \quad \text{Ans.}$$

Multiply numbers first then the letters.

$$ab^2 \times a^3b = a \times a^3 \times b^2 \times b \text{ or } a \times b \times b \times a \times a \times a \times b.$$
$$= \underline{a^4b^3} \quad \text{Ans.}$$

Multiply *a*'s first then *b*'s.
The indices only apply to the one letter.

ab^2 means $a \times b^2$ or $a \times b \times b$

a^2b^2 means $a^2 \times b^2$ or $a \times a \times b \times b$ or $(ab)^2$

EXERCISE 87

Multiply these terms together.

1. $p \times p$
2. $y \times y$
3. $z \times z \times z$
4. $m \times m \times m$
5. $2z \times z$
6. $y \times 2y$
7. $2x \times x$
8. $2a \times 2a$
9. $x^2 \times x$
10. $x^2 \times x^2$
11. $p^2 \times p$
12. $p^2 \times p^2$
13. $2p^2 \times 3p^2$
14. $3a^2 \times a^2$

15. $4a^2 \times a^3$
16. $3a^3 \times a^4$
17. $4t^3 \times t^2$
18. $2t \times t^3$
19. $3t^2 \times 2t^2$
20. $4g^2 \times 2g^3$
21. $a^2 \times b^2 \times 2a^2$
22. $a^3 \times a^2$
23. $(ab)^2 \times ab$
24. $pt \times pt$
25. $fn \times f^2$
26. $(fn)^2 \times f$
27. $(3a)^2 \times (2a)^2$
28. $(2x)^3 \times (2x)^2$
29. $(my)^2 \times (my)^3$
30. $(3ab)^2 \times 2ab$
31. $g^2h^3 \times gh^4$
32. $y^3z^4 \times y^2z^3$
33. $p^5q^3r^2 \times p^2qr^4$
34. $a^3b^2c \times a^4b^3c^3$
35. $3a^2b \times 4ab^2$
36. $5c^3d^2 \times 2c^4d^3$
37. $7m^3n^2 \times 3m^4n^3$
38. $12r^2s^3t^4 \times 5rst^2$
39. $8f^2g^2h^2 \times 3f^3gh$
40. $3b^5c^3d^4 \times 5bc \times 2b^2cd^3$

Just as you added the indices when multiplying the terms, you subtract them when dividing.

EXAMPLE 1

$a^3 \div a^2 = a$ Ans. and $b^5 \div b^3 = b^2$ Ans.

It might be easier to continue to write them as fractions and divide out especially when the divisor has the bigger index.

EXAMPLE 2

$\dfrac{a^2x^3}{ax^2}$ This means multiply a^2 by x^3 and divide by a multiplied by x^2.

$$= \frac{\overset{1}{\cancel{a}} \times a \times \overset{1}{\cancel{x}} \times \overset{1}{\cancel{x}} \times x}{\underset{1}{\cancel{a}} \times \underset{1}{\cancel{x}} \times \underset{1}{\cancel{x}}}$$

On the top we have $a \times x$ while at the bottom there is 1.

$= \underline{ax}$ Ans.

EXAMPLE 3

$$\frac{m^3}{m^5} = \frac{\overset{1}{\cancel{m}} \times \overset{1}{\cancel{m}} \times \overset{1}{\cancel{m}}}{\underset{1}{\cancel{m}} \times \underset{1}{\cancel{m}} \times \underset{1}{\cancel{m}} \times m \times m}$$

$$= \underline{\frac{1}{m^2}} \quad \text{Ans.}$$

EXERCISE 88

1. $a^6 \div a^2$
2. $m^5 \div m^3$
3. $r^7 \div r^4$
4. $s^5 \div s^4$
5. $e^3 \div e$
6. $y^7 \div y^4$
7. $g^4h^3 \div gh^4$
8. $m^3n^2 \div mn$
9. $s^5t^3 \div s^4t^4$
10. $6b^3 \div 3b$
11. $4c^3d \div 2cd^2$
12. $10f^2g^3 \div 5fg$
13. $14h^3k^3 \div 7hk^2$
14. $8y^2z^3 \div 6y^3z$
15. $12r^2s^4 \div 9r^3s^2$
16. $20m^5n^3 \div 12mn^4$
17. $16x^2y^2 \div 10xy^2$
18. $9a^3b^4 \div 12a^4b^3$
19. $6p^4q^3 \div 3p^3q^2$
20. $15v^4w^3 \div 9v^3w^5$

21. $\frac{a \times a^2}{a}$
22. $\frac{b^3 \times b}{b^2}$
23. $\frac{ab^2}{ab}$
24. $\frac{abc^2}{c}$
25. $\frac{a^2bc}{ab}$
26. $\frac{m^3p^2c}{m^2p}$
27. $\frac{nt^2s^3}{nts}$
28. $\frac{x^3y^2}{x^2y}$
29. $\frac{a^4b^2c}{abc}$
30. $\frac{3f^2s^2}{9f^3s^2}$
31. $\frac{8n^3o^2p}{2n^2op}$
32. $\frac{16g^2h}{4gh^2}$
33. $\frac{12q^4p^3t}{3q^2p^2}$
34. $\frac{25n^3m^2}{5mn}$
35. $\frac{18a^2b^3c}{3a^3b^2c}$
36. $\frac{4r^3s^3t^2}{r^2st}$
37. $\frac{20m^5n^2}{4m^3n}$
38. $\frac{7htk}{21hk^2}$
39. $\frac{4p^2at}{2a}$
40. $\frac{6a^3b^2c^4}{3a^4bc^2}$

16 Substitution

John and Jane made up a code so that they could write to one another without people knowing what they were saying. It was built up like this.

A	B	C	D	E	F	G	H	I	J	K	L	M
26	25	24	23	22	21	20	19	18	17	16	15	14
N	O	P	Q	R	S	T	U	V	W	X	Y	Z
13	12	11	10	9	8	7	6	5	4	3	2	1

Can you read their messages?

Jane: 4.19.22.9.22/26.9.22/2.12.6/20.12.18.13.20/21.12.9/2.12.6.9/19.12.15.18.23.26.2.8?

John: 4.22/26.9.22/20.12.18.13.20/8.26.18.15.18.13.20/

Jane: 24.26.13/2.12.6/8.4.18.14?

John: 13.12/25.6.7/14.2/25.9.12.7.19.22.9/24.26.13/19.22/4.12.13/26/24.6.11/21.12.9/8.4.18.14.14.18.13.20/

Have you ever wondered what is the meaning of some of the letters on the price tags on the articles in some shops?

The 36 is the price in pence which you are expected to pay for the article, but what does the *bd* mean? Some shops use a code to show the cost price of the goods. Each letter stands for a number.

If $a = 1, b = 2, c = 3$ and $d = 4$, then $bd = 24$p, $ab = 12$p, $cb = 32$p, etc.

If $a = 7, b = 5, c = 3$ and $d = 1$ then $bd = 51$p, $cd = 31$p, $ba = 57$p, etc.

EXERCISE 89

Find the prices indicated if $a = 6, b = 8, c = 4, d = 3$ and $e = 7$.

1. *ab*
2. *ce*
3. *ed*
4. *ba*
5. *eb*
6. *bc*
7. *ac*
8. *bd*
9. *da*
10. *aa*
11. *cd*
12. *de*
13. *cb*
14. *ea*
15. *dc*
16. *ad*
17. *ca*
18. *be*
19. *db*
20. *cc*

EXERCISE 90

Remember that $4a$ means $4 \times a$ and that $7b$ means $7 \times b$.

If $a = 3$, $b = 5$, $c = 2$ and $d = 9$, find the value of these.

1. $8a$	6. $3c$	11. $5c$	16. $6d$
2. $4b$	7. $2c$	12. $9a$	17. $4a$
3. $7c$	8. $9c$	13. $7b$	18. $5b$
4. $4d$	9. $2a$	14. $8d$	19. $7a$
5. $3b$	10. $11b$	15. $6a$	20. $6c$

This substitution of numbers for letters can also be used to find the value of expressions used in the first chapter.

If $p = 7$, $q = 4$, $r = 10$ and $s = 5$.

EXAMPLE 1

$$
\begin{aligned}
& p + q + r - s \\
= \; & 7 + 4 + 10 - 5 \\
= \; & 21 - 5 \\
= \; & \underline{16} \quad \text{Ans.}
\end{aligned}
$$

EXAMPLE 2

$$
\begin{aligned}
& 3p + 4q \\
= \; & (3 \times 7) + (4 \times 4) \\
= \; & 21 + 16 \\
= \; & \underline{37} \quad \text{Ans.}
\end{aligned}
$$

EXAMPLE 3

$$
\begin{aligned}
& 5r + 2s - 3p \\
= \; & 50 + 10 - 21 \\
= \; & 60 - 21 \\
= \; & \underline{39} \quad \text{Ans.}
\end{aligned}
$$

EXAMPLE 4

$$
\begin{aligned}
& 2p - 5q + r - 4s \\
= \; & 14 - 20 + 10 - 20 \\
= \; & 24 - 40 \\
= \; & \underline{-16} \quad \text{Ans.}
\end{aligned}
$$

EXERCISE 91

If $a = 10$, $b = 2$, $c = 5$ and $d = 7$, find the value of these.

1. $a + b$	11. $3a + 4b + c$
2. $b + c + d$	12. $5c - 2d + 4b$
3. $b + a - d$	13. $7b - 3a + 4c$
4. $c - b$	14. $2a + 3b + 4c$
5. $a - b - d$	15. $6d - 2a + 3b$
6. $b + c - d$	16. $3a - 2b - 4c + 2d$
7. $c - d - b$	17. $6c - d + 4b - 2a$
8. $a + b - c - d$	18. $3b - 3c + 4d - a$
9. $a - b + c + d$	19. $5a - 10c + 2b - d$
10. $d - c - a + b$	20. $3d - 4c + 2a - 5b$

EXAMPLE 5 Using the same values as in Exercise 91.

$$\begin{aligned} a^2c &= a \times a \times c \\ &= 10 \times 10 \times 5 \\ &= \underline{500} \text{ Ans.} \end{aligned}$$

EXAMPLE 6

$$\begin{aligned} \frac{b^4 d}{c} &= \frac{2 \times 2 \times 2 \times 2 \times 7}{5} \\ &= \frac{112}{5} \\ &= \underline{22\tfrac{2}{5}} \text{ Ans.} \end{aligned}$$

EXERCISE 92

If $a = 10$, $b = 2$, $c = 5$ and $d = 7$, find the value of these.

1. ab
2. bc^2
3. abc
4. ad^2
5. bc^2d
6. $bc \div d$
7. $ab \div cd$
8. $a^2b \div c$
9. $3cd^2 \div 2a$
10. $\dfrac{4ab^2c^2}{6cd}$
11. $\dfrac{a^2b}{cd^2}$
12. $\dfrac{abc}{d}$
13. $\dfrac{a^2bd}{c^2}$
14. $\dfrac{a}{bc}$
15. $\dfrac{d^2}{ab}$
16. $\dfrac{3c^2}{2a}$
17. $\dfrac{8bd^2}{4c^3}$
18. $\dfrac{20c}{a^2}$
19. $\dfrac{3b^2c}{4d}$
20. $\dfrac{5a^2b^2}{100c}$

In these multiplication sums a letter stands for a number, can you see what each letter stands for?

1. $\begin{array}{r} \text{C P} \\ \text{O} \\ \hline \text{A N N} \\ \hline \end{array}$

2. $\begin{array}{r} \text{A T} \\ \text{T} \\ \hline \text{T M} \\ \hline \end{array}$

3. $\begin{array}{r} \text{A C} \\ \text{S} \\ \hline \text{M P} \\ \hline \end{array}$

4. $\begin{array}{r} \text{A P} \\ \text{A P} \\ \hline \text{C C P} \\ \hline \end{array}$

17 Simple Equations

Five stamps cost 10p,
so each stamp costs?

$5s = 10$

$s = 10 \div 5$

$s = 2$

Each stamp costs 2p.

Seven boys had 98 marbles,
how many each was this?

$7b = 98$

$b = 98 \div 7$

$b = 14$

Each boy had 14 marbles.

In the second line we divided each side of the equations by the same number.

$5s = 10$
divide by 5
$s = 2$

$7b = 98$
divide by 7
$b = 14$

EXERCISE 93 Solve these equations.

1. $4a = 12$	**6.** $11g = 121$	**11.** $9e = 45$	**16.** $4t = 11$
2. $3h = 15$	**7.** $3r = 45$	**12.** $5q = 40$	**17.** $10c = 35$
3. $7f = 42$	**8.** $6b = 30$	**13.** $8k = 88$	**18.** $7w = 22$
4. $6m = 12$	**9.** $8v = 104$	**14.** $6s = 27$	**19.** $5h = 32$
5. $8j = 32$	**10.** $7p = 84$	**15.** $8d = 20$	**20.** $12n = 50$

EXERCISE 94 Make equations out of these and solve them.

1. The total weight of three children is 126 kg. What is their average weight?
2. The area of my lawn is 200 m^2. If its length is 20 m^2., what is its width?
3. Three numbers multiplied together are equal to 36. If two of the numbers are 2 and 6 what is the third?
4. The total fare for four children from Watford to London is 66p. How much does each pay?
5. Three dozen sandwiches were prepared for four boys to eat on a hike. How many was this for each boy?
6. The weight of three litres of water is 3000 g. What is the weight per litre.

7. Five acts in a concert were to take 40 minutes, how many minutes each were they allowed?
8. On my bookshelf 0·96 m. wide is a set of books of equal thickness. If there are 24 books in the set, how thick is each one?
9. A liner takes 30 days to travel to New Zealand 12 000 nm. away. How many nm. is this per day?
10. In a street of 50 houses live 200 people. What is the average per house?

While on holiday in Holland a few years ago, I went into a shop to buy a book. When I asked the price the shopkeeper replied in Dutch which I did not understand. I gave him 10 guilders after which he gave me the book and 6 guilders change. I had the book and 6 guilders while he had 10 guilders. How much did the book cost?

If $b + 6 = 10$ to find the cost we must get b on its own. If I take off my 6 guilders the shopkeeper must do the same, so $b = 10 - 6$

$$b = 4$$

The book cost 4 guilders.

Normally it takes me 25 minutes to get to school. If I spend 10 minutes walking to the station, for how long am I on the train?

$$t + 10 = 25$$

Take 10 from each side

$$t = 25 - 10$$

$$t = 15$$

I was on the train for 15 minutes.

EXERCISE 95

Make equations from these problems and solve them.

1. From a piece of ribbon 1 m. long, 20 cm. is cut. How much is left?
2. A piece of butter weighing 25 g. was left from a kg., how much had been used?
3. A boy bought some chocolate and had 2p change from 5p. How much did the chocolate cost?

4. From a can containing 42 litres, 15·2 litres were used. What quantity was left?
5. A stick 35 cm. long was held vertically in a stream. If 16 cm. of the stick showed above the surface, what was the depth of the water?
6. John, who is 1·35 m. tall, is 17 cm. taller than Mary. What is her height?
7. When full the petrol tank of my car holds 28 litres. After a journey of 224 km. I have 12 litres left, how much have I used? How many km. is this to the litre.
8. When lowering a 50 m. rope over a cliff, I have 15 m. left after the lower end has reached the bottom, how high is the cliff?
9. When fencing a chicken-run I found that I had 0·5 m. left from a 20 m. roll. What was the perimeter of the run?
10. I think of a number, add on eight and the result is twenty three. What was the number I thought of?

EXAMPLE 1

Putting up a clothes line we found that our 10 m. of line was 0·60 m. short of the second post.

How far apart were the posts?
Call the distance between the posts d, then $d - 0{\cdot}60 = 10$ m.

Add 0·60 to the line, this is the same as adding 0·60 m. to both sides of the equation $d = 10 + 0{\cdot}60$

$d = \underline{10{\cdot}60 \text{ m.}}$ Ans.

EXAMPLE 2

Having given out two dozen bottles of milk we found that we were nine bottles short. How many children were there in the class?

$$c - 9 = 24$$

Add 9 bottles to each side.

$$c = 24 + 9$$
$$c = 33$$

There were 33 children in the class.

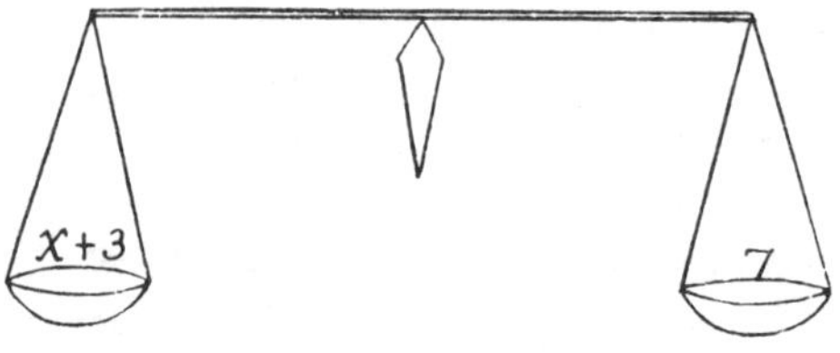

EQUATIONS MUST BALANCE

If you take 3 from the left hand side you must also take 3 from the right hand side or the scales will not balance.

Earlier we had

$b + 6 = 10$ and $t + 10 = 25$

$b = 10 - 6$ $t = 25 - 10$

Then in the second type of question

$d - 2 = 30$ $c - 9 = 24$

$d = 30 + 2$ $c = 24 + 9$

You will see from these examples that the figures have changed sides and the signs in front of them have also changed.

EXERCISE 96

Solve these equations.

1. $x + 5 = 8$
2. $c + 2 = 6$
3. $d + 4 = 9$
4. $p + 3 = 12$
5. $q + 4 = 6$
6. $r + 7 = 10$
7. $s + 4 = 8$
8. $10 + t = 15$
9. $6 + b = 9$
10. $f + 8 = 12$
11. $x - 3 = 4$
12. $d - 8 = 1$
13. $p - 5 = 2$
14. $m - 4 = 6$
15. $r - 9 = 3$
16. $y - 6 = 4$
17. $m - 3 = 7$
18. $z - 5 = 10$
19. $g - 3 = 5$
20. $h - 10 = 8$

EXERCISE 97

1. $2 = x + 3$
2. $13 = 2 + x$
3. $14 + x = 20$
4. $25 = 6 + x$
5. $16 = x - 4$
6. $21 = x - 7$
7. $39 = a - 1$
8. $14\frac{1}{2} = x - 3\frac{1}{2}$
9. $16\frac{1}{4} = x + 7\frac{1}{4}$
10. $7 = x - 3\frac{1}{2}$

I think of a number, multiply it by 5 then add on 7. The answer is 22, what was my number? Black Magic? No, just start at the end and work backwards.

The last thing I did was to add seven, so subtract 7.

$$22 - 7 = 15$$

Before that I multiplied by 5, so divide by 5.

$$15 \div 5 = 3$$

Writing this as an equation with x as number I first thought of we should have

$$5x + 7 = 22$$
$$5x = 22 - 7$$
$$5x = 15$$
$$\underline{x = 3} \quad \text{Ans.}$$

EXAMPLE 3

I think of another number, multiply it by 3 then take off 9. The answer now is 15, what was my number?

Suppose my number is n

$$3n - 9 = 15$$
$$3n = 15 + 9$$
$$3n = 24$$
$$\underline{n = 8} \quad \text{Ans.}$$

EXERCISE 98

Solve these equations.

1. $2b + 3 = 9$
2. $4n + 5 = 13$
3. $8v + 3 = 35$
4. $9d + 6 = 15$
5. $5m + 4 = 19$
6. $3y + 5 = 29$
7. $6p + 8 = 38$
8. $9s + 2 = 47$
9. $8m + 6 = 70$
10. $10a + 3 = 43$
11. $3g - 5 = 19$
12. $8r - 2 = 46$
13. $2w - 9 = 5$
14. $7s - 5 = 72$
15. $3r - 2 = 34$
16. $5x - 6 = 14$
17. $9d - 2 = 25$
18. $4j - 5 = 31$
19. $6s - 3 = 39$
20. $7f - 3 = 25$
21. $5m + 4 = 49$
22. $5h + 14 = 44$
23. $8e - 5 = 27$
24. $3h + 4 = 31$
25. $4c - 6 = 4$
26. $9g + 5 = 17$
27. $6f - 4 = 23$
28. $3a + 4 = 11$
29. $2t - 5 = 8$
30. $5z - 3 = 13$

These examples have unknown quantities on both sides.

EXAMPLE 4

$3x - 4 = x + 8$

$3x - x = 8 + 4$ Collect unknown quantities on 1 side and numbers on the other.

$2x = 12$

$\underline{x = 6}$ Ans.

EXAMPLE 5

$4x - 2 = 9x - 12$

$12 - 2 = 9x - 4x$ Collect terms.

$10 = 5x$

$2 = x$

or $\underline{x = 2}$ Ans.

You could collect the terms on the other side.

$4x - 2 = 9x - 12$

$4x - 9x = 2 - 12$

$-5x = -10$

$-x = -2$

$\underline{x = 2}$ Ans.

EXERCISE 99

1. $2x + 3 = x + 8$
2. $5x + 2 = 3x + 10$
3. $3x - 1 = x + 7$
4. $5x - 2 = 3x + 10$
5. $4x - 5 = x + 10$
6. $12x + 4 = 5x + 25$
7. $10x - 3 = 4x + 21$
8. $3x - 12 = x - 2$
9. $22x - 20 = 16x - 8$
10. $7x + 7 = 2x + 42$
11. $3x + 10 = 4x + 2$
12. $x - 5 = 3x - 15$
13. $12x - 3 = 4x + 29$
14. $25 - 7x = 2x - 2$
15. $13 - 5x = 1 - x$
16. $10x - 2 = 8 - 5x$
17. $14x + 3 = 10x + 11$
18. $7x - 2 = 5x + 14$
19. $6 - x = 3x - 10$
20. $2x + 4 = 5x - 2$

If I buy $\frac{1}{2}$ kg. of butter for 40p, what is the cost of a kg?
If I cycle 2 km. in $\frac{1}{4}$ of an hour what is my speed in km/h?

$$\frac{b}{2} = 40$$

multiply both sides by 2

$$b = 80\text{p}.$$

The butter cost 80p per kg.

$$\frac{d}{4} = 2 \text{ km.}$$

multiply both sides by 4

$$d = 8$$

My speed was 8 km/h.

EXAMPLE 6

$$\frac{x}{4} = \frac{3}{8}$$

$$x = \frac{3}{\cancel{8}_2} \times \frac{\cancel{4}^1}{1}$$

$x = 1\frac{1}{2}$ Ans.

EXAMPLE 7

$$\frac{p}{3} = \frac{4}{9}$$

$$p = \frac{4}{\cancel{9}_3} \times \frac{\cancel{3}^1}{1}$$

$p = 1\frac{1}{3}$ Ans.

EXERCISE 100

1. $\frac{x}{3} = 4$

2. $\frac{p}{5} = 2$

3. $\frac{r}{7} = 3$

4. $\frac{m}{8} = \frac{3}{4}$

5. $\frac{t}{6} = \frac{2}{9}$

6. $\frac{n}{5} = \frac{14}{15}$

7. $\frac{b}{10} = \frac{3}{5}$

8. $\frac{h}{4} = \frac{5}{6}$

9. $\frac{e}{11} = \frac{20}{33}$

10. $\frac{k}{8} = \frac{7}{16}$

11. $\frac{s}{6} = \frac{7}{12}$

12. $\frac{y}{4} = \frac{5}{6}$

13. $\frac{a}{2} = \frac{7}{8}$

14. $\frac{d}{9} = \frac{8}{15}$

15. $\frac{g}{12} = \frac{3}{4}$

16. $\frac{f}{4} = 1\frac{1}{2}$

17. $\frac{c}{7} = 3\frac{1}{7}$

18. $\frac{g}{9} = 2\frac{1}{3}$

19. $\frac{z}{8} = 3\frac{1}{4}$

20. $\frac{j}{6} = 3\frac{1}{3}$

Another use for Equations

In the drawing you see Jack and Jill on a see-saw. What would be the position of the plank if they were both the same weight?

The question of whether or not the boys in the above drawing will fall into the sea depends upon two facts:

(*a*) the point of balance of the plank and

(*b*) the weight of the body on the other end.

Bore a hole in the centre of a uniform piece of wood or metal and mark off your rod in equal units from the centre. Put a peg through the hole and fix it to the wall. Your rod should balance. If it does not, stick a small piece of plasticine to the lighter end.

If we now suspend a weight from the right hand side it should balance a similar weight suspended from the left hand side providing the weights are the same distance from the middle of the rod. We call the point of balance the pivot.

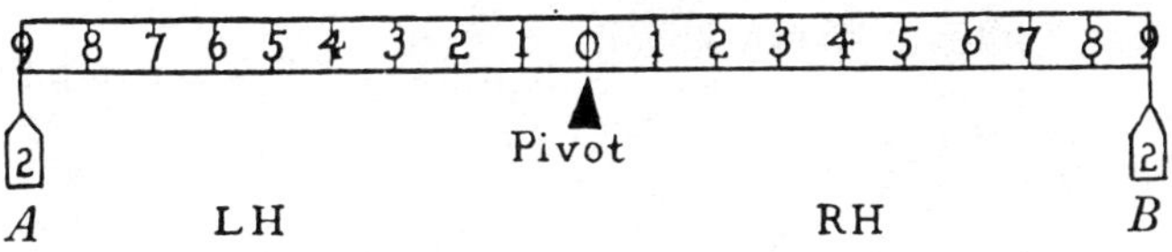

What happens if we slide weight A towards the pivot?

How can the rod be made to balance once more?

It will balance if we move B nearer to the pivot, but can you think of another way?

We can do it by adding more weight to the left hand side but where? This is where algebra comes to our aid.

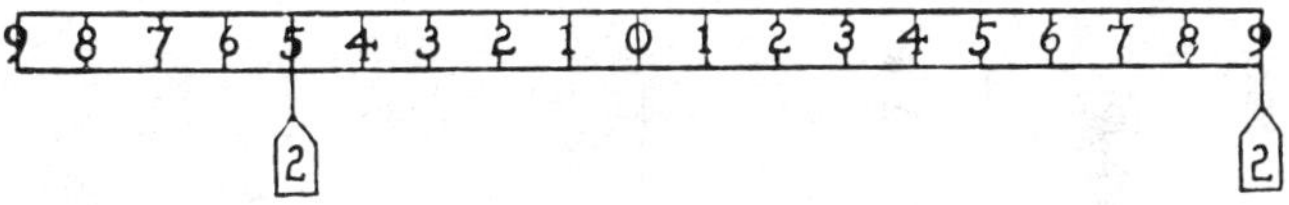

We have moved the left hand weight to a point which is 5 units from the pivot while the right hand weight is still 9 units from the pivot.

Where can we put a 1 kg. weight so that the rod again will balance?

EXAMPLE 8

Call the unknown distance from the pivot x units, then since we know that L.H. = R.H. to maintain balance we can solve the problem like this

$$\begin{aligned}
&\text{L Handside} &&= \text{R Handside}\\
&(2\text{ kg.} \times 5\text{ units dist.}) + (1\text{ kg.} \times x\text{ units dist.}) &&= (2\text{ kg.} \times 9\text{ units dist.})\\
&(2 \times 5) \quad + \quad (1 \times x) &&= (2 \times 9)\\
&10 \quad + \quad x &&= 18\\
& && x = 8
\end{aligned}$$

The weight of 1 kg. must be hung at 8 units from the pivot on the left hand side to maintain balance.

EXAMPLE 9

Now consider the next problem.

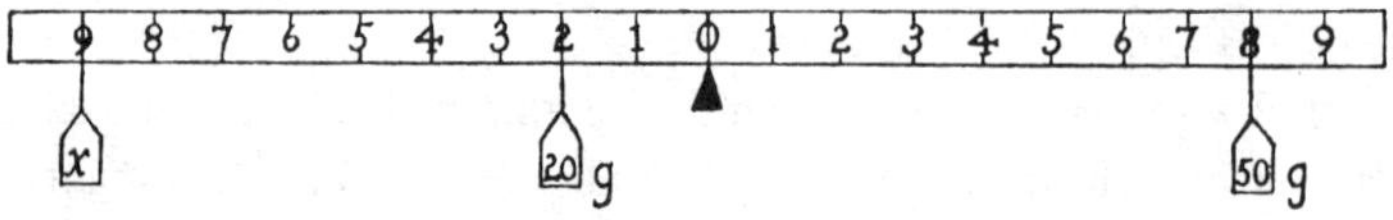

What weight would be required at a point 9 units to the left of the pivot to make this rod balance?
Let x represent the unknown weight.

$$\text{L.H.} = \text{R.H.}$$
$$(x \times 9) + (20 \times 2) = (50 \times 8)$$
$$9x + 40 = 400$$
$$9x = 400 - 40$$
$$9x = 360$$
$$x = 40 \text{ g.}$$

We should have to place a weight of 40 g. at 9 units from the pivot in order to maintain balance.

EXAMPLE 10

What weight at x will maintain the balance?

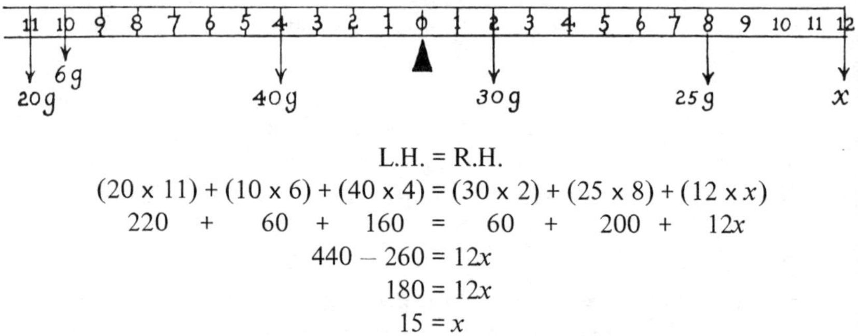

$$\text{L.H.} = \text{R.H.}$$
$$(20 \times 11) + (10 \times 6) + (40 \times 4) = (30 \times 2) + (25 \times 8) + (12 \times x)$$
$$220 + 60 + 160 = 60 + 200 + 12x$$
$$440 - 260 = 12x$$
$$180 = 12x$$
$$15 = x$$

A weight of 15 g. must be hung at 12 units on the right hand side of the pivot.

EXERCISE 101

In this exercise find the value of x.

1. 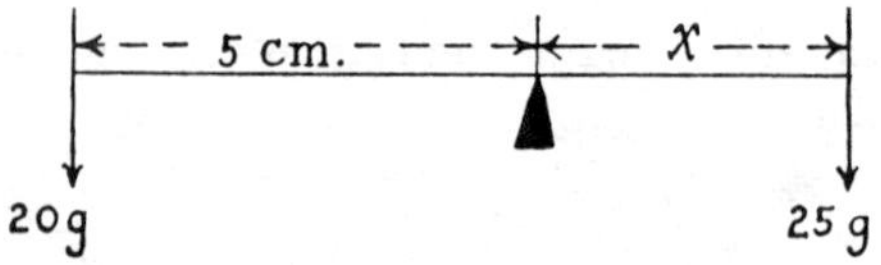

(2)

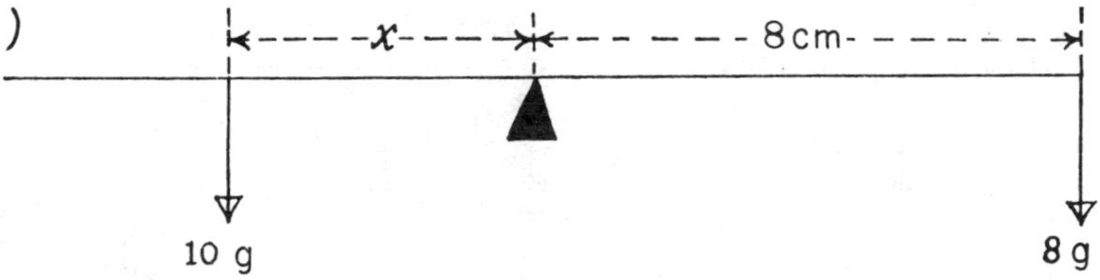
x
8 cm
10 g
8 g

(3)

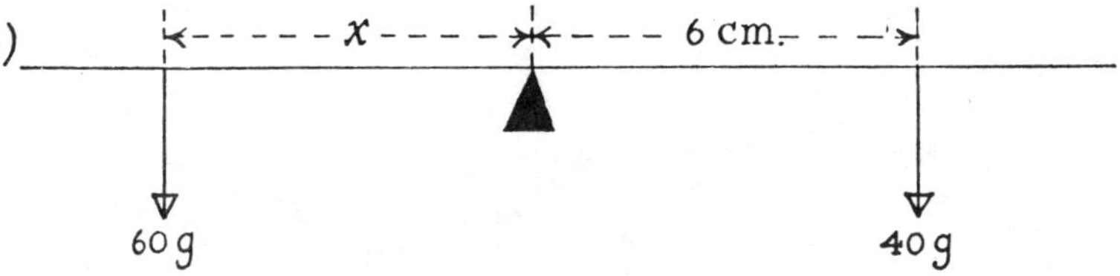
x
6 cm.
60 g
40 g

(4)

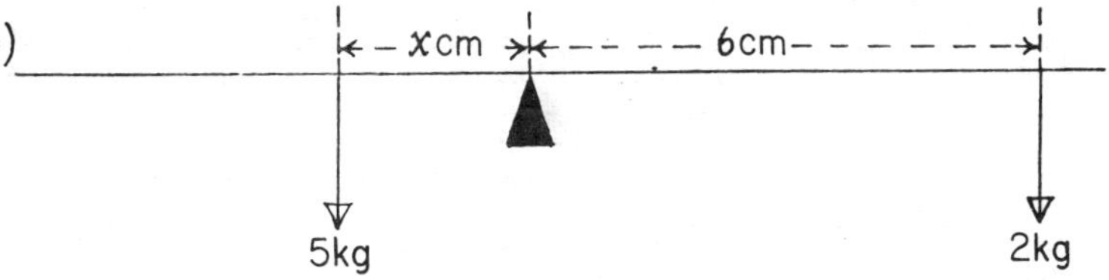
x cm
6cm
5kg
2kg

(5)

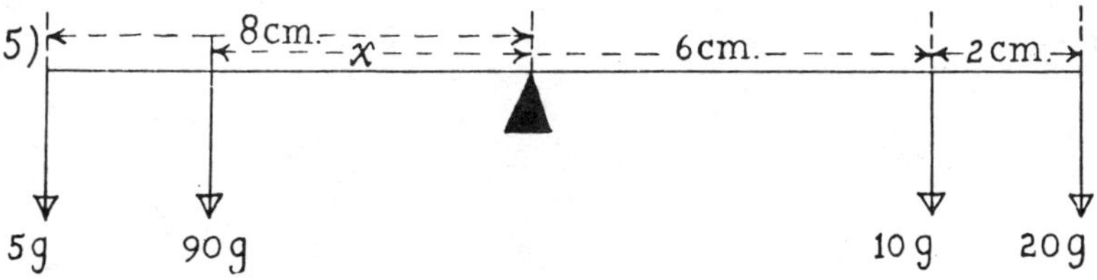
8cm.
x
6cm.
2cm.
5g
90g
10g
20g

(6)

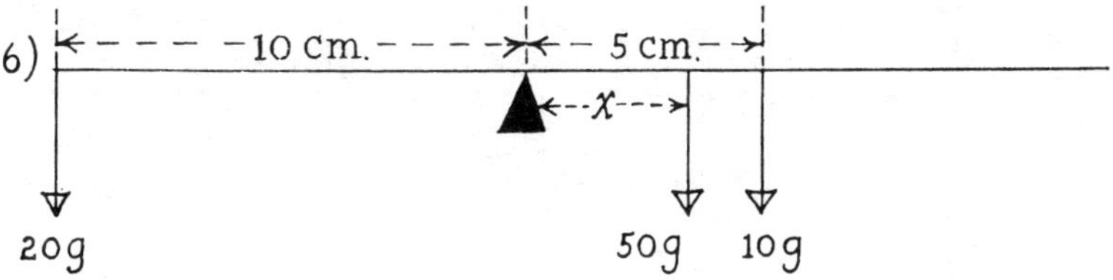
10 cm.
5 cm.
x
20g
50g
10g

(7)

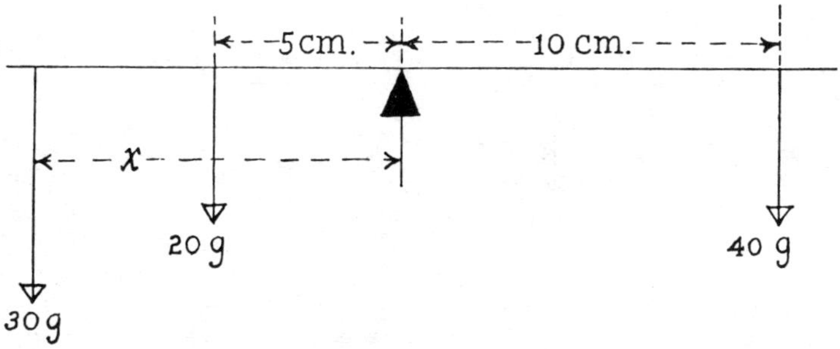
5cm.
10 cm.
x
20 g
40 g
30 g

(8)

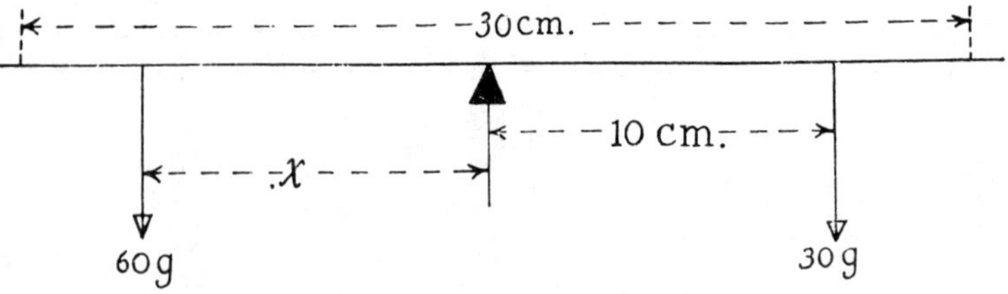
30cm.
10 cm.
x
60g
30g

(9)

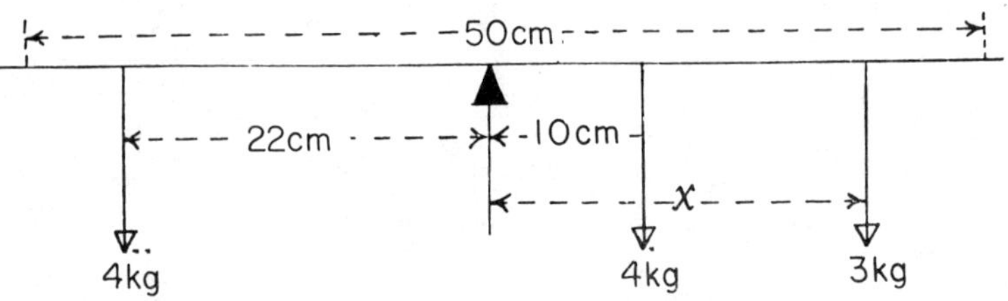
50cm
22cm
10cm
x
4kg
4kg
3kg

(10)

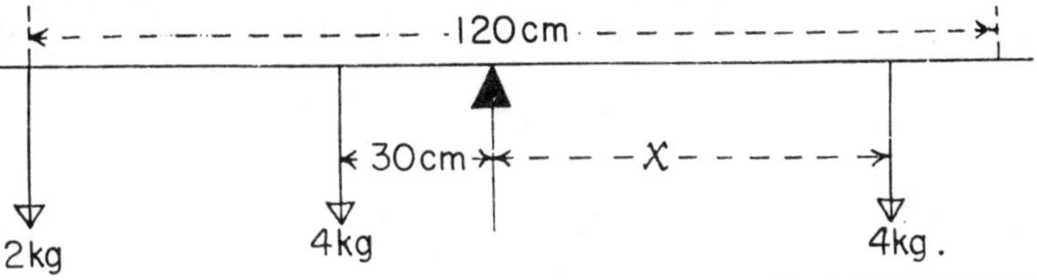
120cm
30cm
x
2kg
4kg
4kg.

(11)

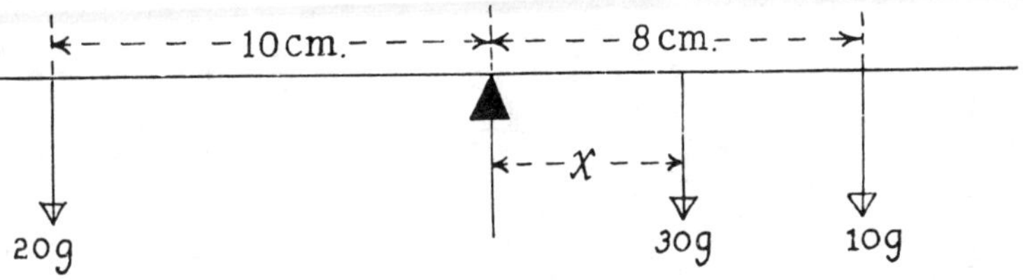
10cm.
8cm.
x
20g
30g
10g

(12)

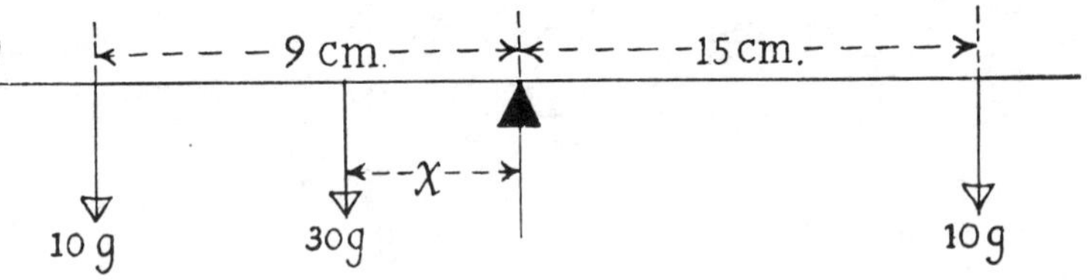
9 cm.
15cm.
x
10 g
30g
10g

(13)

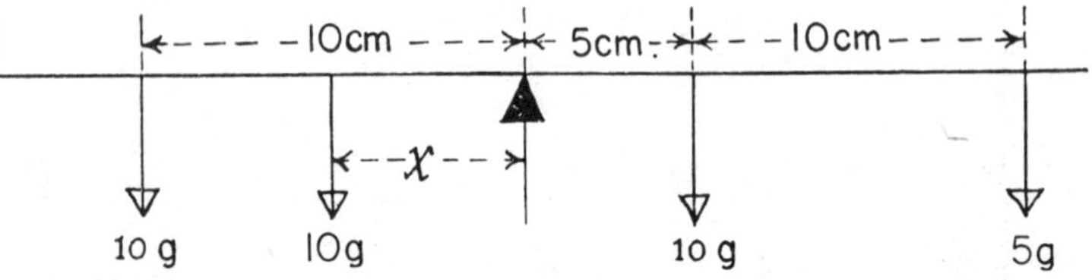
10cm
5cm.
10cm
x
10 g
10g
10 g
5g

(14)

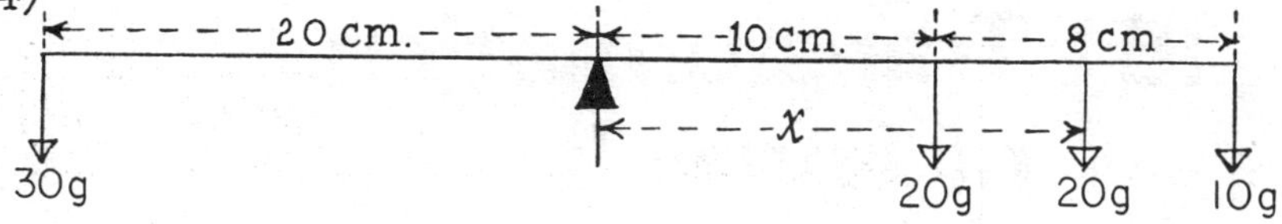

(15)

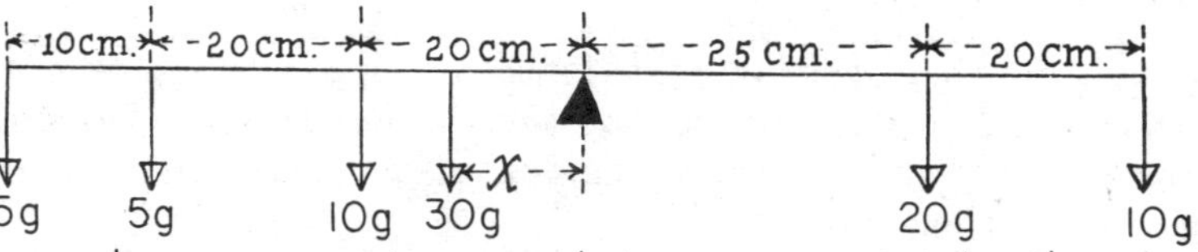

(16)

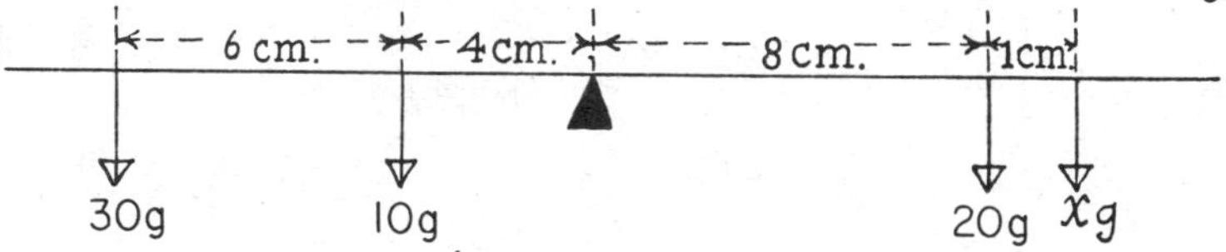

(17)

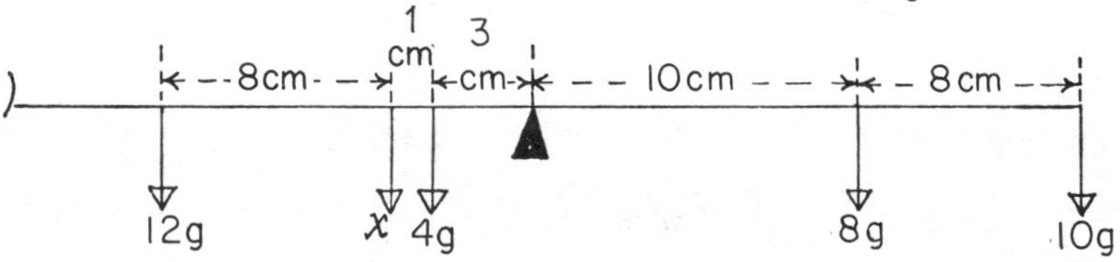

(18)

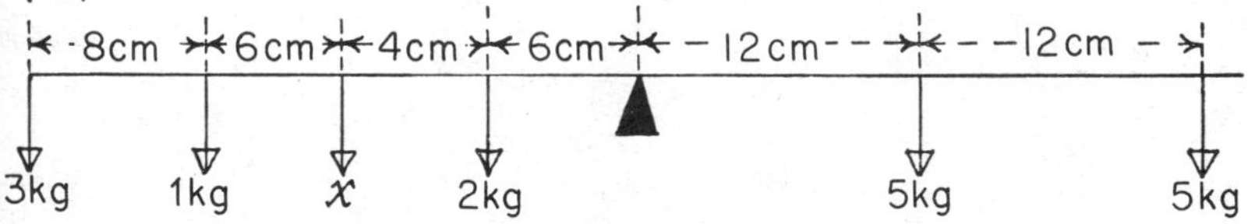

(19)

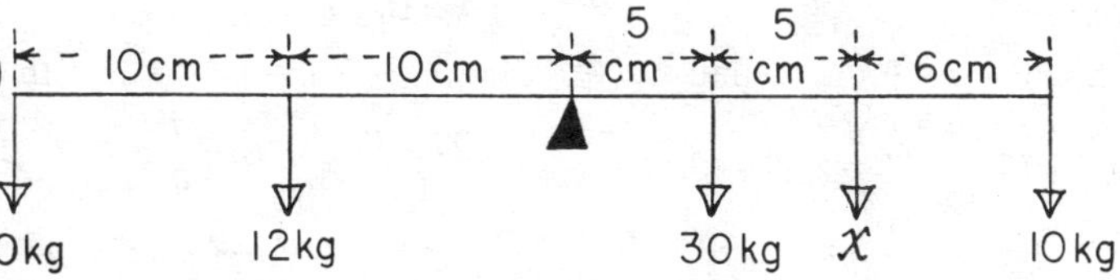

(20)

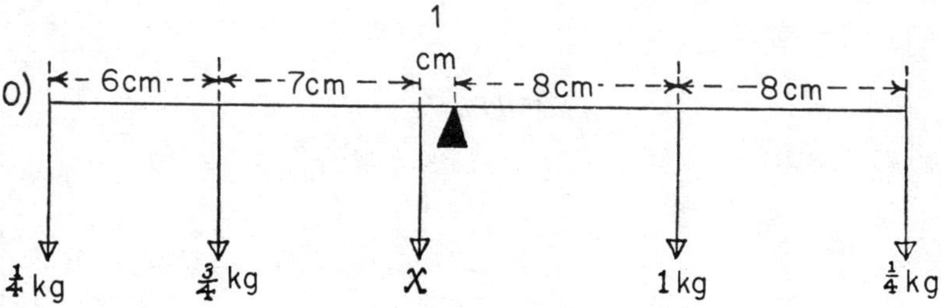

18 Transformation of Formulae

In the last few pages you have read a lot about changing the sign when you change the side, and you have solved many equations finding a numerical value for the letters. We are now going to juggle with the letters and express one letter in terms of others.

EXAMPLE 1

$a + b = c$ find *a*.

$a = c - b$

EXAMPLE 2

$r + s - t = p$, find *r*.

$r = p - s + t$

EXAMPLE 3

$\frac{m}{n} = q$, find *m*

$m = n \times q$

$m = qn$

EXAMPLE 4

$\frac{a}{x} = b$, find *x*.

$a = bx$

$x = \frac{a}{b}$

EXAMPLE 5

$x - y - z = h$, find *y*.

$x - z - h = y$

EXAMPLE 6

$3p + q = s$, find *p*.

$3p = s - q$

$p = \frac{s - q}{3}$

EXAMPLE 7

$d = \frac{5e}{f}$, find *e*.

$df = 5e$

$e = \frac{df}{5}$

EXERCISE 102

Change the subject of these expressions.

1. $d + e = f$,	find e	11. $a = lb$,	find b
2. $p + q = r$,	find p	12. $c = pd$,	find d
3. $m - n = t$,	find m	13. $t = \frac{ph}{2}$,	find h
4. $a - b - c = d$,	find a	14. $v = lbh$,	find l
5. $r - s = t + w$,	find w	15. $p = \frac{lbh}{3}$,	find b
6. $x + y = a - b$,	find b	16. $\frac{m}{p} = r$,	find m
7. $2m + n = r$,	find m	17. $\frac{rs}{t} = u$	find t
8. $a = b - 3c$,	find b	18. $ab = \frac{c}{d}$,	find b
9. $2p - q = 3r - s$,	find s	19. $\frac{g}{h} = \frac{j}{k}$,	find h
10. $4g + 2h - 3j = 2k$,	find j	20. $\frac{pq}{r} = st$,	find q

EXERCISE 103

Change the following formulae before substituting the numbers for the letters.

1. $c = 2\pi r$,	find r	when $c = 14$, $\pi = 3\frac{1}{7}$
2. $s = \frac{m}{v}$,	find v	when $s = 6$, $m = 30$
3. $pv = k$,	find v	when $k = 8$, $p = 2$
4. $m = \frac{wo}{d}$,	find d	when $m = 6$, $w = 8$, $o = 9$
5. $u = \frac{f}{r}$,	find f	when $u = 0{\cdot}2$, $r = 10$
6. $v = u + at$,	find t	when $v = 44$, $u = 0$, $a = 5$
7. $a = \frac{L}{e}$,	find L	when $e = 50$, $a = 0{\cdot}5$
8. $r = \frac{D}{d}$	find r	when $D = 40$, $d = 4$

9. $c = \dfrac{v}{u + t}$,	find v	when $u = 14, t = 0{\cdot}2, c = 8$
10. $f = \dfrac{r}{2}$,	find r	when $f = 10{\cdot}5$
11. $s = (u + v)t$	find t	when $s = 80, u = 2, v = 38$
12. $a = p(1 + r)$,	find p	when $a = 150, r = 0{\cdot}5$
13. $p = x(L + b)$,	find x	when $p = 28, L = 4, b = 2$
14. $I = \dfrac{prt}{100}$,	find p	when $r = 3, t = 4, I = 12$
15. $I = \dfrac{prt}{100}$,	find t	when $I = 15, p = 300, r = 2\frac{1}{2}$

19 Graphs

Picture and Column Graphs

This diagram shows the number of ices bought at a small shop in a week in February. It is easy enough to see on which day they sold the most and on which days they sold the least ices, but how many did they sell? I suppose that all of you can count up to 40 so that question will soon be answered.

The picture shows the number of ices bought one day in a week in July. Would you like to count them? We could show the actual number by using a large cone to represent 100 ices, a smaller one to represent 10 and a little one like the ones on the previous page for the odd ones. Then our diagram for the week would look like this.

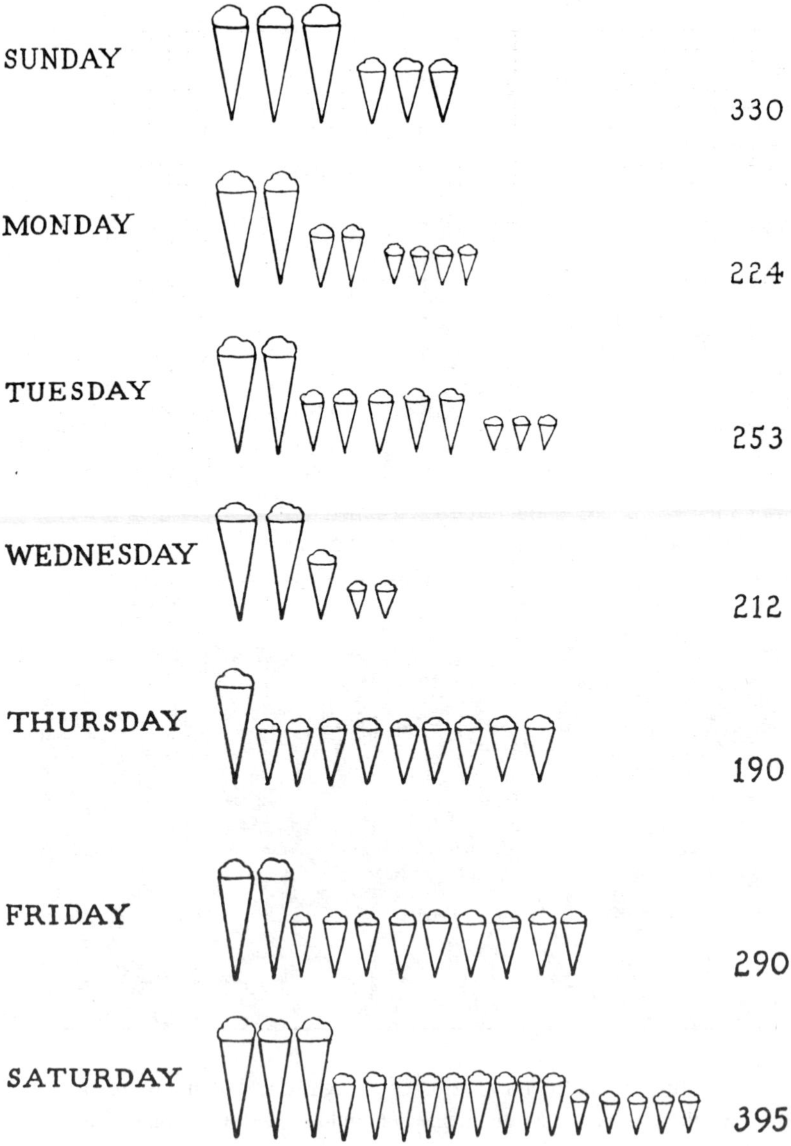

There is a simpler way of illustrating the consumption of ices by using a squared paper. To show the February figures we use 1 square to represent 1 ice on the vertical scale and make a column 5 squares wide for each day's sales.

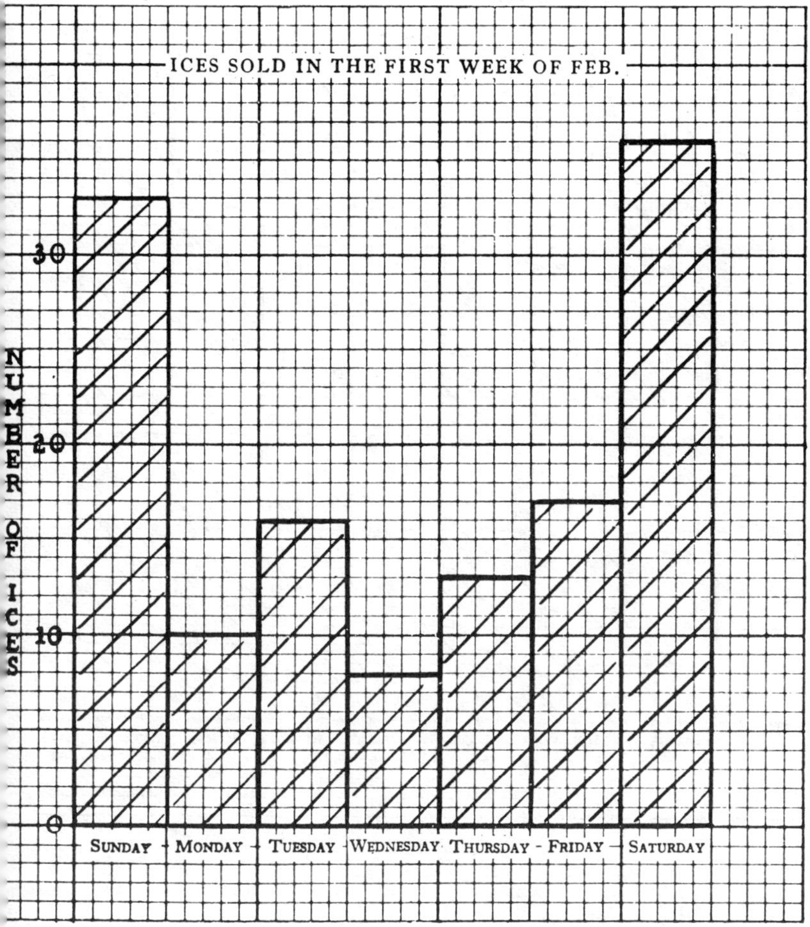

We would need an awfully long sheet of paper if we tried to illustrate the July figures to the same scale, so just as we used the larger cone to represent 10 ices we use each small square for ten. 10 squares will then represent 100 ices and the odd ones will have to be marked in between the lines.

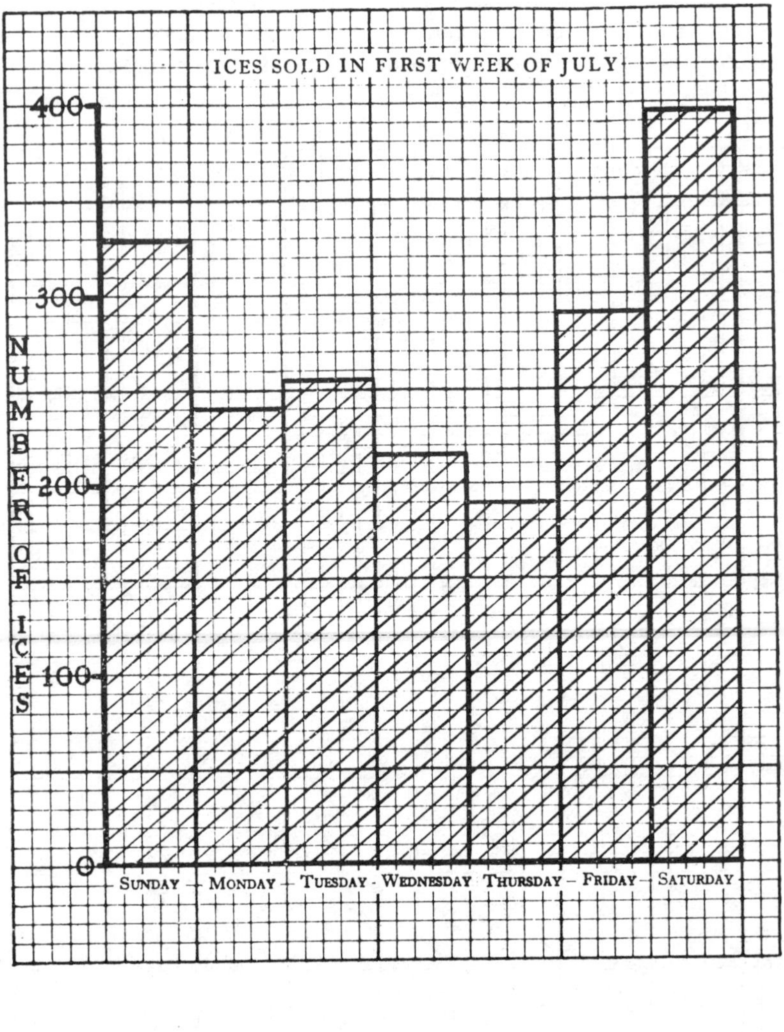

Looking at the two graphs answer these questions.

1. On which 2 days were most ices sold in February?
2. On which 2 days were most ices sold in July?
3. Can you think of any reasons for this?
4. Which day in February do you think was the coldest? Why?
5. How does the fact that many children receive their pocket money on Friday affect sales?

In a series of tests a boy gained the following marks:

1. 73%	5. 40%
2. 65%	6. 56%
3. 32%	7. 62%
4. 45%	8. 70%

What scale should we use to make a graph of these figures?

One small square to represent 1 mark is an easy scale to use if your paper is big enough. If not, use 1 small square to represent 2 marks instead, as we have done here.

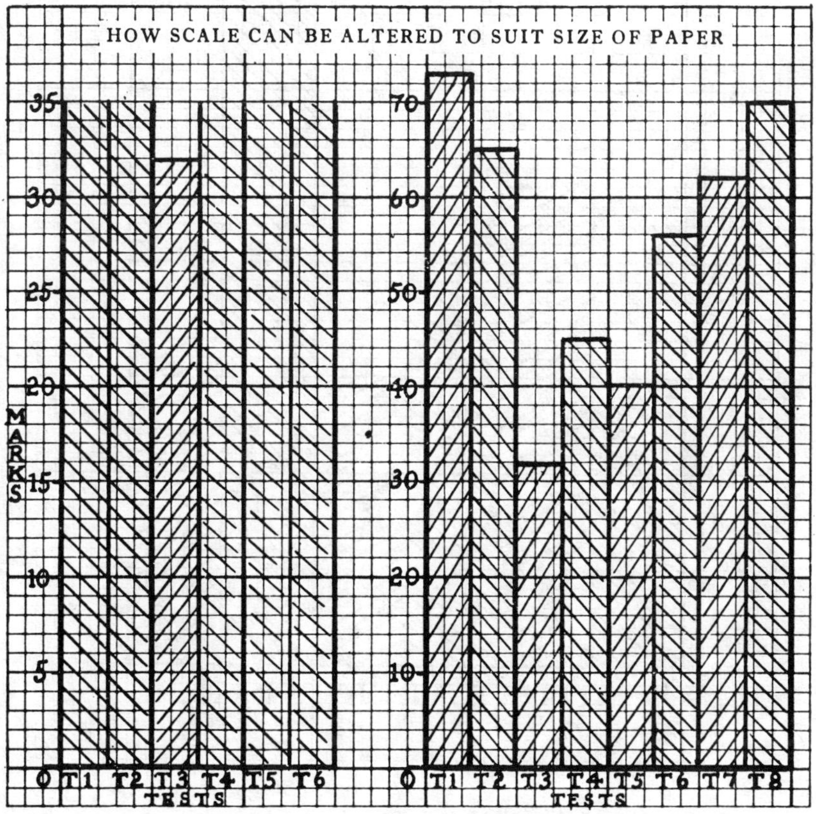

From the graph you will get a good idea of his progress. Can you think of a reason for his low mark in Test 3? Did he make steady improvement afterwards?

You can quite easily keep a record of your own test marks like this and so keep a check on your own progress.

This graph shows the amount of pocket money which 6 boys are given each week.

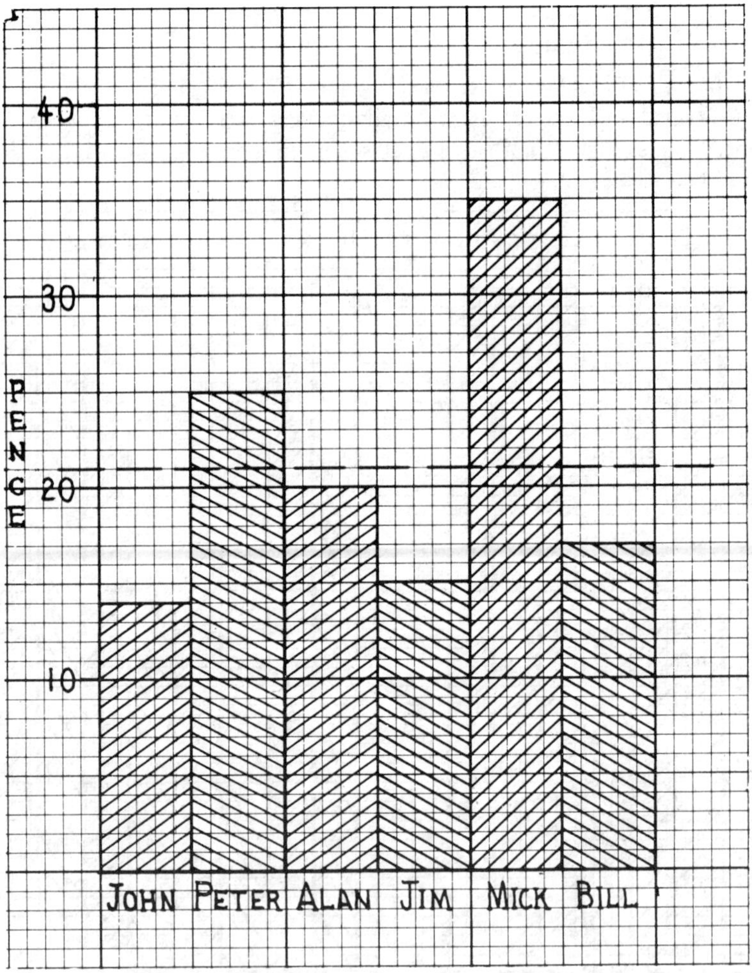

How much are they given altogether? Suppose this amount was divided equally among them, how much each would they receive? The dotted line across the graph actually shows this amount and is called the average.

You will learn more about averages when you study Chapter 11 if you have not already done so.

EXERCISE 104

1. Draw the following graph using a scale of 1 small square to represent 1 kg. vertically, and a scale of 5 small squares horizontally to represent each child.

Sheila	34 kg.	Jack	31 kg.
Margaret	30 kg.	Phil	29 kg.
Pat	35 kg.	Simon	33 kg.

When you have drawn the graph work out the average weight of these children and mark it with a dotted line.

2. Choose your own scale to draw this graph to show the heights of these eight children.

Joy	122 cm.	Peter	122 cm.
Mary	118 cm.	Michael	118 cm.
Betty	115 cm.	John	123 cm.
Jane	120 cm.	Mark	116 cm.

Show their average height by a dotted line.

3. Draw a graph to show number of runs scored by a cricketer.

May	7	34	June	4	56
	14	72		11	92
	21	40		18	86
	28	24		25	68

What was his average score for these matches.

4. Draw a graph to show the amount of rain which fell in a week in March.

Sunday	2 mm.
Monday	5 mm.
Tuesday	3 mm.
Wednesday	7 mm.
Thursday	8 mm.
Friday	4 mm.
Saturday	6 mm.

To illustrate these figures clearly you will have to use 1 small square to represent 1 mm. Show the average.

Many graphs like these can be drawn from information which you find in your own classroom. You can find the ages, heights and weights of your friends, the attendance and dinner numbers for your form, the number of house-points gained by each house or the amount of rain which falls each week.

5. Draw the following graph to show the rainfall in mm. in a certain town in this country.

Jan	Feb	Mar	Apl	May	June	July	Aug	Sept	Oct	Nov	Dec
65	45	42	35	55	38	25	8	15	67	97	63

(*a*) In which month did most rain fall?
(*b*) Which was the driest month?
(*c*) February is often called February-fill-dyke, do these figures show that this title is well earned or not?

6.

Jan	Feb	Mar	Apl	May	June	July	Aug	Sept	Oct	Nov	Dec
2	4	12	35	92	63	72	118	53	32	47	8

These figures were recorded in a country south of the Equator, in mm. Plot a graph to a convenient scale, and answer these questions.

(*a*) In which months did most rain fall?
(*b*) In which months did least fall?
(*c*) In what country do you think this place may be situated?
(*d*) Is the average greater or smaller than the average of the graph in question 5?

7. Savings collected in Ridge Road.

Jan	Feb	Mar	Apl	May	June	July	Aug	Sept	Oct	Nov	Dec
£9	£12	£19	£20	£15	£18	£12	£3	£6	£5	£11	£8

Draw a column graph to illustrate these figures and from it answer the following questions.

(*a*) What was the total collected for the year?
(*b*) What was the average monthly saving?
(*c*) In which month was most saved?
(*d*) Why do you think people saved less in Aug, Sept, and Oct, and why was there another drop in Dec and Jan?

8. Number of people who had afternoon tea in Brown's Cafe.

Monday	Tuesday	Wednesday	Thursday	Friday	Saturday
48	88	62	12	75	160

(*a*) Which day would you say was half day closing in the town?
(*b*) Why do you think so many people had tea out on Saturday?
(*c*) Why do you suppose fewer people had tea in town on Monday?

Statistics

All the graphs that you have so far drawn have been column graphs. Because there was 10 mm. of rain on Monday and 20 mm. on Wednesday, there was not necessarily 15 mm. on Tuesday. There may not have been any rain at all on that day. These figures are discontinuous and must therefore be illustrated by blocks, or columns.

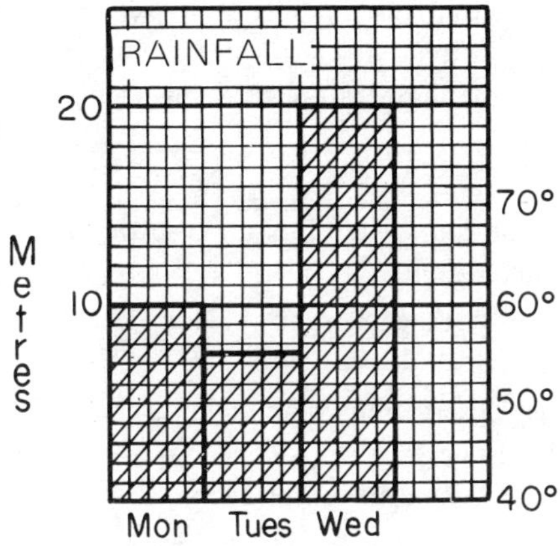

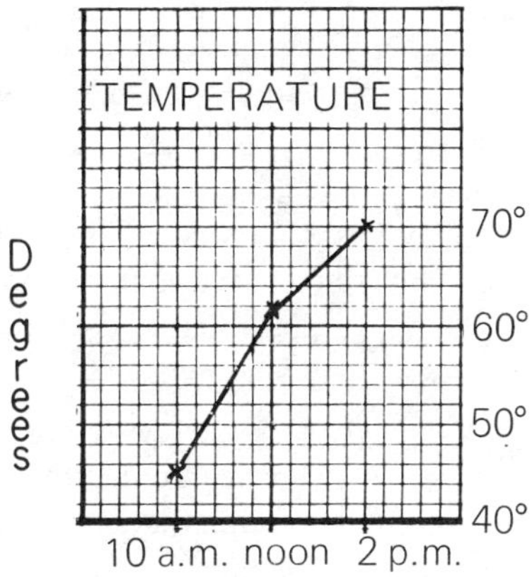

The temperature figures, however, are continuous and the line shows that the temperature rose between 10 a.m. and noon, and again between noon and 2 p.m.

(*a*) How many degrees did the temperature rise between 10 a.m. and 12 noon?
(*b*) How many degrees did it rise between 12 noon and 2 p.m.?
(*c*) Between what times was the greater rise in temperature?
(*d*) How do the slopes of the graphs show you this?
(*e*) Why should we not join the points if the temperature was read at noon each day?

EXERCISE 105

1. The following table shows temperature taken at school during one day in February in degrees celsius.

9 a.m.	10 a.m.	11 a.m.	12 noon	1 p.m.	2 p.m.	3 p.m.	4 p.m.
3°	4°	6°	7°	7°	8°	7·5°	7°

Plot these points on a graph and join them by a straight line.

(*a*) At what time of the day was the temperature lowest?
(*b*) At what time was the highest temperature recorded?

2. The average temperatures for each month in London were as follows;

Jan :	Feb :	Mar :	Apl :	May :	June :	July :	Aug :	Sept :	Oct :	Nov :	Dec:
3	7	8·5	10	11	15	17	16	15	12	8	6

(*a*) Draw a graph and put in a dotted line to indicate the average temperature for the year.
(*b*) For how many months was the temperature above average?

3. In this table is shown the number of hours when the sun is above the horizon. Draw a graph to show these results.

Jan :	Feb :	Mar :	Apl :	May :	June :	July :	Aug :	Sept :	Oct :	Nov :	Dec:
$8\frac{1}{2}$	10	$11\frac{1}{2}$	$13\frac{1}{2}$	$15\frac{1}{2}$	$16\frac{1}{2}$	16	$14\frac{1}{2}$	$12\frac{1}{2}$	9	8	$7\frac{1}{2}$

(*a*) Can you say whether these observations were taken in the Northern or Southern Hemisphere?
(*b*) During what months should this particular place have its Summer?

4. A baby's weight is recorded on the days after its birth as follows.

Days after birth	3	10	14	21	28	35
Weight in kg.	3·6	3·7	3·8	4·0	4·1	4·3

(*a*) What do you notice about your graph between the 3rd and 14th days?
((*b*) Is the increase in weight a uniform figure or does it vary?

5. This graph shows average speeds made by a racing driver in ten consecutive laps of the racing circuit.

Laps	1	2	3	4	5	6	7	8	9	10
Speed in km/h.	136,	128,	140,	138,	120,	144,	139,	140,	134,	144,

(*a*) Draw a dotted line on your graph to show the average speed.
(*b*) For how many laps was his average speed exceeded?

6. Boil a pint of water and note the temperature at intervals of 5 minutes while it cools. Your results may look like this.

Time in minutes	3	5	10	15	20	25	30
Temperature °C	90	76	68	62	58	54	54

Draw the graph with a vertical scale of 1 cm. to 10° and a horizontal scale of 1 cm. to 10 minutes. Join up the points of your graph by a smooth curve.

Is the rate of cooling constant?

7. The greatest distance that one can see varies with the height of the spectator. This table shows the greatest distance you can see from various heights, across a plain or out to sea.

Height in m. of spectator	1·5	6·0	15	30	
Distance spectator can see	4·6	9·4	14·9	21·1	in km.

Draw a smooth curve to pass through these points.

(*a*) What distance would you expect to see from a height of 3m.

(*b*) At what height would you need to be in order to see a distance of 10 km?

8. The following table gives the distance in metres in which a vehicle can be stopped on dry roads at different speeds.

Speed in km/h.	80	64	48	30
Distance in m.	55	36	18	14

(*a*) How far would the car travel in order to come to a stop from a speed of 72 km/h?

(*b*) How fast is a car travelling which can stop in 30 m?

9. This table shows the velocity of a stone at different times after release from a height of 200 m.

Time in seconds	1	2	3	4	5	6
Speed in m/sec.	9·8,	19·6,	29·4,	39·2,	49·0,	58·8

(*a*) What is the speed after $3\frac{1}{2}$sec., $4\frac{1}{2}$sec.

(*b*) At what time is the speed 54 m/sec.

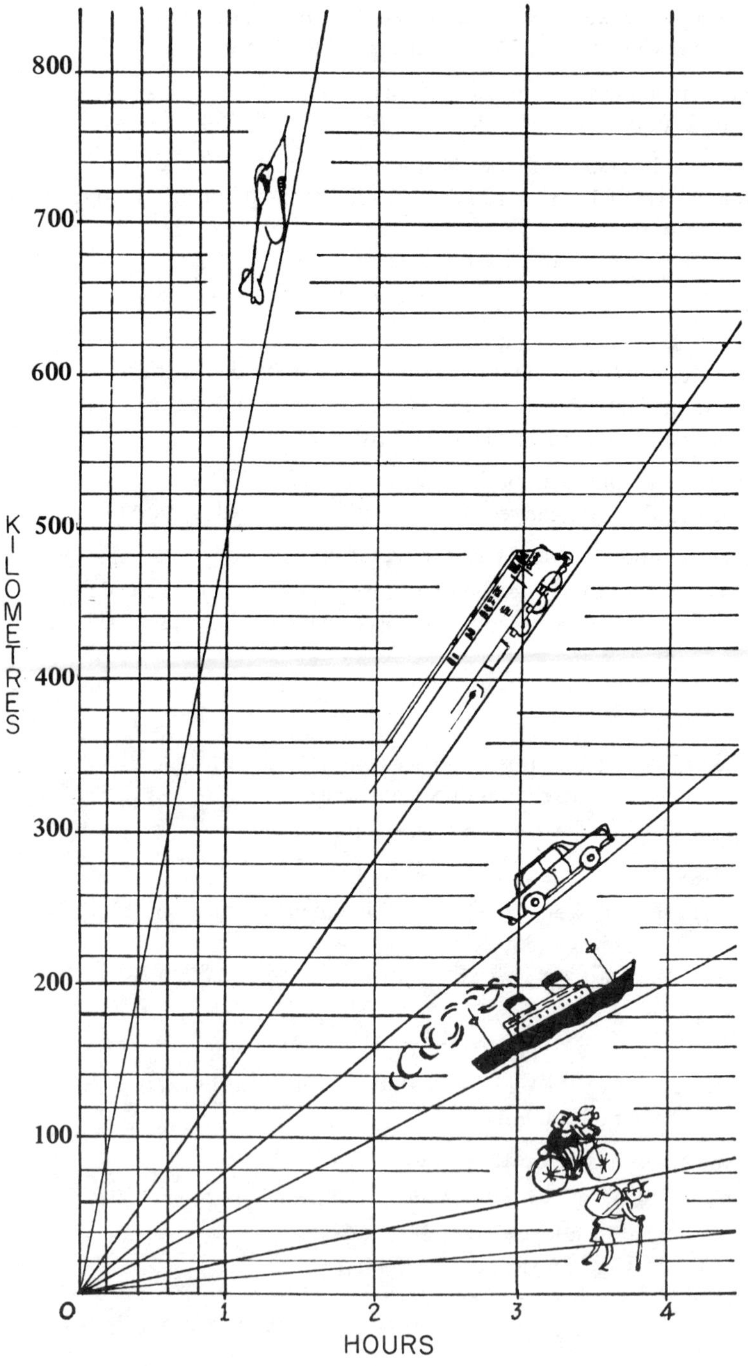
800
700
600
500
400
300
200
100
O
KILOMETRES
1
2
3
4
HOURS

Travel

You will have noticed when studying your temperature graphs that when there was a sharp rise in temperature, the graph was steeper, the angle of the line to the horizontal was greater.

Study the graph opposite the page.

(*a*) Which is the fastest means of transport shown?
(*b*) Which is the slowest?
(*c*) How many kilometres do they each go in one hour?
(*d*) How far from port was the ship after 4 hr.?
(*e*) How much farther than the ship did the car travel in 4 hr.?

Graphs like this will help us to solve other problems in involving travel.

EXAMPLE 1

Dennis was travelling from Stoke-on-Trent to Shrewsbury, a distance of 56 km. He started out on his bicycle at 10 a.m. and had travelled a distance of 24 km. by 11 a.m., when he stopped for a drink of coffee. He rested for half of an hour before carrying on with his journey, and arrived in Shrewsbury at 1 p.m. Geoffrey started from Shrewsbury at 11.30 a.m. and arrived by car in Stoke at 1 p.m.

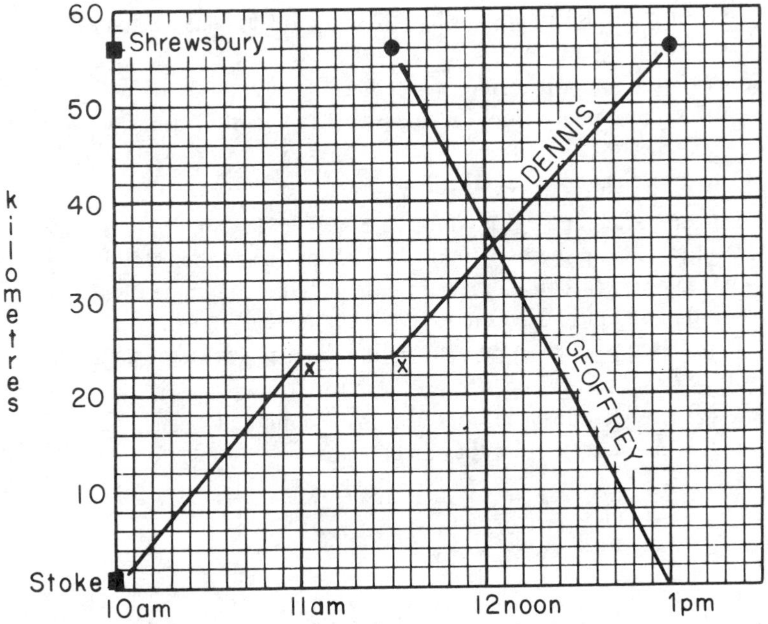

The point where the two lines intersect will tell us the distance from either Shrewsbury or Stoke where these two travellers meet. We can also read off Geoffrey's speed from the graph.

EXERCISE 106

1. An express starts from Manchester for London 300 km. away at 8.30 a.m. Its speed is 100 km/h. At 7.30 a.m. a train travelling at 75 km/h starts out from London for Manchester. At what distance from Manchester would they pass?
2. Alana started to cycle from Northampton to Cambridge a distance of 80 km., at 7.30 a.m. In the first hour she did 20 km. when she had a puncture. With help, she was on her way again after 45 minutes delay. She then made good time covering 24 km. in the next hour, but realising that she would be late getting to her destination, she abandoned her bicycle and proceeded by bus, arriving in Cambridge at 11.30 a.m. Lesley started from Cambridge for Northampton at 9.30 a.m. using public transport. She arrived in Northampton at 12 noon.
   (*a*) At what distance from Northampton did they pass each other?
   (*b*) Find Alana's average speed.
3. Georgina set off to walk to Rickmansworth a distance of 10 km. at a steady speed of 5 km/h. Alice started an hour later on her cycle, and arrived in Rickmansworth at 12 noon. If Georgina started at 10.30 a.m. what would be
   (*a*) the time when Alice passed her?
   (*b*) the distance from Rickmansworth where they passed (to nearest km.)?
   (*c*) average speed of Alice?
4. Doreen started out at 1 p.m. for Beaconsfield, 80 km. away, to see her mother. She did the first 16 km. in 20 minutes and then stopped to buy some.cakes, continuing her journey after a 10 minute break, at a speed of 48 km/h. Olive anxious to meet her daughter started out from Beaconsfield at 1.20 p.m. She travelled at a constant speed of 60 km/h.
   (*a*) At what distance from Beaconsfield did they meet?
   (*b*) At what time did they meet?

(*c*) For what time was Doreen travelling?
(*d*) If they spent 20 minutes in chatter before going on to Beaconsfield, at what speed must they travel in order to be there by 3 p.m.?

5. The local Youth Club planned to visit the Club in a town 40 km. away. The energetic members set off on cycles at 2 p.m. planning to average 16 km/h. The rest were to go on the 3 p.m. bus which was scheduled to arrive at 4 p.m.

(*a*) How long did the cyclists take?
(*b*) At what time did they arrive?
(*c*) At what time did the bus pass them?
(*d*) How far had they gone when the bus passed them?
(*e*) How long did the party that went by bus have to wait before the cyclists arrived?

6. Alan started out from Manor Park for South Oxhey, 48 km. away, at 8.30 on one crisp, November evening, but realised after 16 minutes that he had forgotten some exhibition pottery. He returned to Manor Park, picked up the pottery and once more set out for South Oxhey. After continuing for half an hour he stopped to call on a friend. This delayed him for 24 minutes. If he kept a constant speed of 48 km/h:

(*a*) How long did the journey take him including the return home?
(*b*) How many km. did he travel?
(*c*) How long should the journey have taken him?

7. A bank was robbed at 10 a.m. on Friday morning. The thieves made off in a small van which they abandoned 15 minutes later in a village 8 km. away. Here they stole a car and drove north at 64 km/h.

The Police were on the scene at 10.5 a.m. and within 20 minutes had found the abandoned van. They spent 5 minutes questioning people in the village then followed the car at 80 km/h. At what time did they catch up with it? How far had the thieves got from the scene of their crime?

Conversion

This graph can be used for quick conversion from one unit to another.

EXAMPLE 2

To convert km/h. to m/sec. given 90 km/h. = 25 m/sec.

On the horizontal scale let 2 cm. represent 10 m/sec. and on the vertical scale let 1 cm represent 10 km/h. Plot the point where the 90 km/h. vertical line cuts the 25 m/sec. horizontal line. Join this point to where the axes meet.

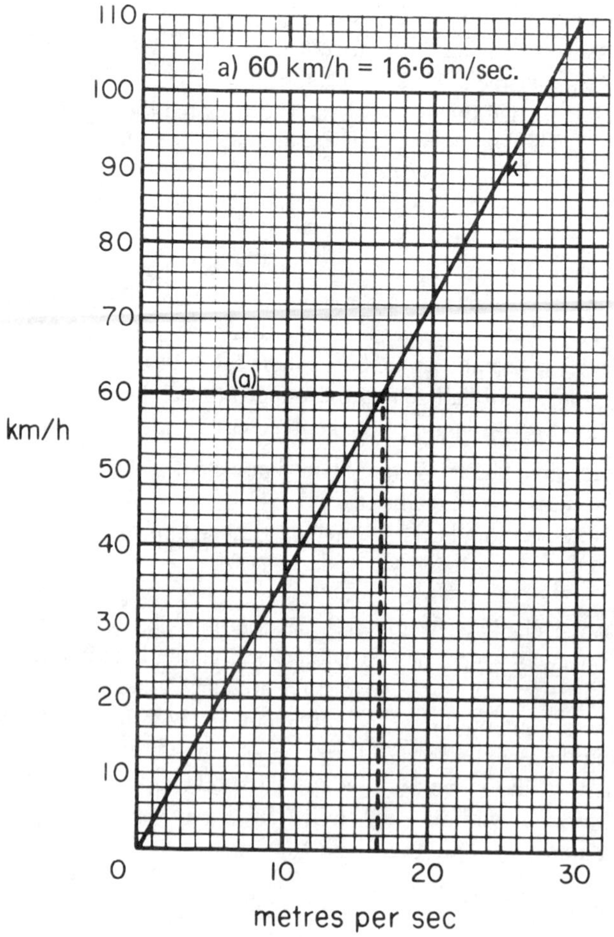

To convert 60 km/h. to m/sec., follow the 60 km/h. horizontal line until it meets the conversion line. From this point move vertically and read off speed in m/sec. In this case 16·6 m/sec.

Measure a distance of 30 m. on your playground. Measure the time in seconds it takes you to cover this distance running, walking, etc. If you took 10 seconds to cover this distance it would mean that you had travelled 3 m. in one second, by referring to your graph you will see that this is equivalent to a speed of 10·1 km/h.

EXERCISE 107

1. Use your graph to find the following:
   (*a*) John runs 16 m. in 10 seconds; find his speed in km/h.
   (*b*) Mavis can maintain a speed of 24 km/h. when cycling. What speed is this in m/sec.?
   (*c*) In order to run 1 km. in 4 minutes an athlete has to cover 4·2 m. in 1 sec. What speed is this in km/h?
   (*d*) Speed limits in a country are 50 km/h and 70 km/h. What are the speeds in m/sec?

2. The following are boiling and freezing points on the Fahrenheit and Celsius thermometers:

	Fahrenheit	*Celsius*
Boiling Point	212°	100°
Freezing Point	32°	0°

   Draw a graph for converting Fahrenheit to Celsius and vice versa.
   (*a*) The temperature of the body is 98·4° F. What is the equivalent on the Celsius scale?
   (*b*) A glass plate measures 400 cm^2 at 10° C and 400·36 cm^2 at 60° C. Rewrite this statement putting the temperatures in degrees Fahrenheit.

3. Milk costs 9p a litre. Draw a graph to illustrate this then use this to answer the following questions.
   (*a*) How much milk would I get for $40\frac{1}{2}$p?
   (*b*) A family drank 14 litres in a week. What did this cost?
   (*c*) The following week they only drank 9 litres. How much did this cost?
   (*d*) What was the difference in their milk bills for these two weeks?

4. The relationship between the circumference and diameter of a circle can be shown as a straight-line graph, since $C = \pi \times d$.

Mark the vertical scale to represent the circumference and the horizontal scale to represent the diameter in centimetres. Join the point where diameter = 10 cm. and circumference = 31·4 cm. to the origin.

(*a*) Find the circumference of circles having diameter 3, 7, 8, 11 cm.

(*b*) Find the diameters of circles having circumferences 6·28 cm., 28·26 cm., 15·7 cm., 21·98 cm.

We have used a straight line for converting one unit to another. A straight line may also be used to illustrate an algebraic expression.

In earlier exercises we have met equations such as $5 = x + 4$ where we have had to find the value of x. In this equation we can see that x is equal to 1 and it cannot in this particular case have any other value.

Instead of putting $x + 4 = 5$, put it equal to y.

Our equation now reads $y = x + 4$ and you will see from the following that for each value we give to x then there is a correspondingly different value for y.

EXAMPLE 3 If $x = 2$ then from $y = x + 4$

$y = 2 + 4$

$\underline{y = 6}$

If $x = 1$ then $y = 1 + 4$

$\underline{y = 5}$

If $x = 0$ then $y = 0 + 4$

$\underline{y = 4}$

If $x = -1$ then $y = -1 + 4$

$\underline{y = 3}$

If $x = -2$ then $y = -2 + 4$

$\underline{y = 2}$

Tabulate results:

x	−2	−1	0	1	2
y	2	3	4	5	6

You will notice that as x decreases uniformly then so does y. How are we going to show these points graphically? It would have been simpler if all the values had been positive, but here we have some negative values as well. Proceed as follows;

Draw the x and y axes at right angles to each other. Notice that the highest value for y is 6 and the lowest is 2, while x goes from 2 to -2. This should help you with your choice of scale.

To plot the point where $x = 2$ and $y = 6$ follow the vertical line from where x is 2 until it cuts the horizontal line where y is 6. I have shown the points on the graph and labelled them, e.g. (2, 6), x values always being given first.

What do you notice about all these points?

$$\text{If } x = -4 \text{ then from } y = x + 4$$
$$y = -4 + 4$$
$$\underline{y = 0}$$

Does this point fall on our straight line?

No matter how many points you work out and plot using this equation, they will always fall on this straight line, or this line produced.

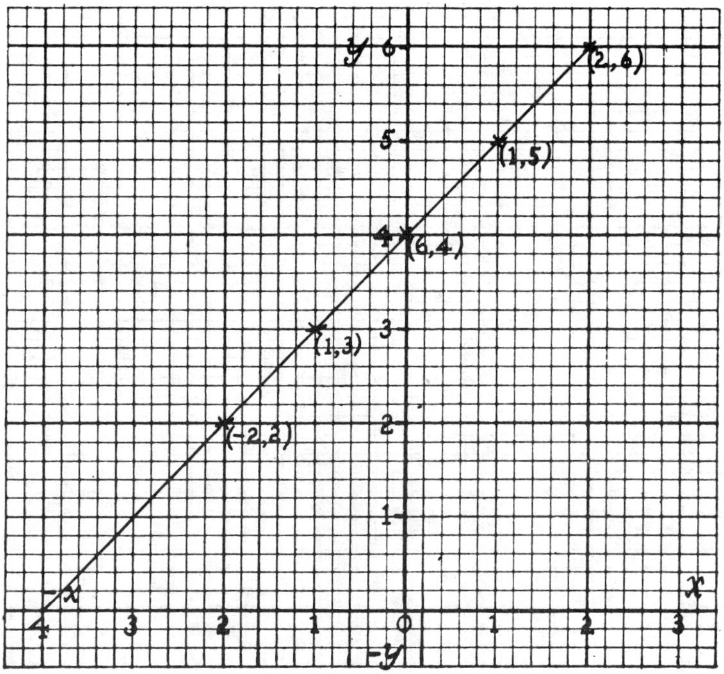

EXAMPLE 4

It is not necessary to have the same scale on the y axis as you have on the x axis. But you must have the same scale in $+x$ as $-x$, and on $+y$ as $-y$.

Using the equation $y = 2x + 2$.

If $x = 2$ then $y = (2 \times 2) + 2$ — $\underline{y = 6}$

If $x = -2$ then $y = (2 \times -2) + 2$ — $\underline{y = -2}$

Our table is:

x	-2	-1	0	1	2
y	-2	0	2	4	6

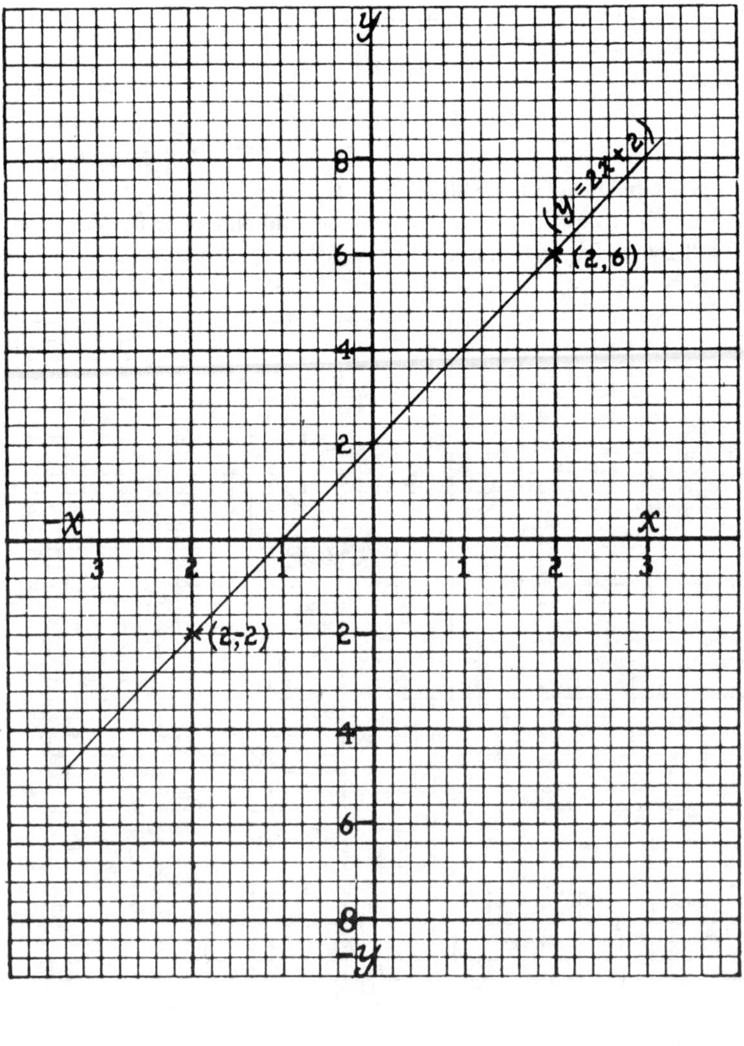

EXAMPLE 5

You can draw more than one graph using the same axes providing the scales are appropriate to each.

$y = 2 - x$	x	-2	-1	0	1	2	3
	y	4	3	2	1	0	-1
$y = 2x - 4$	x	-2	-1	0	1	2	3
	y	-8	-6	-4	-2	0	2

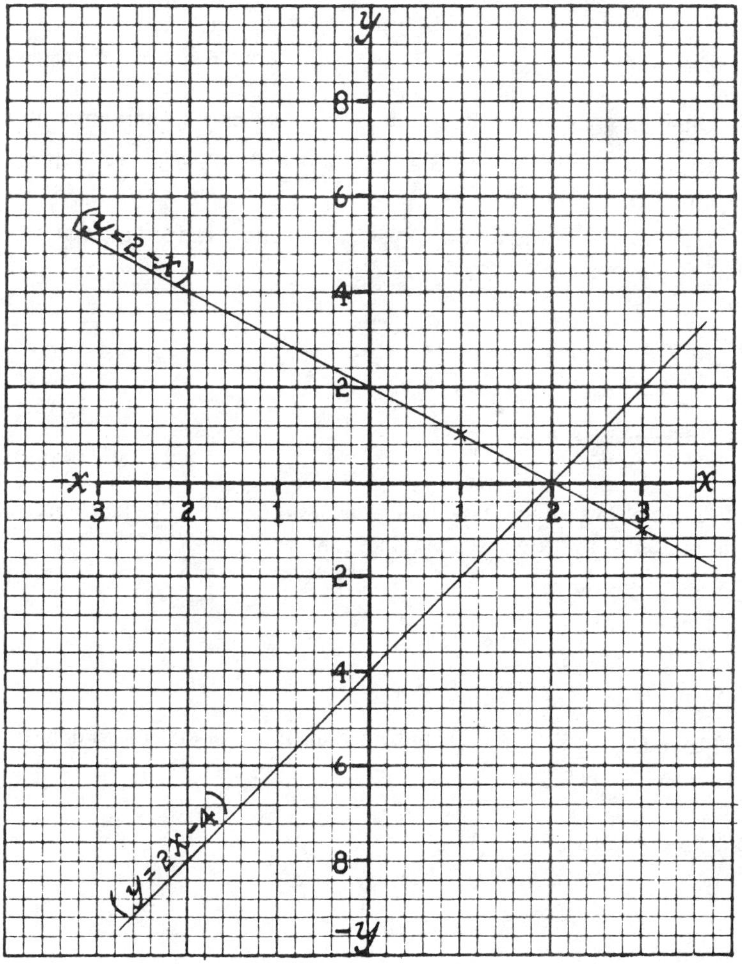

EXERCISE 108

Work out the respective values for x and y in these equations and plot their points joining them with a straight line.

Do you need to work out more than two values for each question?

1. $y = x + 2$
2. $y = x - 1$
3. $y = 6 - x$
4. $y = 2x + 2$
5. $y - x = 4$
6. $y = 2x - 2$
7. $y + x = 5$
8. $y = 3x - 1$
9. $y = 3x + 1$
10. $y = 3 - x$
11. Draw the graph of 1 and 3 on the same page and see where they cross.
12. Draw the graph of $y = x + 3$ and $y = 7 - x$. At what points do they cross?

Draw a graph of the following pairs of equations and say at what point they intersect.

13. $y = 2 + 2x$ and $y = x + 1$
14. $y = 3x - 2$ and $y = 2x + 2$
15. $y = 2x$ and $y = 3 - x$

In Example 3 I showed how to plot values for x and y. We labelled each point plotted with the respective values for x and y, e.g. (3, 7) means $x = 3, y = 7$.

EXAMPLE 6

Plot these points and join them by a straight line in the groups they are given.

(*a*) $(\frac{1}{2}, 4\frac{1}{2}), (3\frac{1}{2}, 1), (\frac{1}{2}, 1)$
(*b*) $(-\frac{1}{2}, 4), (-\frac{1}{2}, 1), (-2\frac{1}{2}, 1)$
(*c*) $(4, 0), (3\frac{1}{2}, -\frac{1}{2}), (-2\frac{1}{2}, -\frac{1}{2}), (-3, \frac{1}{2})$

What picture do we get?

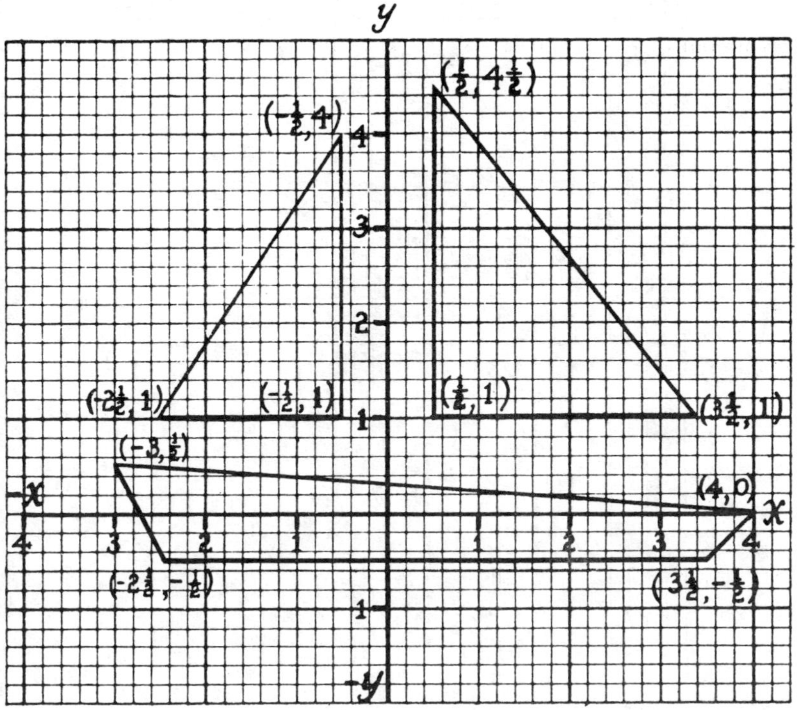

EXAMPLE 7

Plot the points (2, 2), (1, −2), (−2, 1) and join them by straight lines. What shape do we get? See figure over page.

In this triangle I have taken the side (−2, 1), (1, −2) as the base and by measuring I find it to be approximately 21 tenths. Perpendicular to the base I have drawn a line passing through (2, 2) which measures 18 tenths. I now have a triangle with base and height known, the area of which is

$$= \tfrac{1}{2} \times b \times h$$

$$= \frac{1}{\cancel{2}_1} \times \frac{21}{1} \times \frac{\overset{9}{\cancel{18}}}{1} \quad \text{sq. units.}$$

$$= \underline{189 \text{ sq. units.}}$$

You can make a rough check of this area by counting the number of squares it is covered by, don't forget to make up whole squares from the parts around the edge.

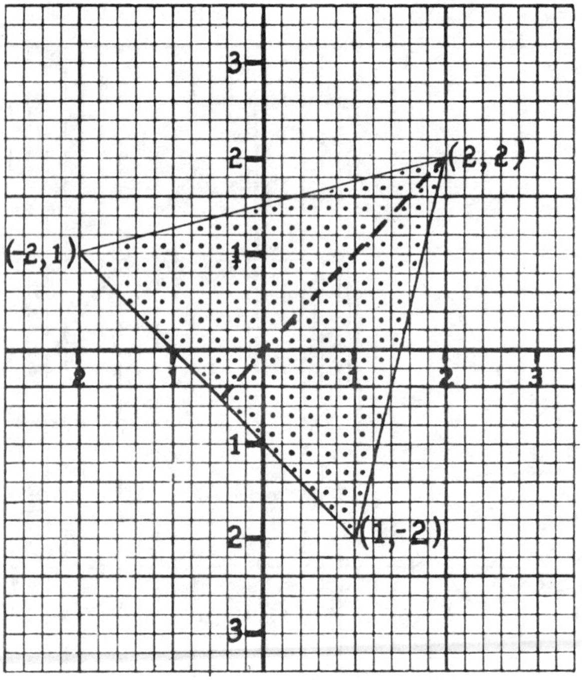

EXERCISE 109

1. Plot these points in the groups given.

   (*a*) $(4, 3\frac{1}{2}), (4, -1), (1\frac{1}{2}, -1), (1\frac{1}{2}, 3\frac{1}{2})$
   (*b*) $(3\frac{1}{2}, 3), (3\frac{1}{2}, 1\frac{1}{2}), (2, 1\frac{1}{2}), (2\frac{1}{2}, 3)$
   (*c*) $(1, 2), (1, -1), (-4, -1), (-4, 2)$
   (*d*) $(-3, 3\frac{1}{2}), (-3, 2\frac{1}{2}), (-3\frac{1}{2}, 2\frac{1}{2}), (-3\frac{1}{2}, 3\frac{1}{2})$
   (*e*) $(-2\frac{1}{2}, -1\frac{1}{2}), (-2, -2), (-2\frac{1}{2}, -2\frac{1}{2}), (-3, -2)$. Join these points with curve.
   (*f*) $(3, -1\frac{1}{2}), (3\frac{1}{2}, -2), (3, -2\frac{1}{2}), (2\frac{1}{2}, -2)$. Join these points with curve.

The following points when plotted will form triangles. Choose one of their sides as a base and draw a perpendicular to it passing through the third point. Calculate their area. (Use the same scale as example 7).

2. $(1, 2), (1, -1), (-2, 0)$
3. $(2, 1), (-1, 2), (-2, 0)$
4. $(1, 1), (-2, 0), (-1, 2)$
5. $(2, -1), (-2, -2), (), 2)$
6. $(1, -1), (-1, -1), (0, 2)$
7. $(1, 1), (1, -2), (-2, 0)$
8. $(0, 2), (2, -1), (-2, 0)$

Things that you have discovered.

1. Add

$$\begin{array}{r} 3a + 4b + 2c \\ 2a - 3b - c \\ \hline 5a + b + c \\ \hline \end{array}$$

2. Multiply $a \times b \times c$

$= \underline{abc}$

3. $2a \times 3b \times 4c$

$= \underline{24abc}$

4. $a^3 \times a^4 = a^{\overline{3+4}}$

$= \underline{a^7}$

5. $a^4 \div a^3 = a^{4-3}$

$= \underline{a}$

6. $(2a)^2 \times (4a)^2 = 2 \times 2 \times a \times a \times 4 \times 4 \times a \times a$

$= \underline{64a^4}$

7. $\dfrac{x^3p^2y}{x^2y} = \dfrac{x \times x \times x \times p \times p \times y}{x \times x \times y}$

$= \underline{xp^2}$

8. If $n = 4$, $p = 2$, and $q = 3$

$3n - 2p + q$

$= 12 - 4 + 3$

$= 15 - 4$

$= \underline{11}$

9. If $a = 10$, $b = 2$, $c = 5$

$\dfrac{a^2b^2c}{2ab}$

$= \dfrac{\overset{10}{\cancel{100}} \times \overset{\overset{1}{2}}{\cancel{4}} \times 5}{\underset{1}{\cancel{2}} \times \underset{1}{\cancel{10}} \times \underset{1}{\cancel{2}}}$

$= \underline{50}$

10. $2x + 4 = 3x + 8$

$-x = 4$

$\underline{x = -4}$

11. $2x + 4 = 5x - 2$

$4 + 2 = 5x - 2x$

$6 = 3x$

$\underline{x = 2}$

12. $\dfrac{a}{16} = \dfrac{4}{5}$

$5a = 64$

$\underline{a = 12\frac{4}{5}}$

13. If $a + b = c$

$\underline{b = c - a}$

14. If $v = \dfrac{lbh}{3}$

$\underline{h = \dfrac{3v}{lb}}$

15.

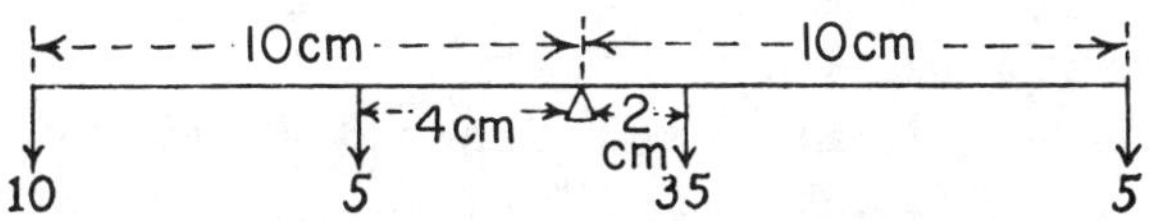

$(10 \times 10) + (5 \times 4) = (35 \times 2) + (5 \times 10)$

20 Revision

In Exercises 110 and 111, plot the paths indicated to scale. State direction and distance you must travel to reach base.

EXERCISE 110

1. Walk 10 paces North and then 10 paces to the East. Now walk towards the South West for 7 paces.
2. Walk East for 10 paces, turn towards the North West and walk 14 paces.
3. Take 10 paces South, turn to North East and walk 14 paces.
4. Walk South East 14 paces, then 10 paces North followed by 5 paces to the West.
5. After walking 10 paces North turn to the North East, walk 14 paces, then 10 paces South.

EXERCISE 111

1. Walk 14 paces South West and then walk 20 paces East.
2. 14 paces South East from your base turn North and take 5 paces, then proceed in a Westerly direction for 10 paces.
3. When you have walked 10 paces East, take 5 paces to the North and then 14 paces to the North West.
4. Take 10 paces South, 10 paces East, 7 paces North West, 7 North East and 5 West.
5. Walk 10 paces South, 7 East and 14 paces North West.

EXERCISE 112

Choose a suitable scale and plot the following bearings.

1. On a bearing of 120° take 20 paces. From this point move 13 paces on a bearing of 000°. What bearing must you take and what distance must you travel in order to reach base?
2. Walk 17 paces on a bearing of 060°, turn on to 180° and walk 9 paces. What is the distance and bearing of base.

3. Walk 10 paces on a bearing of 090°, turn on to 220° and walk $15\frac{1}{2}$ paces.
   (*a*) How far are you from base?
   (*b*) What bearing are you from base?
4. On a bearing of 330° walk for 20 paces. Turn on to a heading of 180° and walk for $17\frac{1}{2}$ paces. What is your bearing and distance from base?
5. Walk 15 paces on a bearing 270°, turn on to a heading of 045°, walk for 21 paces.
   (*a*) What is your distance from base?
   (*b*) What is the bearing of base?

EXERCISE 113

1. Walk in direction of 015° for 18 paces, turn on to 150° and walk 20 paces. What is your bearing and distance from base?
2. When you have moved along a bearing of 070° for 20 paces, turn on to a heading of 180° for 17 paces. Give the bearing and distance of base from you.
3. Turn on to 201° after walking along 060° for 17 paces, walk on this heading for 23 paces. What bearing and distance is base?
4. Walk along 020° for 15 paces, then turn on to a heading of 130° for 15 paces. Find the bearing and distance of base from your present position.
5. Walk 20 paces in a direction of 050°, then walk along 090° for 15 paces, make a final turn on to 180° and walk for 13 paces. What is the bearing and distance of base from your present position?

EXERCISE 114

1. From an open area the following were observed. Using a scale of 1 cm. to 10 m. plot their positions.

	OBJECT	BEARING	DISTANCE
(*a*)	Tree	040°	60 metres
(*b*)	Well	090°	40 metres
(*c*)	Gate	110°	20 metres
(*d*)	Chimney stack	180°	80 metres
(*e*)	River bridge	200°	70 metres
(*f*)	Hall	270°	80 metres
(*g*)	Hay-rick	300°	30 metres
(*h*)	Farm	000°	50 metres

2. Plot the following positions to scale of 1 cm. to 10 m.

	OBJECT	BEARING	DISTANCE
(*a*)	Copse	025°	100 metres
(*b*)	Viaduct	050°	80 metres
(*c*)	Five bar gate	100°	40 metres
(*d*)	Pond	120°	50 metres
(*e*)	Road Junction	230°	70 metres
(*f*)	House	270°	90 metres
(*g*)	Church	295°	60 metres
(*h*)	Farm	320°	40 metres

3. Choose your scale and plot these positions.

	OBJECT	BEARING	DISTANCE
(*a*)	Village	020°	2 kilometres
(*b*)	Lighthouse	087°	6 kilometres
(*c*)	Castle	135°	$2\frac{1}{2}$ kilometres
(*d*)	Bay	184°	3 kilometres
(*e*)	Ship	240°	4 kilometres
(*f*)	Island	260°	2 kilometres
(*g*)	Inlet	293°	3 kilometres
(*h*)	Mountain peak	000°	8 kilometres

4. Choose your own scale for plotting these.

	OBJECT	BEARING	DISTANCE
(*a*)	Railway station	012°	6 kilometres
(*b*)	Nearest point to sea	065°	4 kilometres
(*c*)	Town	090°	3 kilometres
(*d*)	Pleasure beach	104°	3 kilometres
(*e*)	Caravan site	160°	3 kilometres
(*f*)	Lighthouse	234°	6 kilometres
(*g*)	Town	251°	3 kilometres
(*h*)	Golf course	275°	2 kilometres

EXERCISE 115

1. *ABC* is straight line. Find y

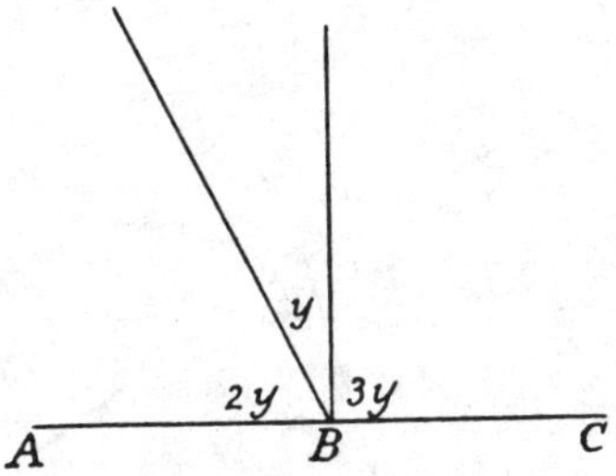

2. Find a and size of each angle.

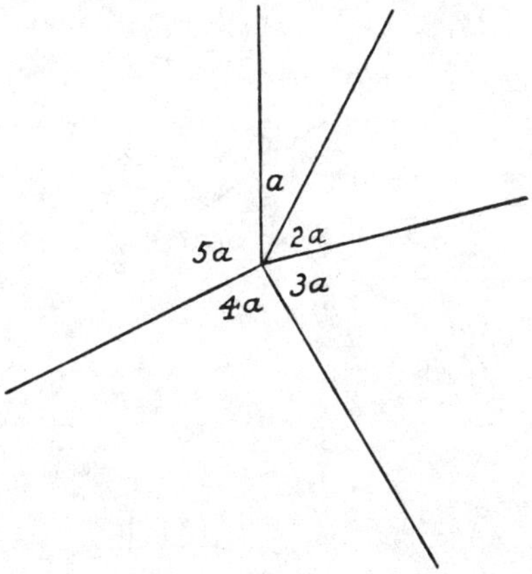

3. *ABC* is right angle.

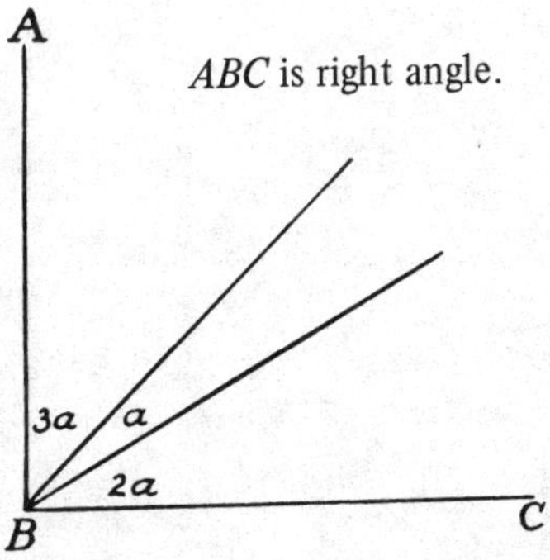

Find angle a and value of each angle.

4.

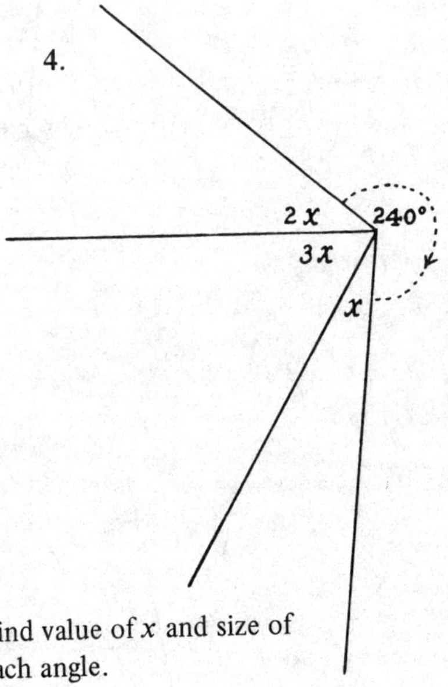

Find value of x and size of each angle.

5.

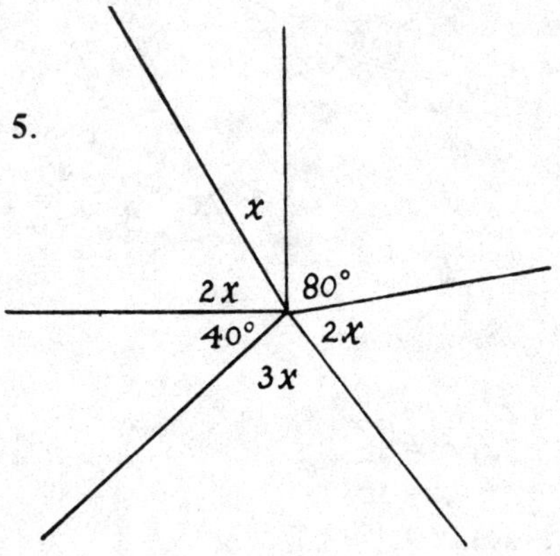

EXERCISE 116

1. Find value for a

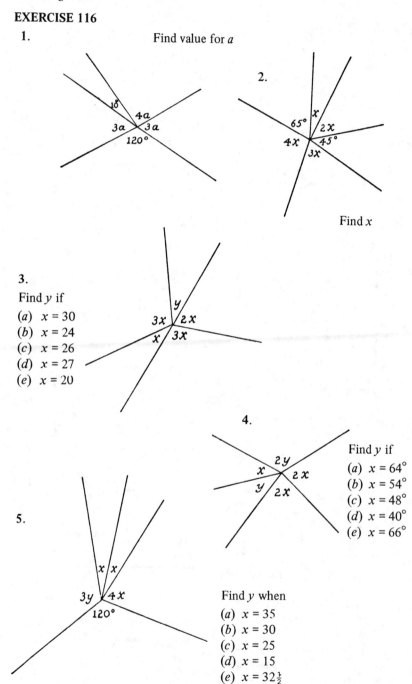

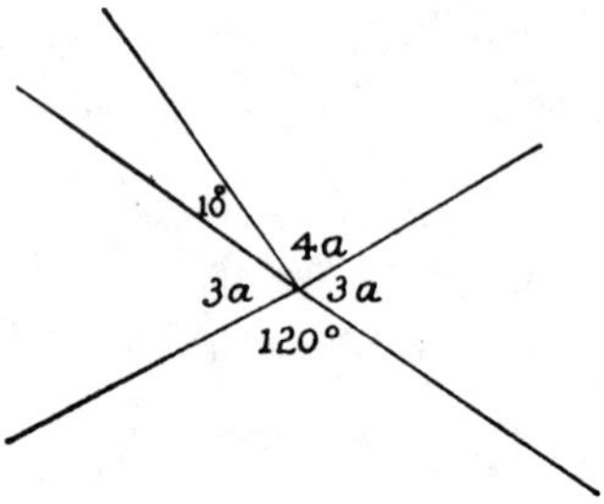

2. Find x

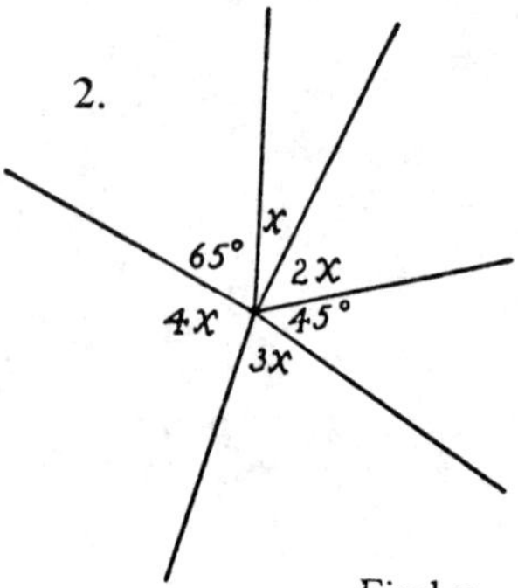

3. Find y if

(*a*) $x = 30$
(*b*) $x = 24$
(*c*) $x = 26$
(*d*) $x = 27$
(*e*) $x = 20$

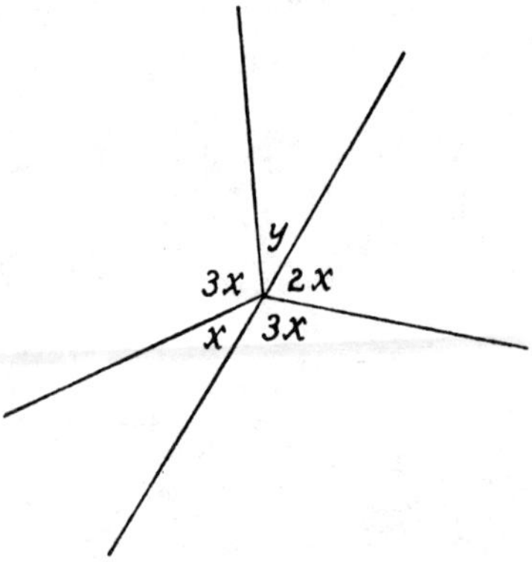

4. Find y if

(*a*) $x = 64°$
(*b*) $x = 54°$
(*c*) $x = 48°$
(*d*) $x = 40°$
(*e*) $x = 66°$

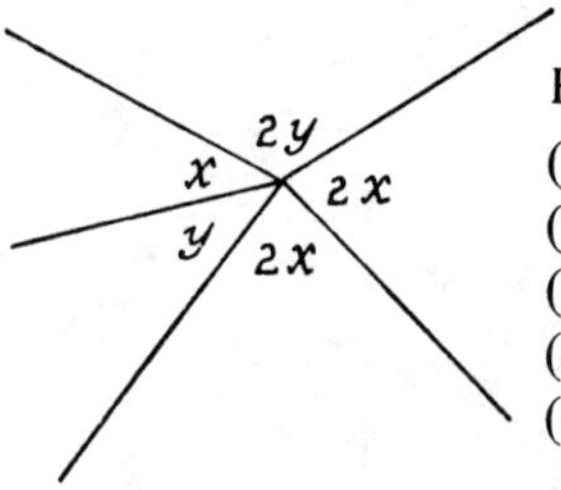

5. Find y when

(*a*) $x = 35$
(*b*) $x = 30$
(*c*) $x = 25$
(*d*) $x = 15$
(*e*) $x = 32\frac{1}{2}$

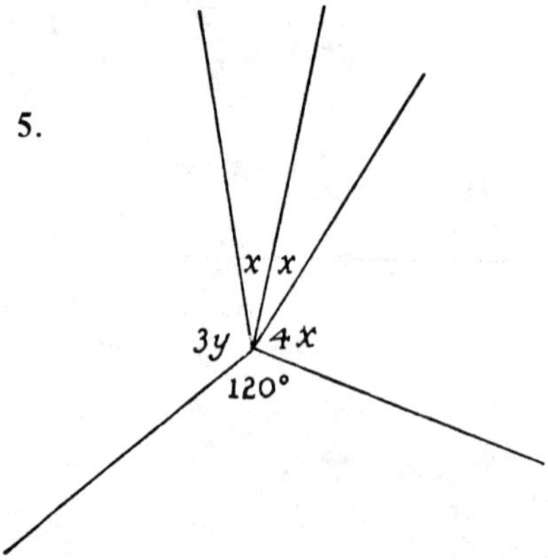

EXERCISE 117

1.
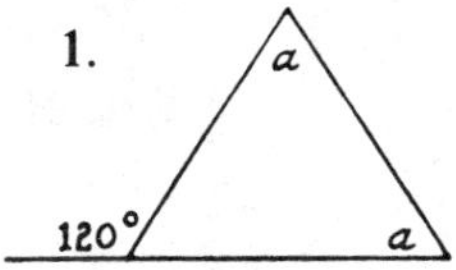

Find *a*.

2.
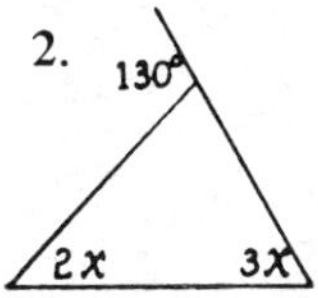

Find *x*.

3.
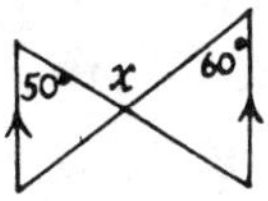

Find *x*.

4.
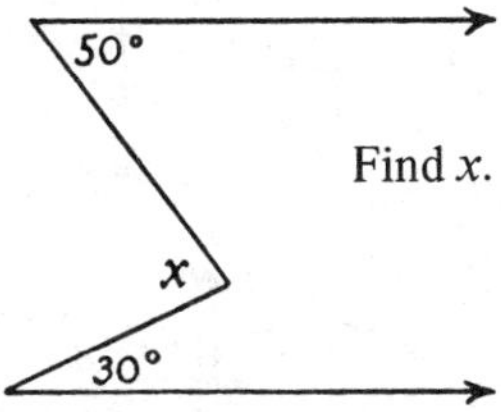

Find *x*.

5.
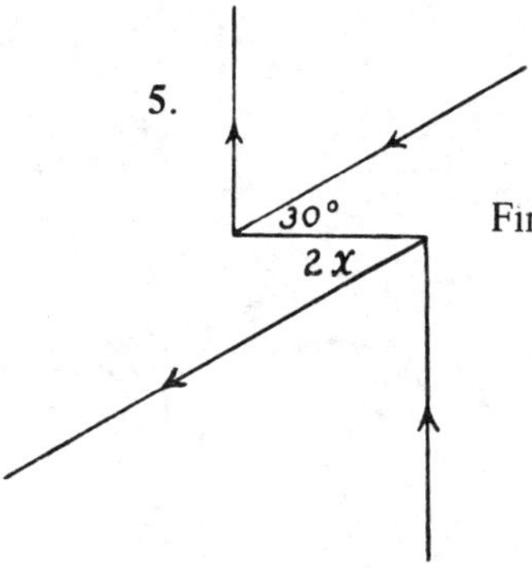

Find *x*.

EXERCISE 118

1.
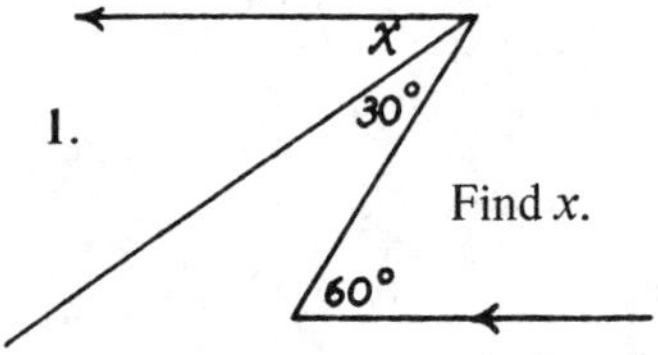

Find *x*.

2.
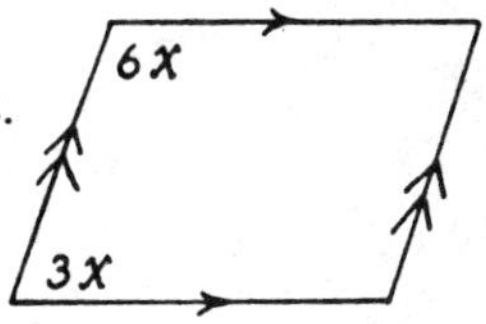

Find *x*.

3.
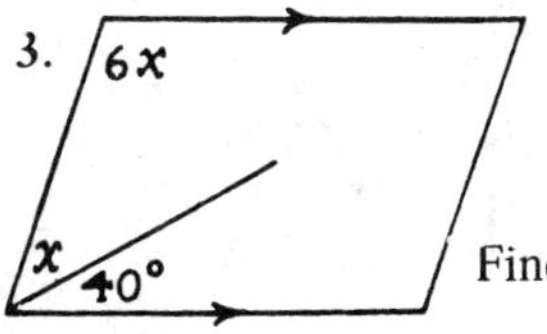

Find *x*.

4.
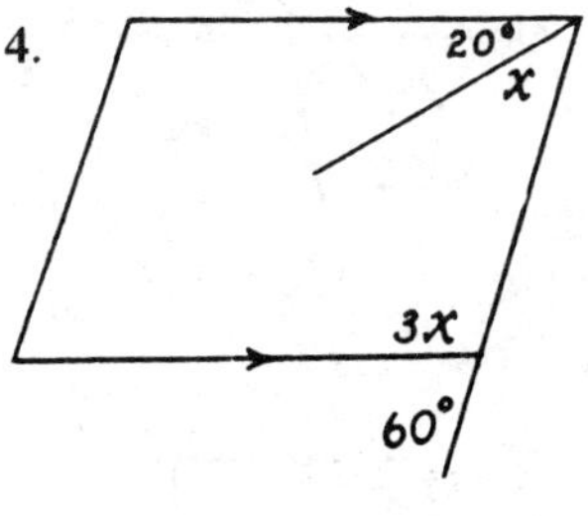

Find *x*.

5.
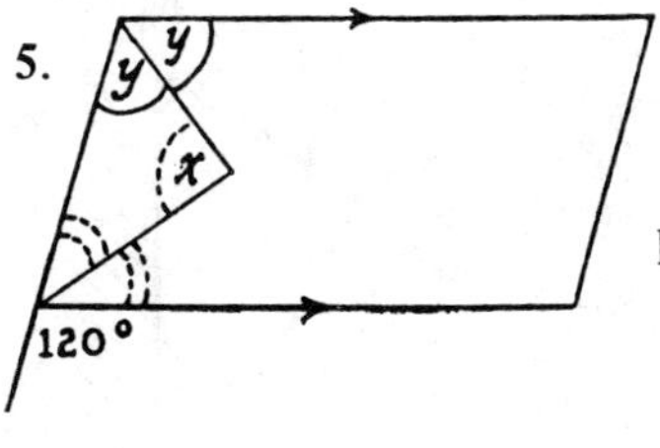

Find *x*, *y*.

M – M ' indicate an axis of reflection. Copy the diagrams in Ex. 119 and 120. Use a mirror to help you to draw the reflected images.

EXERCISE 119

1.

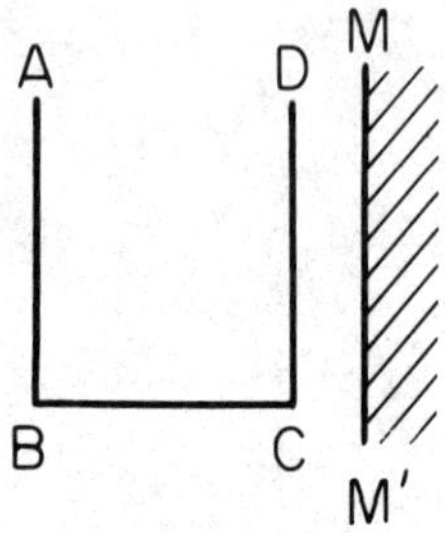

2.

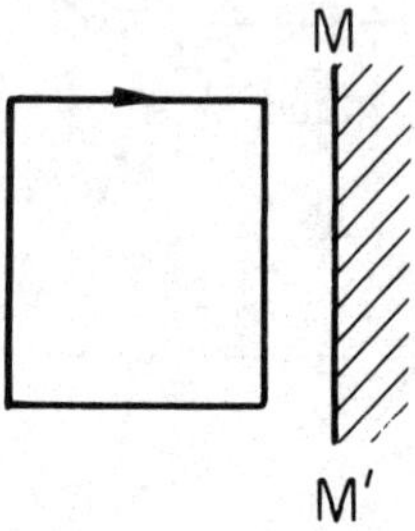

3.

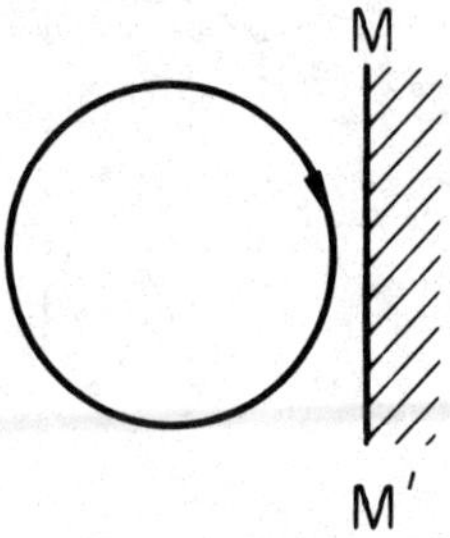

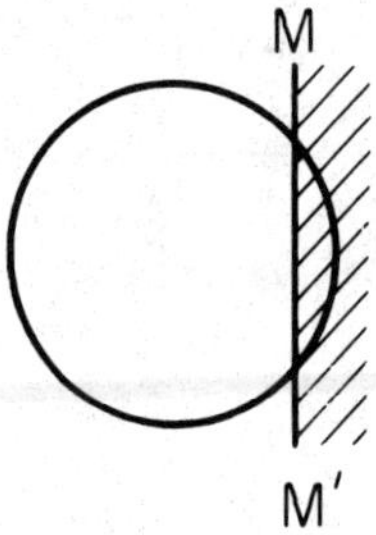

EXERCISE 120

1.

2.

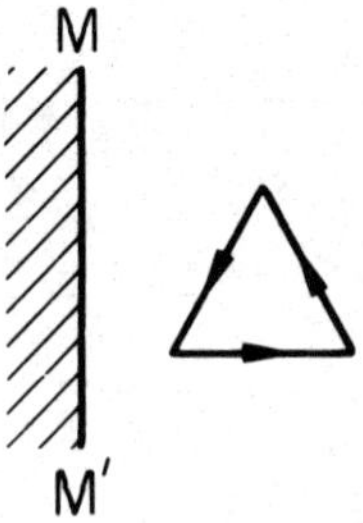

3.

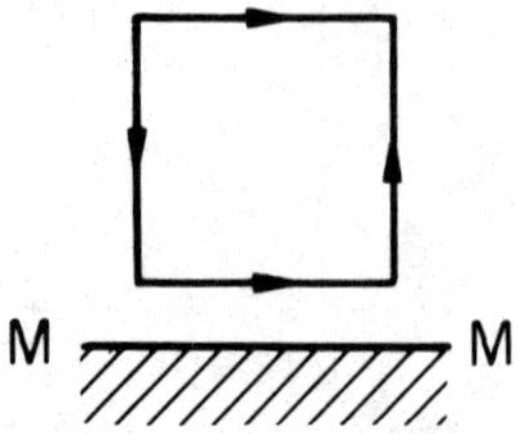

4.

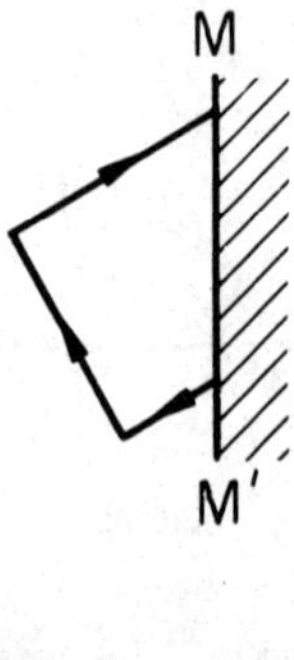

EXERCISE 121

Study these figures, and discuss whether or not they have reflective symmetry.

1.

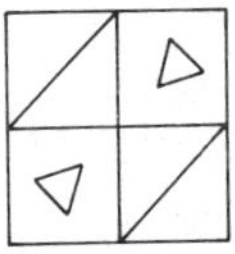

2.

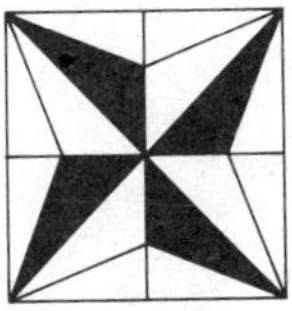

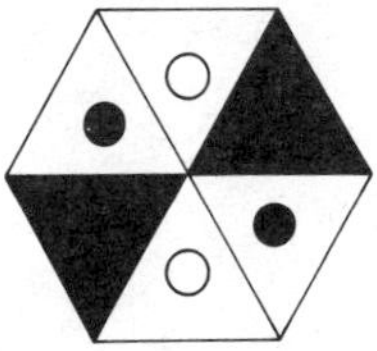

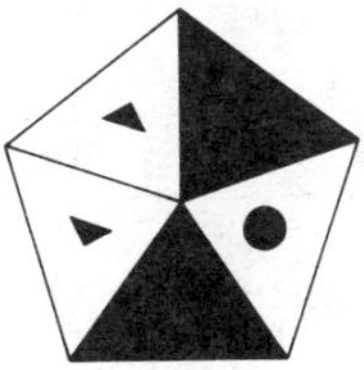

EXERCISE 122

(*a*) Draw this shape when reflected in the mirror M–M′

(*b*) Copy the shape and test to see if it tessellates.

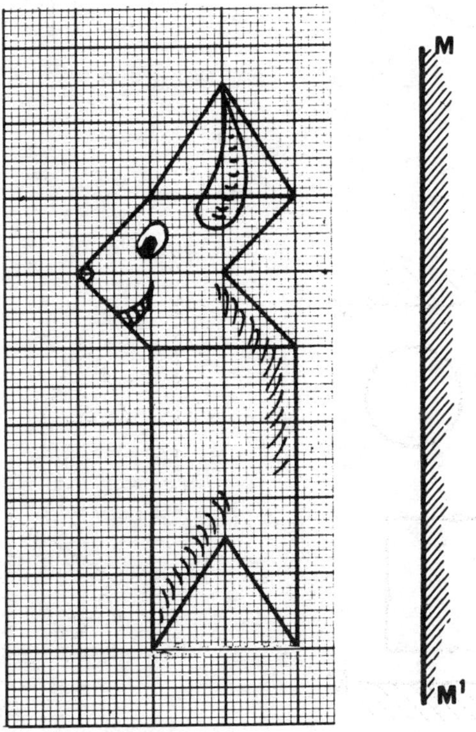

EXERCISE 123

1. $2\frac{1}{4} + 1\frac{1}{3} + \frac{5}{6}$
2. $1\frac{3}{5} + \frac{7}{10} + \frac{1}{2}$
3. $1\frac{5}{8} + 2\frac{1}{2} + 1\frac{3}{4}$
4. $1\frac{1}{3} + 2\frac{3}{4} - \frac{7}{12}$
5. $2\frac{1}{2} + 1\frac{3}{4} - 1\frac{5}{8}$
6. $3\frac{3}{5} - (1\frac{1}{4} - \frac{9}{10})$
7. $3\frac{1}{9} - (1\frac{2}{3} + \frac{5}{6})$
8. $2\frac{1}{3} - (2\frac{1}{4} - 1\frac{1}{2})$
9. $\frac{1}{2} - \frac{7}{8} + \frac{3}{4}$
10. $1\frac{7}{8} - (\frac{2}{3} - \frac{1}{4})$

EXERCISE 124

1. $1\frac{1}{3} + 2\frac{1}{4} + 1\frac{1}{6}$
2. $2\frac{1}{2} + 1\frac{1}{3} + 2\frac{1}{5}$
3. $2\frac{1}{8} + 1\frac{1}{3} + 1\frac{3}{4}$
4. $2\frac{1}{8} + 1\frac{1}{2} - 2\frac{3}{4}$
5. $4\frac{1}{2} - (2\frac{1}{3} - \frac{2}{5})$
6. $1\frac{1}{2} + (2\frac{1}{3} - 1\frac{1}{5})$
7. $2\frac{1}{4} - \frac{2}{3} + \frac{1}{2}$
8. $3\frac{1}{3} - 2\frac{1}{4} + \frac{1}{2}$
9. $2\frac{2}{3} + \frac{1}{2} - 1\frac{1}{4}$
10. $2\frac{7}{8} - 1\frac{1}{4} + 3\frac{1}{3}$

EXERCISE 125

1. $\frac{2}{3} \div \frac{3}{5} \times \frac{5}{6}$
2. $2\frac{3}{4} \div (1\frac{1}{2} + \frac{1}{8})$
3. $(1\frac{1}{2} + \frac{1}{8}) \div 2\frac{3}{4}$
4. $3\frac{1}{4} \div (1\frac{3}{4} - \frac{1}{8})$
5. $(\frac{5}{6} \times \frac{3}{5}) \div \frac{1}{4}$
6. $(\frac{3}{4} \times \frac{4}{5}) \div \frac{3}{5}$
7. $(2\frac{1}{2} \times 3\frac{3}{4}) \div 2\frac{3}{8}$
8. $(6\frac{2}{7} \times \frac{2}{11}) \div \frac{3}{14}$
9. $(2\frac{1}{3} \times \frac{4}{7}) \div 4\frac{1}{2}$
10. $(3\frac{1}{4} \times \frac{1}{3}) \div 1\frac{1}{2}$

EXERCISE 126

1. $(\frac{1}{2} + 1\frac{3}{8}) \div 2\frac{1}{2}$
2. $(2\frac{1}{4} - \frac{3}{8}) \div 1\frac{3}{4}$
3. $(3\frac{3}{8} - 1\frac{1}{2}) \div 2\frac{1}{4}$
4. $(2\frac{1}{3} + 1\frac{5}{12}) \div 2\frac{7}{8}$
5. $(1\frac{1}{4} + 1\frac{3}{5}) \div 1\frac{9}{10}$
6. $2\frac{1}{4} \div (2\frac{3}{8} - 1\frac{1}{2})$
7. $3\frac{3}{4} \div (2\frac{1}{3} + 1\frac{5}{12})$
8. $2\frac{1}{2} \div (3\frac{1}{4} - \frac{1}{2})$
9. $(1\frac{2}{3} + 2\frac{3}{4}) \div (2\frac{3}{4} - 1\frac{2}{3})$
10. $(1\frac{1}{4} + 2\frac{1}{3}) \div (2\frac{1}{3} - 1\frac{1}{4})$

EXERCISE 127

1. 2·05 x 1·13
2. 0·11 x 2·04
3. 11·2 x 3·01
4. 0·025 x 2·012
5. 13·6 x 5·2
6. 0·778 ÷ 0·24
7. 5·76 ÷ 4·8
8. 3·604 ÷ 1·06
9. 55·44 ÷ 15·4
10. 3·8488 ÷ 113·2

EXERCISE 128

1. 3·5 x 1·02
2. 14·4 x 2·1
3. 0·034 x 3·102
4. 10·01 x 25·4
5. 1·002 x 31·3
6. 26·88 ÷ 8·4
7. 32·64 ÷ 2·4
8. 0·366 ÷ 12·2
9. 2·3562 ÷ 1·02
10. 19·264 ÷ 3·01

EXERCISE 129

Change to units indicated.

1. 2 km. 3 hm. 4m. in metres
2. 3 dam. 4 m. 2 dm. 6 cm. in dm.
3. 3 hm. 4 dam. 1 m. in dam.
4. 2 dam. 1 dm. 3 mm. in cm.
5. 1 km. 2 hm. 6 dam. 3 dm. in km.
6. 2 m. 5 cm. 4 mm. in m.
7. 3 dam. 4 m. 1 dm. in dm.
8. 2 hm. 7 cm. 1 mm. in hm.
9. 1 m. 5 dm. in dam.
10. 2 km. 6 m. 4 dm. in m.

EXERCISE 130

Change to metric units.

1. 37·51 hm.
2. 24·37 dm.
3. 1420·2 cm.
4. 3·42 hm.
5. 21·05 m.
6. 10·74 dam.
7. 3014 cm.
8. 217 mm.
9. 3·102 km.
10. 20·034 dam.

EXERCISE 131

Find area of shaded part except where indicated in the next two exercises.

1.

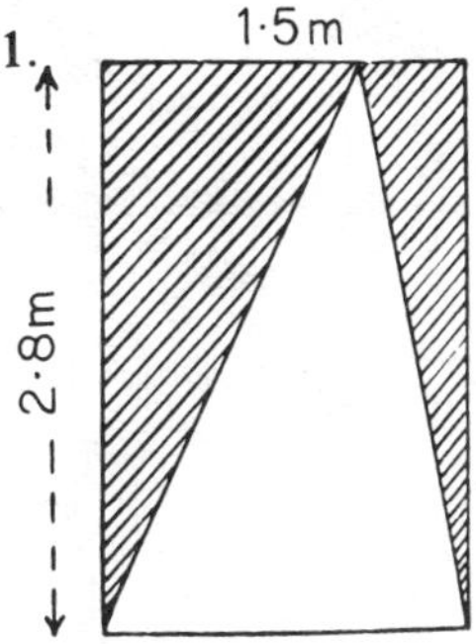

2.

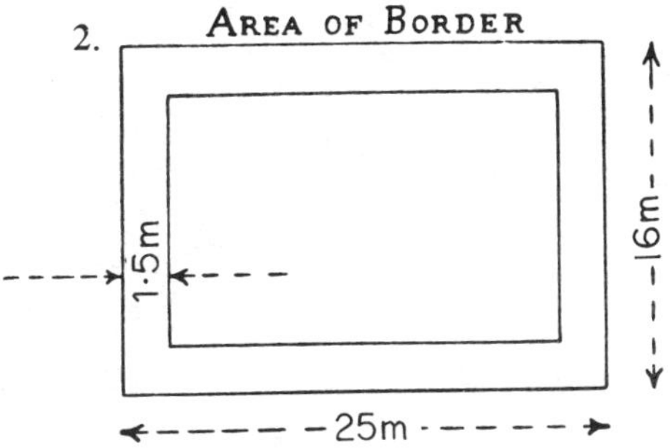

AREA SHADED

3.

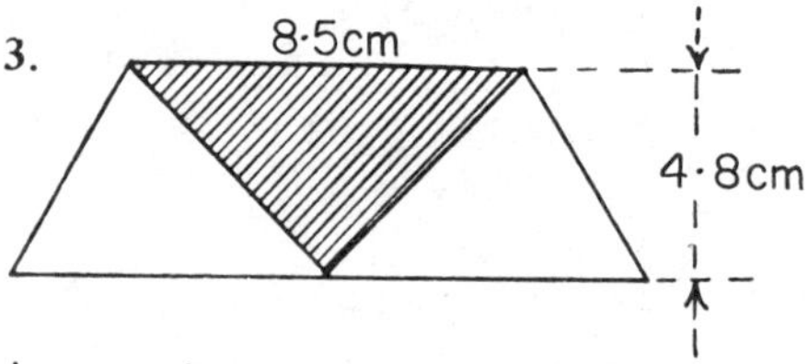

4.

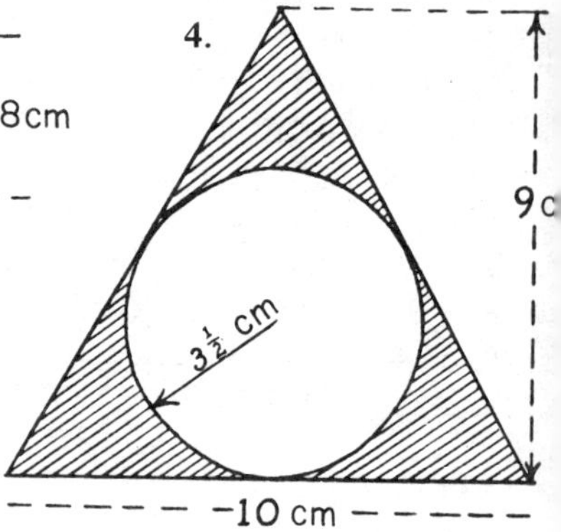

5.

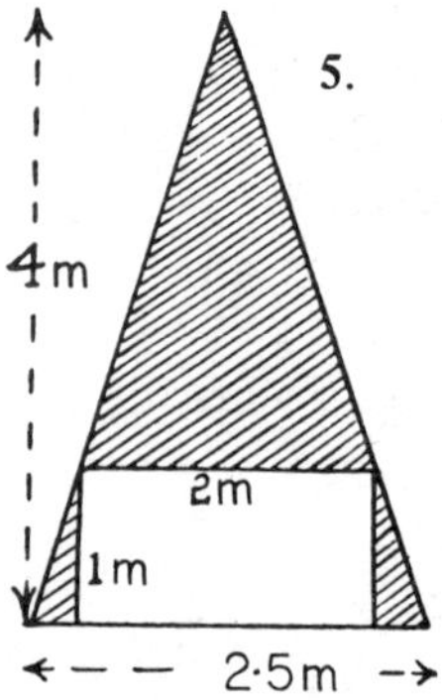

2.

AREA OF WHOLE

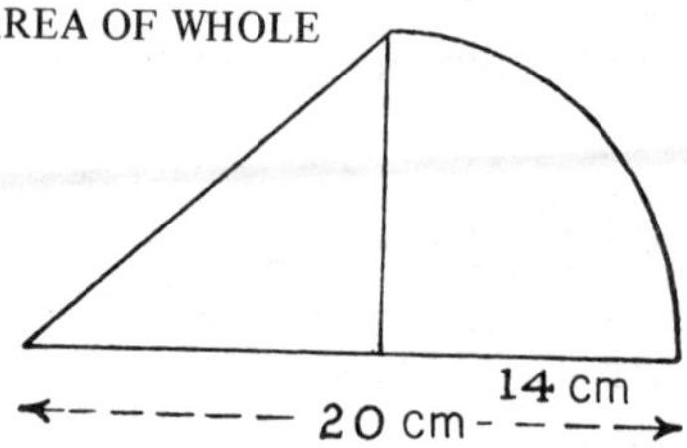

EXERCISE 132

1.

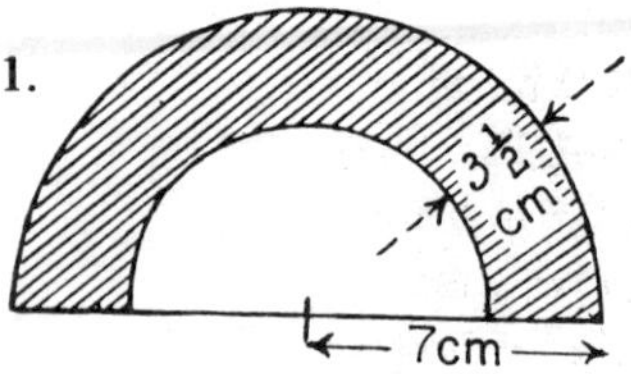

3.

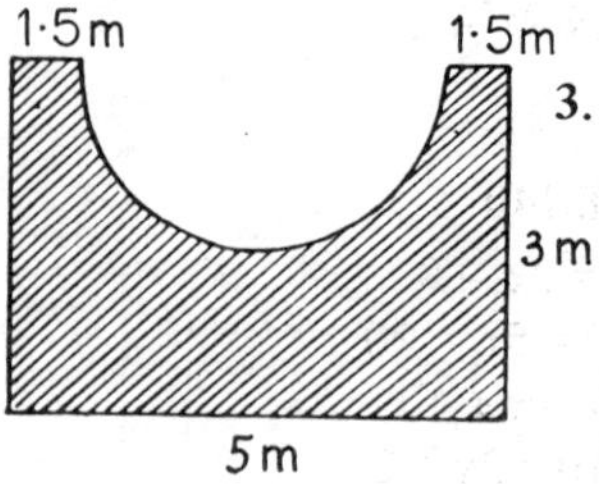

4.

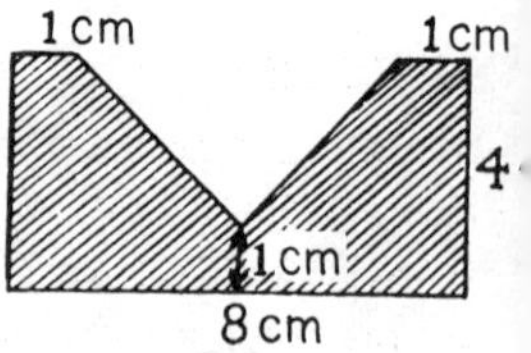

5.

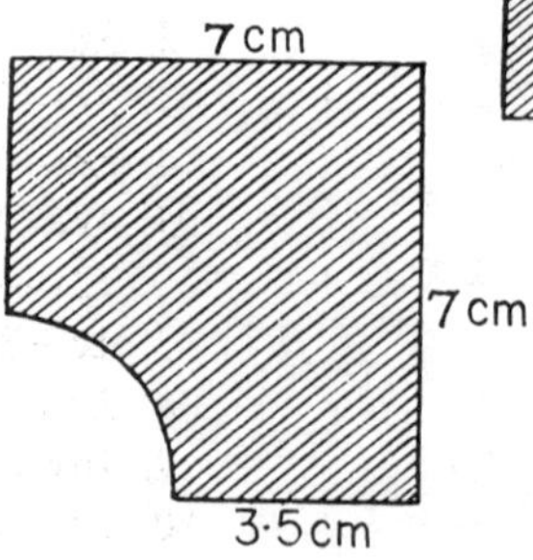

EXERCISE 133

1. $4x + 2y - 3x - y$
2. $-3a + 2b - 4a - b$
3. $6a - 2b + 3a + b$
4. $2a + 3b - 4a + b$
5. $5a + 2b - 3a + 4c$
6. $2y + 3x - 5y - x$
7. $2a + b - 4a + 3b$
8. $2r - 3s + 4r + 2s$
9. $-4n + 2s - 3p + 2p - s$
10. $-2y - 2z + 3y + z$

EXERCISE 134

1. $2n + 3p - 4n$
2. $6p - 3q + 2p - 4q$
3. $4e + 2j - 3e - j$
4. $2t - 4s + 3t - s$
5. $5v - 4f - 2v + f$
6. $-5a + 2b + 3a - 4b$
7. $-x - 2y + 3x - c + y$
8. $-2y - 3a + 2y + 4a$
9. $3b - 2c + b + 4c$
10. $-4x - 2y + 3x + 3y + z$

EXERCISE 135

1. $3a^2bc^3 \times 2a^3b^2c$
2. $4p^4q \times 2pq^3$
3. $12x^3y^2 \times 3xy^4$
4. $4m^3n^2 \times 2m^4n$
5. $5r^3s^2t \times 2rs^3t^2$

EXERCISE 136

1. $2a^3b^2c \times 3b^3c^2$
2. $3x^4y \times 2xy^3$
3. $4f^3gh^2 \times 5fg^2h$
4. $2t^4s \times 3st^2u$
5. $4c^4d^2 \times 2d^3c^2$

EXERCISE 137

1. $4c^4d^2 \div 2d^3c^2$
2. $8f^4gh \div 2fg^2h$
3. $4m^4n^2 \div 2m^3n$
4. $8x^3y \div 4x^2$
5. $10p^3r^2 \div 5pr$

EXERCISE 138

1. $25m^3n^2 \div 5m^2n$
2. $5c^3d^2 \div 2c^4d^3$
3. $12r^4s^2 \div 6r^3$
4. $4a^3x^2 \div 2ax^2$
5. $8g^4h^3 \div 2g^3h$

EXERCISE 139

1. $3x - 2 = 2x + 4$
2. $3x - 1 + 2x = 4x - 2$
3. $5x - 2 = 2x + 4$
4. $14x - 11 = 10x - 3$
5. $5x - 2 = 8 - 10x$
6. $x + 13 = 1 + 5x$
7. $-3x - 5 = x + 15$
8. $2x - 25 = 2 - 7x$
9. $3x - 2 = 4x - 10$
10. $3x - 6 = 10 - x$

EXERCISE 140

1. $x - 3 = 2x - 8$
2. $3x - 2 = 5x - 10$
3. $3x - 7 = x + 1$
4. $5x - 10 = 3x + 2$
5. $4x - 10 = x + 5$
6. $12x - 25 = 5x - 4$
7. $4x + 3 = 10x - 21$
8. $3x + 2 = x + 12$
9. $16x + 20 = 22x + 8$
10. $7x - 42 = 2x - 7$

EXERCISE 141

Find value for x.

1.

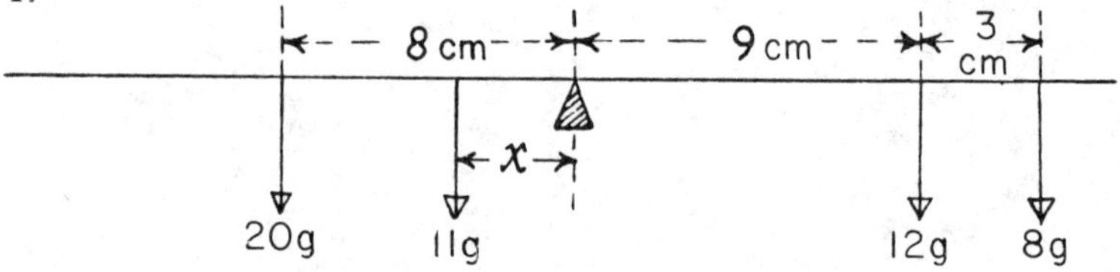

2.

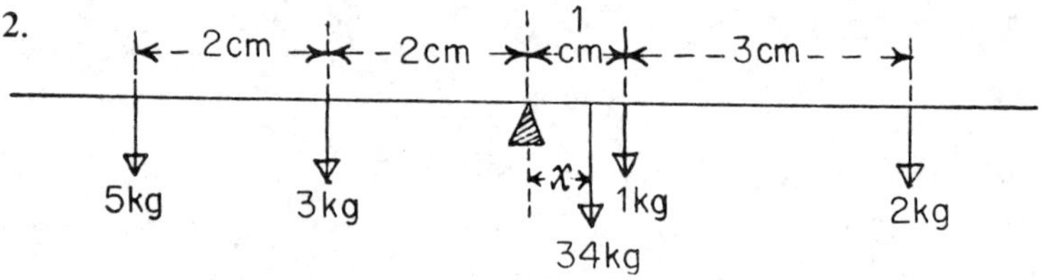

3.

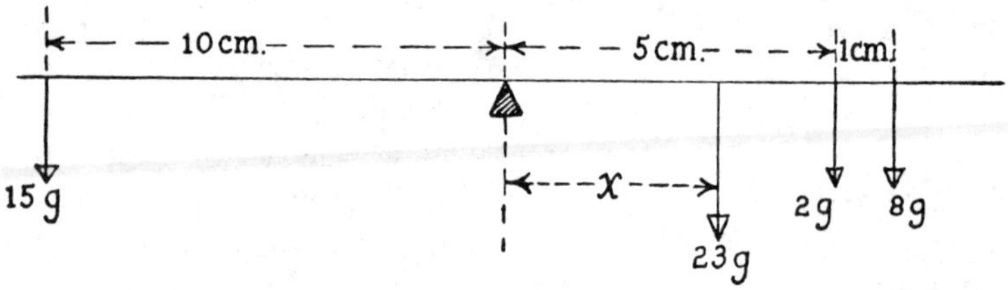

4.

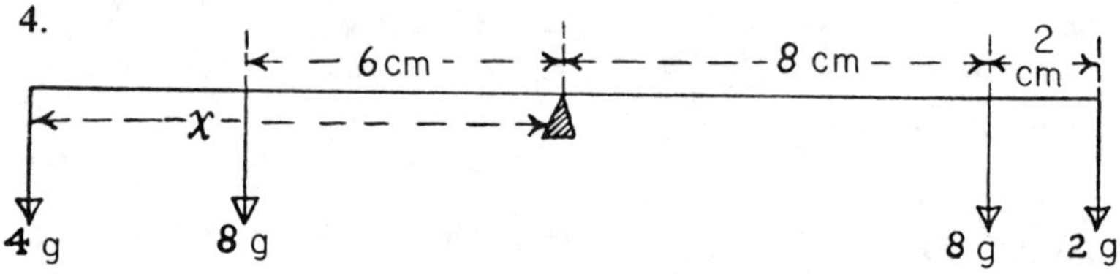

5.

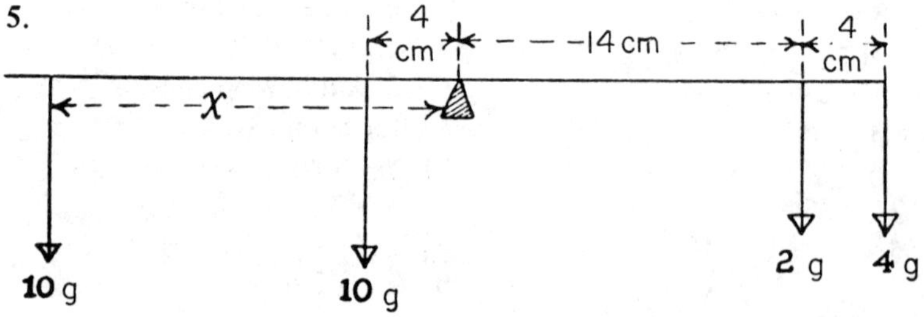

TEST 1

1. Arrange in order, largest first
   $\frac{3}{4}$ $\frac{7}{8}$ $\frac{5}{6}$ $\frac{1}{2}$ $\frac{5}{12}$ $\frac{2}{3}$.
2. $4\frac{3}{4} + 2\frac{5}{8} - 3\frac{1}{2}$.
3. $\frac{7}{8} \times \frac{3}{5} \times \frac{10}{21}$.
4. $6\frac{1}{4} \div 2\frac{1}{2}$.
5. Which is the bigger
   $\frac{1}{2}$ of $2\frac{2}{3}$ or $\frac{2}{3}$ of $1\frac{1}{2}$
6. Find x.

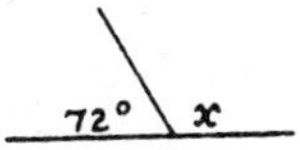

7. Draw rectangle *ABCD* and write down 2 facts about its sides and 2 facts about its angles.
8. What is the angle between the hands of a clock at 7 p.m.?
9. $3a + 4a - 2b - 3a - 4b$.
10. If $a = 4$, $b = 6$, $c = 5$, find $3a + 4b - 2c$.

TEST 2

1. Arrange in order, greatest first
   $\frac{7}{10}$ $\frac{4}{5}$ $\frac{1}{2}$ $\frac{3}{4}$ $\frac{9}{10}$ $\frac{17}{20}$
2. $7\frac{1}{3} - 2\frac{3}{4} + 3\frac{1}{6}$.
3. $\frac{9}{16} \times \frac{4}{5} \times \frac{5}{3}$.
4. $3\frac{5}{9} \div 5\frac{1}{3}$.
5. A boy drew a line 5·75 cm. long. It was 1·88 cm. too long. How long should it have been?
6. Find x.

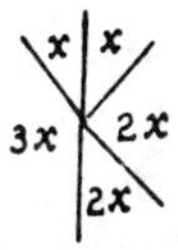

7. What is the angle between S.S.E. and S.W.?
8. How many degrees are there in $3\frac{1}{2}$ right angles?
9. $11p - 2q - 3p + 4q + 5r - 3p$.
10. If $a = 4$, $b = 6$, $c = 5$, $d = 10$, find $5ab + 2c - 7d$.

TEST 3

1. $5\frac{1}{2} + 2\frac{2}{5} - 3\frac{7}{10}$
2. $1\frac{1}{4} \times 2\frac{2}{3} \times 3\frac{1}{5}$
3. $7\frac{1}{2} \div 1\frac{1}{4}$.
4. 3·75 + 12·9 + 4·386 – 5·8.
5. 36·3 × 5.
6. 48·4 ÷ 1·1.
7. Find x.

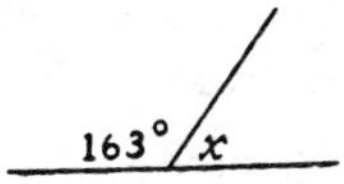

8. Plot the course of a plane flying 060° at 300 km/h. in a 20 km/h. wind blowing from the South.
9. $7m - 4n - 3m - 4n - 3m + 2n + 5m$.
10. $a \times a \times a \times b \times b$.

TEST 4

1. $3\frac{3}{8} - 5\frac{3}{4} + 3\frac{2}{3}$.
2. $2\frac{1}{2} \times 1\frac{1}{7} \div 3\frac{1}{3}$.
3. $14{\cdot}98 - 2{\cdot}764 + 3{\cdot}521$.
4. $7{\cdot}9 \times 1{\cdot}2$.
5. Change $\frac{3}{4}$ and $\frac{2}{3}$ to decimals.
6. Plot the position of a plane after one hour, flying at 450 km/h. on a course 240° in a 50 km/h. wind from the North.
7. Find x.

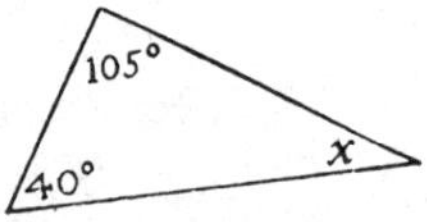

8. Draw a triangle with sides 3 cm., $3\frac{1}{2}$ cm., $4\frac{1}{2}$ cm.
9. If $a = 4, b = 6, c = 5$, find $\frac{a+b}{c}$.
10. $a^4 \times a^2$.

TEST 5

1. $6\frac{2}{3} + 4\frac{2}{5} - 5\frac{5}{6}$.
2. $5\frac{1}{3} \times 4\frac{1}{2} \div 1\frac{3}{5}$.
3. $44{\cdot}21 \times 1{\cdot}5$.
4. $6{\cdot}4 \div 0{\cdot}08$.
5. Find $\frac{2}{3}$ of £7·35 correct to the nearest penny.
6. Find x.

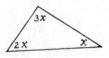

7. Give bearings for these directions: N 30° E, S 40° W, S 65° E and N 27° W.
8. $5c + 2d - 8e - 6c + 2d + 4e$.
9. If $a = 4, b = 6, c = 5$ and $d = 10$ find $\frac{a^2bc}{bd}$.
10. $a^2b \times ab^2$

TEST 6

1. $4\frac{1}{5} \div 1\frac{2}{5} \times 1\frac{1}{3}$.
2. Change to decimals: $\frac{7}{8}$ and $\frac{5}{12}$.
3. Who is the taller, Suzanne 125 cm. or Mary 1·36 m.? By how many cm.?
4. Find $\frac{3}{5}$ of £4·45 correct to the nearest penny.
5. Tom gained the following marks 52, 39, 81, 66, 72. What was his average?
6. Find x.

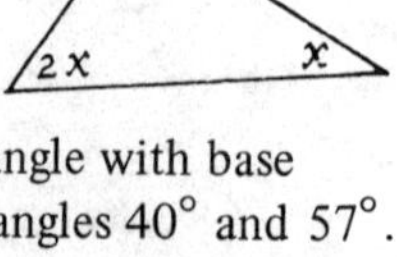

7. Draw a triangle with base 5 cm. and angles 40° and 57°.
8. Draw and label pairs of alternate, corresponding and vertically opposite angles.
9. $p^3q \times pq^2$.
10. Solve $3x + 2 = 8$.

TEST 7

1. Arrange in order, greatest first
   $\frac{2}{3}$ $\frac{5}{6}$ $\frac{7}{10}$ $\frac{8}{15}$ $\frac{1}{2}$ $\frac{4}{5}$.
2. 7·45 + 3·491 – 8·5.
3. 58·32 ÷ 1·2.
4. Who is the heavier, Karen who is 28 kg. or Betty who is 30 000 g. By how many g?
5. Find the average weight of these children:

John	34 kg.
Martin	32 kg.
Stephen	30 kg.
Jane	33 kg.
Sally	28 kg.
Susan	29 kg.

6. Give directions for these bearings; 030°, 197°, 142°, 264°.
7. Copy and put in all angles.

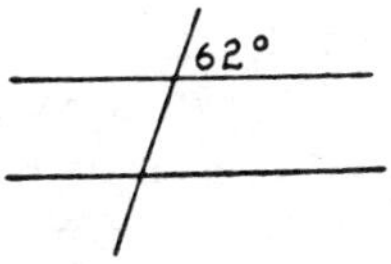

8. Draw the image when reflected in M-M'

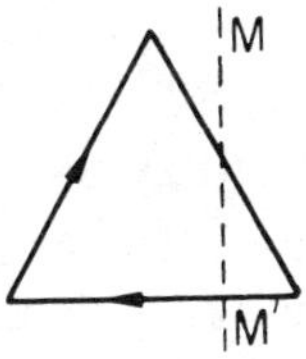

9. $4m - 3 = 2m + 5$.
10. I bought 2 rubbers and received 7 pence change from a ten-penny piece. Make an equation to find out how much each cost.

TEST 8

1. $6\frac{1}{2} + 3\frac{3}{4} - 4\frac{7}{8}$.
2. Mary has 1·5 m. of ribbon, Joan has 1·7 m. and Pam has 163 cm. Who has the longest piece? What is the total length in cm?
3. 2·6 x 0·04.
4. Which is travelling faster, an Anglia doing 55 km. in 30 minutes or a Volkswagen doing 90 km/h?
5. Find the area of this rectangle.

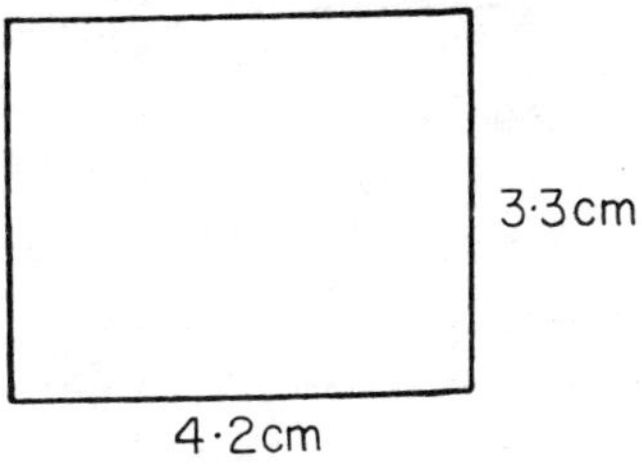

6. Copy and fill in the other angles.

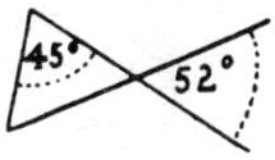

7. Draw the image you would see if the mirror is on M-M'.

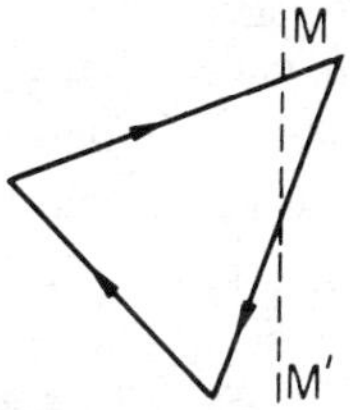

8. In one room are 3 Dutchmen, 4 Englishmen and 2 Scots; in another room are 4 Scots, 6 Welshmen and 2 English. Write shortly and collect terms.
9. $8 = 12 - 4x$.
10. $p = q + r - s$ find r.

TEST 9

1. $5\frac{1}{2} \times 3\frac{1}{3} \times 1\frac{1}{5}$.
2. Change to decimals: $\frac{3}{16}$ and $\frac{1}{6}$.
3. The average height of 3 boys is 135 cm. One of them is 142 cm. and another is 136 cm. How tall is the third?
4. Find the area of the border.

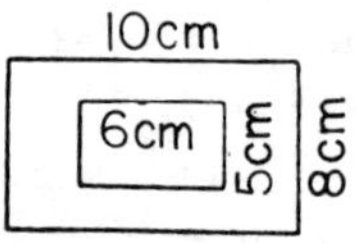

5. Find the area.

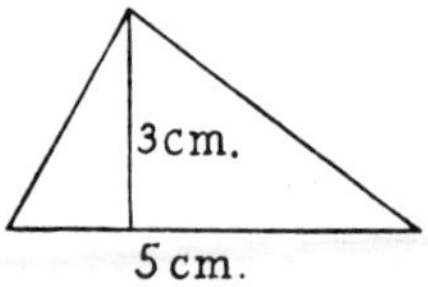

6. Find x.

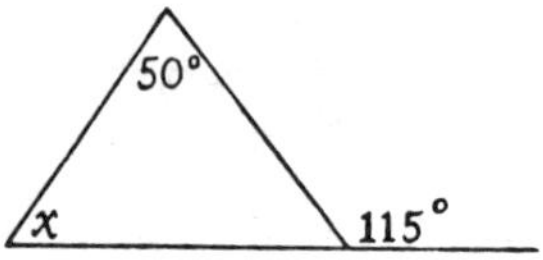

7. Draw to scale a room 5 m. by 4·5 m. Put a 4 m. square carpet on the floor and find the area of the border.
8. 30 m. from the bottom of a tree the angle of elevation is 36°. Draw to scale and find the height of the tree.
9. $ab = \frac{c}{d}$ find b.
10. Draw graphs $y = 3 + x$
    $y = 5 - x$.
    Say where graphs intersect.

TEST 10

1. $3\frac{1}{3} \div 1\frac{2}{3} \times 3\frac{1}{2}$.
2. 3·851 − 2·9 + 1·63.
3. Find 0·28 × £3·11 correct to the nearest penny.
4. A cloth 45 dm. square is put on a round table with 35 dm. diameter. Find
   (*a*) the area of the cloth.
   (*b*) the area of the table.
   (*c*) the area of the cloth hanging over the edge.
5. Find the circumference of a circle with radius 21 cm.
6. Find x.

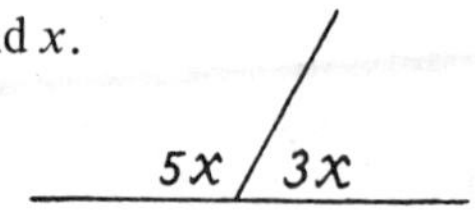

7. Draw to scale a garden 120 m. × 70 m. Put in 20 m. square lawn, fruit 70 m. × 50 m. vegetables 50 m × 30 m., shed 8 m. × 6 m. Fill in spaces with flower borders and paths and put in all dimensions.
8. $3r - 2s - 2t + s - 3t - 5r + s$.
9. $17 - 3x = 25 - 5x$.
10. $c = pd$, find d.

Answers

Page 15 Exercise 1

(1) E.–W. (2) N.–S. (3) N.W.–S.E. (4) N.E. (5) S.E.
(6) E. (7) S.W. (8) S.E. (9) W. (10) N.E.
(11) N.W. (12) N.E. (13) S.E. (14) E. (15) N.
(16) S.W. (17) S. (18) S.W. (19) W. (20) S.

Page 16 Exercise 2

(1) 45. (2) 45. (3) $112\frac{1}{2}$. (4) 225. (5) 135.
(6) 135. (7) $247\frac{1}{2}$. (8) 270. (9) $67\frac{1}{2}$. (10) $112\frac{1}{2}$.
(11) $157\frac{1}{2}$. (12) $67\frac{1}{2}$. (13) $247\frac{1}{2}$. (14) 225. (15) $112\frac{1}{2}$.
(16) 45. (17) 180. (18) $112\frac{1}{2}$. (19) 135. (20) 135.

Page 18 Exercise 3

(1) N.10°E. (2) N.80°E. (3) E. (4) S.70°E. (5) S.50°E.
(6) S.10°E. (7) S.10°W. (8) S.30°W. (9) S.60°W. (10) W.
(11) N.80°W. (12) N.60°W. (13) N.1°W. (14) N.15°E. (15) N.86°E.
(16) S.85°E. (17) S.65°E. (18) S.48°E. (19) S.6°E. (20) S.68°W.

Page 23 Exercise 6

(1) 100°. (2) 30°. (3) 50°. (4) 30°. (5) 141°.

Page 23 Exercise 7

(1) 45°. (2) 65°. (3) 80°. (4) 75°. (5) 138°.
(6) 10°. (7) 72°. (8) 74°. (9) 98°. (10) 87°.
(11) 59°. (12) 88°. (13) 77°. (14) 90°. (15) 115°56′.
(16) 39° 30′. (17) 37° 34′. (18) 50° 40′. (19) 76° 29′. (20) 140° 31′.

Page 24 Exercise 8

(1) 120° (2) 130°. (3) 85°. (4) 65°. (5) 74°.
(6) 108°.

Page 27 Exercise 11

(1) 25 m. (2) 16 m. (3) 10m. (4) 13 m. (5) 10 m.
(6) 18 m. (7) ·277 m. (8) 5·8 m. (9) 14 m. (10) 0·88 km.

Page 28 Exercise 12

(1) 133 m. (2) 28 m. (3) 260 m. (4) 137 m.

Page 46 Exercise 20

(1) Vert. opp., corr., supp., alt. (2) $b = 140^\circ$, $c = 140^\circ$, $d = 40^\circ$, $e = 40^\circ$.
(3) 90°. (4)

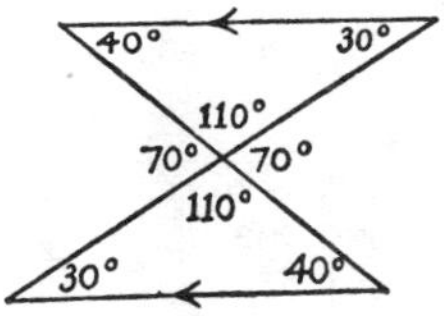

(5) $a = 70^\circ, b = 135^\circ$.
(6) $a = 150^\circ, b = 30^\circ, c = 30^\circ, d = 150^\circ$.
(7) $a = 120^\circ, b = 70^\circ$.
(8)

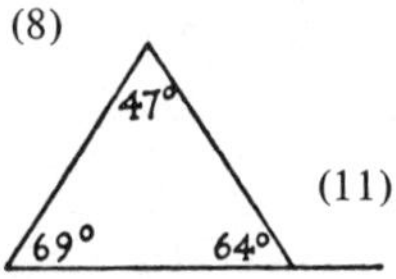

(9) $p = 77^\circ$.

(10)

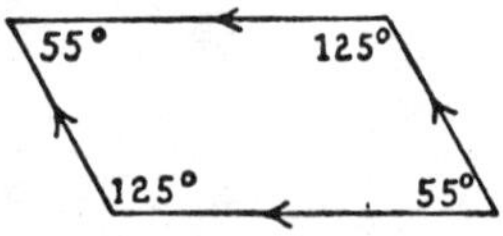

(11) $a = b = c = d = 90^\circ$.

(12) $a = d$ (opp. ∠s ||gram)
$b = e$, $c = f$ } alt. ∠s.

(13) $a = 145^\circ$. (14) $b = 68^\circ$. (15) $x = 36^\circ$.

(16)

(17)

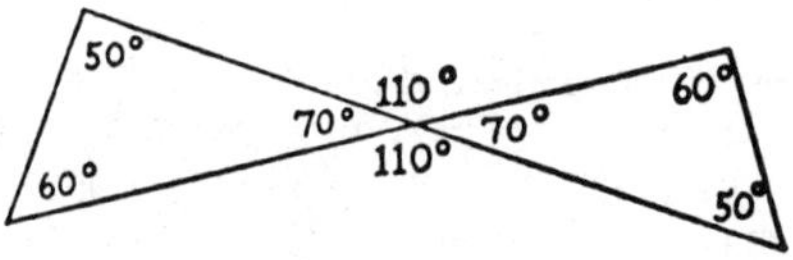

(18) $x = 40^\circ$.
$a = 140^\circ$.
$b = 10^\circ$.
$c = 30^\circ$.

(19) $x = 24^\circ$.
$3x = 72^\circ$.
$5x = 120^\circ$.

(20) $x = 22\frac{1}{2}^\circ$.
$2x = 45^\circ$.
$5x = 112\frac{1}{2}^\circ$.

Page 48
REVISION CROSSWORD

9	0	■	3	6	0	■	■
5	4	0	■	■	2	1	■
4	5	■	1	■	5	7	6
0	■	1	3	2	■	0	■
■	2	■	1	7	5	■	3
1	1	4	■	0	■	2	6
■	5	2	■	■	2	1	0
■	■	2	8	0	■	6	0

Page 55 Exercise 22
(1) 600 m^2. (2) 695 m^2. (3) 1070 m^2.

Page 58 Exercise 23
(1) 17 m. (2) 23 m. (3) 2·5 m. (4) 8 m. (5) 8·7 m.
(6) 58·4 m. (7) 29 m. (8) 2 m. (9) 39 m. (10) A
(11) Church from Pylon is 0·9 km., Power Station to Church is 1·0 km. approx.
(12) AC = 37 m, BC = 33 m, DC = 28 m. (13) 48 m. (14) 254 m.
(15) 2·6 m.

Page 62 Exercise 24
(1) 15·6 m. (2) 2·8 m. (3) 125 m. (4) 0·4 km. (5) 26 m.
(6) Yes. Approx. 4 metres.

Page 64 Exercise 25
(1) 108 m. (2) 167 m. (3) 203 m. (4) 78 m. (5) 280 m.

Page 77 Exercise 26
(1) 374. (2) 1476. (3) 1674. (4) 1632. (5) 1064.
(6) 1134. (7) 3072. (8) 1690. (9) 1230. (10) 1404.
(11) 18. (12) 27. (13) 36. (14) 46. (15) 84.
(16) 63. (17) 57. (18) 66. (19) 31. (20) 56.

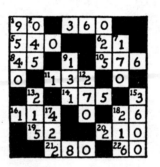

Page 78 Exercise 27

	HCF	LCM		HCF	LCM
(1)	3	45	(2)	6	84
(3)	8	120	(4)	9	108
(5)	26	156	(6)	3	90
(7)	4	24	(8)	7	490
(9)	5	700	(10)	22	1320

Page 79 Exercise 28

[1]2	4		[2]1		[3]1	0	[4]8
5		[5]2	6	0			1
6			9			[6]1	
	[7]4	[8]4		[9]2	[10]2	4	0
[11]2	4	9	0		[12]5	4	
	1			[13]1			[14]1
[15]6			[16]2	2	5		9
[17]4	0	0		1		[18]3	6

Page 80 Exercise 29

[1]1	[2]2	1			[3]1	[4]7	[5]5
[6]5	5		[7]3	6		[8]7	5
6		[9]1	2		[10]1		0
	[11]8	0		[12]1	0	[13]8	
	[14]1	0	4		[15]2	4	
[16]1		0		[17]5	4		[18]1
[19]6	[20]3		[21]2	7		[22]5	0
[23]9	0	0			[24]3	6	0

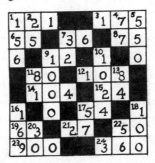

Page 81 Exercise 30

[1]2	[2]1		[3]1	7	[4]6	0	
	[5]4	4	1		6		[6]8
[7]1	8		[8]2	[9]5		[10]6	6
0		[11]1		[12]1	[13]2	2	4
[14]1	[15]1	5	[16]5		7		0
[17]2	4		[18]5	[19]2		[20]6	0
0		[21]3		[22]4	4	0	
	[23]2	2	4	0		[24]5	6

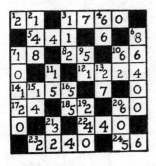

Page 82 Exercise 31

[1]1	[2]4	[3]1	6		[4]1	[5]2	[6]1
[7]4	0	0		[8]3		[9]5	6
3		0		[10]6	3	0	0
	[11]1	0	[12]8	0			0
[13]6			[14]1	0	[15]1	2	
[16]3	[17]4	5	6		1		[18]5
[19]3	4		0		[20]3	[21]7	5
[22]6	4	0		[23]3	1	1	5

Page 87 Exercise 32

(1) $\frac{2}{3}, \frac{1}{4}$. (2) $\frac{3}{4}, \frac{1}{2}$. (3) $\frac{5}{6}, \frac{5}{12}, \frac{1}{6}$. (4) $\frac{2}{3}, \frac{1}{3}, \frac{1}{4}$. (5) $\frac{3}{10}, \frac{1}{5}, \frac{1}{10}$.
(6) $\frac{2}{5}, \frac{7}{20}$. (7) $\frac{1}{2}, \frac{3}{8}, \frac{5}{16}$. (8) $\frac{3}{4}, \frac{11}{16}, \frac{1}{4}$. (9) $\frac{3}{4}, \frac{2}{3}, \frac{1}{2}, \frac{1}{4}$.
(10) $\frac{3}{5}, \frac{2}{5}, \frac{7}{20}, \frac{3}{10}, \frac{1}{10}$. (11) $\frac{5}{6}, \frac{2}{3}, \frac{1}{2}, \frac{5}{12}, \frac{1}{4}, \frac{1}{6}$. (12) $\frac{1}{3}, \frac{1}{4}, \frac{1}{5}$.
(13) $\frac{2}{3}, \frac{1}{16}$. (14) $\frac{7}{10}, \frac{3}{5}, \frac{4}{15}$. (15) $\frac{5}{6}, \frac{3}{4}, \frac{2}{3}$. (16) $\frac{1}{2}, \frac{7}{16}, \frac{2}{9}$. (17) $\frac{4}{5}, \frac{2}{3}, \frac{7}{30}$.
(18) $\frac{1}{2}, \frac{4}{11}, \frac{7}{22}$. (19) $\frac{7}{14}, \frac{3}{7}, \frac{4}{21}$. (20) $\frac{1}{5}, \frac{1}{6}, \frac{1}{7}$.

Page 88 Exercise 33

(1) $\frac{3}{4}$. (2) $\frac{3}{16}$. (3) $\frac{1}{3}$. (4) $1\frac{1}{4}$. (5) $\frac{5}{8}$.
(6) $\frac{4}{5}$. (7) $3\frac{1}{4}$. (8) $2\frac{1}{8}$. (9) $2\frac{1}{8}$. (10) $3\frac{11}{16}$.
(11) $3\frac{9}{10}$. (12) $4\frac{1}{2}$. (13) $\frac{1}{8}$. (14) $\frac{9}{16}$. (15) $1\frac{5}{6}$.
(16) $\frac{1}{4}$. (17) $\frac{1}{8}$. (18) $\frac{1}{5}$. (19) $1\frac{3}{4}$. (20) $1\frac{3}{8}$.
(21) $\frac{3}{8}$. (22) $3\frac{5}{16}$. (23) $\frac{3}{10}$. (24) $\frac{3}{4}$. (25) $2\frac{5}{16}$.
(26) $\frac{1}{2}$. (27) $1\frac{3}{4}$. (28) $1\frac{5}{8}$. (29) $1\frac{5}{16}$. (30) $\frac{3}{8}$.

Page 88 Exercise 34

(1) $\frac{5}{6}$, (2) $\frac{3}{16}$, (3) $\frac{1}{3}$ (4) $1\frac{1}{4}$ (5) $\frac{5}{8}$.
(6) $1\frac{1}{10}$. (7) $\frac{7}{8}$. (8) $1\frac{1}{8}$ (9) $1\frac{1}{8}$. (10) $\frac{13}{21}$.
(11) $\frac{23}{44}$. (12) $\frac{31}{35}$. (13) $1\frac{1}{9}$. (14) $\frac{23}{56}$. (15) $\frac{11}{12}$.
(16) $1\frac{5}{12}$. (17) $1\frac{5}{12}$. (18) $\frac{3}{5}$. (19) 1. (20) $1\frac{3}{16}$.

Page 89 Exercise 35

(1) $\frac{5}{12}$. (2) $\frac{1}{16}$. (3) $\frac{1}{6}$. (4) $\frac{1}{4}$. (5) $\frac{1}{8}$.
(6) $\frac{1}{5}$. (7) $\frac{3}{8}$. (8) $\frac{1}{44}$. (9) $\frac{11}{35}$. (10) $\frac{2}{9}$.
(11) $\frac{9}{56}$. (12) $\frac{8}{45}$. (13) $\frac{17}{32}$. (14) $\frac{7}{24}$. (15) $\frac{1}{6}$.
(16) $\frac{7}{24}$. (17) $\frac{2}{7}$. (18) $\frac{1}{6}$. (19) $\frac{13}{30}$. (20) $\frac{7}{30}$.

Page 90 Exercise 36

(1) $3\frac{10}{21}$. (2) $5\frac{7}{12}$. (3) $4\frac{1}{24}$. (4) $5\frac{11}{15}$. (5) $7\frac{5}{18}$.
(6) $8\frac{10}{21}$. (7) $3\frac{11}{28}$. (8) $4\frac{5}{12}$. (9) $6\frac{3}{8}$. (10) $3\frac{19}{30}$.
(11) $5\frac{7}{16}$. (12) $7\frac{13}{15}$. (13) $5\frac{5}{12}$. (14) $9\frac{11}{16}$. (15) $5\frac{1}{12}$.
(16) $6\frac{1}{8}$. (17) $10\frac{7}{36}$ (18) $7\frac{19}{40}$. (19) $7\frac{17}{36}$. (20) $7\frac{3}{5}$.

Page 90 Exercise 37

(1) $\frac{11}{24}$. (2) $1\frac{1}{12}$. (3) $1\frac{17}{24}$. (4) $1\frac{1}{15}$. (5) $1\frac{13}{18}$.
(6) $4\frac{2}{63}$. (7) $1\frac{7}{8}$. (8) $1\frac{1}{12}$. (9) $2\frac{1}{8}$. (10) $\frac{4}{5}$.
(11) $\frac{15}{16}$. (12) $\frac{8}{13}$.

Page 91 Exercise 38

(1) $\frac{1}{6}$. (2) $\frac{2}{15}$. (3) $\frac{3}{8}$. (4) $\frac{15}{32}$. (5) $\frac{6}{50}$.
(6) $\frac{3}{20}$. (7) $\frac{15}{88}$. (8) $\frac{8}{35}$. (9) $\frac{6}{21}$. (10) $\frac{12}{45}$.
(11) $\frac{4}{10}$. (12) $\frac{6}{21}$. (13) $\frac{6}{12}$. (14) $\frac{2}{36}$ (15) $\frac{15}{80}$.

Page 93 Exercise 39

(1) $\frac{2}{9}$, (2) $\frac{5}{48}$. (3) $\frac{1}{8}$. (4) $\frac{5}{8}$. (5) $\frac{3}{25}$.
(6) $\frac{1}{4}$, (7) $\frac{2}{11}$. (8) $\frac{1}{2}$. (9) $1\frac{1}{9}$. (10) $3\frac{1}{3}$.
(11) $3\frac{1}{8}$. (12) $\frac{1}{2}$. (13) $3\frac{3}{4}$. (14) $\frac{57}{64}$. (15) 4.
(16) 1. (17) 9. (18) $\frac{2}{27}$. (19) $\frac{1}{12}$. (20) $\frac{3}{40}$.
(21) $\frac{9}{80}$. (22) $\frac{3}{40}$. (23) $\frac{1}{8}$. (24) $1\frac{1}{9}$. (25) $3\frac{19}{27}$.
(26) $6\frac{1}{4}$. (27) $2\frac{11}{32}$. (28) $\frac{17}{18}$. (29) 20. (30) $1\frac{2}{11}$.

Page 96 Exercise 40

(1) $\frac{1}{2}$. (2) 4. (3) $1\frac{11}{16}$. (4) $\frac{1}{4}$. (5) $\frac{3}{25}$.
(2) $1\frac{1}{4}$. (6) $\frac{1}{2}$. (8) $\frac{5}{8}$. (9) $\frac{10}{13}$. (10) $\frac{8}{9}$.
(11) $1\frac{1}{4}$. (12) $3\frac{1}{5}$. (13) $\frac{1}{12}$. (14) $\frac{2}{27}$. (15) $\frac{1}{12}$.
(16) $\frac{9}{80}$. (17) $\frac{8}{15}$. (18) $\frac{1}{2}$. (19) 2. (20) $4\frac{4}{9}$.

Page 96 Exercise 41

(1) $\frac{5}{12}$. (2) $\frac{19}{24}$. (3) $\frac{2}{3}$. (4) $\frac{1}{12}$. (5) $\frac{13}{24}$.
(6) $\frac{1}{3}$. (7) $\frac{1}{16}$. (8) $\frac{3}{8}$. (9) $\frac{1}{6}$. (10) $\frac{25}{42}$.
(11) $\frac{37}{42}$. (12) $18\frac{1}{2}$. (13) $1\frac{29}{30}$. (14) $7\frac{4}{9}$. (15) 0.
(16) $\frac{1}{18}$. (17) $\frac{1}{8}$. (18) $\frac{1}{8}$. (19) $\frac{7}{36}$. (20) $\frac{21}{32}$.

Page 97 Exercise 42

(1) 26. (2) 250. (3) $\frac{3}{4}$. (4) $\frac{11}{20}$. (5) 12p.
(6) $\frac{7}{25}$. (7) £0·44. (8) £1·50. (9) 75p. (10) 24 l.
(11) 4321. (12) 225 l. (13) Mary 6 km., Betty 9 km., Total 15 km.
(14) $\frac{1}{8}$. (15) 140 rows., 10 cm.

Page 100 Exercise 43

(1) 5·9. (2) 10·5. (3) 10·0. (4) 8·15. (5) 21·06.
(6) 10. (7) 21·316. (8) 19·021. (9) 2·352. (10) 7·223.
(11) 6·763. (12) 18·1 (13) 65·931. (14) 26·159. (15) 15·2283.
(16) 66·08. (17) 26·413. (18) 20·267. (19) 107·383. (20) 20·032.
(21) 10·659. (22) 4·874. (23) 43·8846. (24) 0·65152. (25) 28·0115.
(26) 6·51. (27) 3·9432. (28) 14·365. (29) 22·43. (30) 28·825.

Page 101 Exercise 44

(1) 4·3. (2) 2·24. (3) 3·22. (4) 8·23. (5) 7·22.
(6) 1·22. (7) 4·12. (8) 1·98. (9) 2·57. (10) 2·468.
(11) 0·31. (12) 1·79. (13) 0·205. (14) 0·350. (15) 8·2.
(16) 2·6. (17) 2·3. (18) 24·75. (19) 4·988 (20) 0·999.
(21) 4·3. (22) 0·94. (23) 0·99. (24) 6·38. (25) 3·91.
(26) 2·569. (27) 0·019. (28) 0·129 2 (29) 5·76. (30) 0·09.

Page 101 Exercise 45.

(1) 4·7. (2) 3·35. (3) 6·81. (4) 5·8. (5) 2·989.
(6) 13·098. (7) 0·63. (8) 11·458. (9) 2·967. (10) 9·889.

Page 102 Exercise 46

(1) 31·4. (2) 17·321. (3) 14·14. (4) 103. (5) 1601.
(6) 230. (7) 1103. (8) 1600·1. (9) 2. (10) 101·21.
(11) 44. (12) 96. (13) 60. (14) 510. (15) 7·2.
(16) 0·84. (17) 168·24. (18) 723·6. (19) 808. (20) 331·8.
(21) 12448. (22) 6500·5. (23) 0·6. (24) 1477. (25) 912.
(26) 1836·9. (27) 866. (28) 8682. (29) 3·6. (30) 102.
(31) 12·3. (32) 311400. (33) 174. (34) 11000. (35) 6301.
(36) 543000. (37) 2136. (38) 100·24. (39) 1812·8. (40) 311·42.

Page 103 Exercise 47

(1) 2·7. (2) 3·4. (3) 5·7. (4) 5·2. (5) 5·3.
(6) 3·3. (7) 3·025. (8) 0·465. (9) 4·1. (10) 0·51.
(11) 1·001. (12) 0·36. (13) 1·25. (14) 4·5. (15) 0·025.
(16) 7·5. (17) 0·007. (18) 0·09. (19) 0·009. (20) 0·0009.
(21) 0·25. (22) 0·3. (23) 0·04. (24) 0·08. (25) 1·507.
(26) 4·74. (27) 0·0753. (28) 0·3375. (29) 0·5102. (30) 0·0006.

Page 103 Exercise 48

(1) 0·314.	(2) 0·17321.	(3) 0·1414.	(4) 10·04.	(5) 0·2001.
(6) 0·000032.	(7) 0·1103.	(8) 0·004.	(9) 0·0157.	(10) 0·0866.
(11) 2·335.	(12) 0·73.	(13) 0·6103.	(14) 0·000204.	(15) 0·02022.
(16) 0·731.				

Page 103 Exercise 49.

(1) 0·308.	(2) 0·082.	(3) 0·62.	(4) 0·405.	(5) 0·405.
(6) 0·024.	(7) 3·09.	(8) 0·21.	(9) 1·172.	(10) 0·081.
(11) 0·143.	(12) 0·72.	(13) 0·045.	(14) 0·302.	(15) 5·001.

Page 104 Exercise 50

(1) 0·22.	(2) 0·2.	(3) 0·06.	(4) 0·2.	(5) 0·42.
(6) 0·56.	(7) 0·72.	(8) 0·9.	(9) 0·04.	(10) 0·09.
(11) 0·16.	(12) 0·25.	(13) 0·36.	(14) 0·49.	(15) 0·64.
(16) 0·81.	(17) 2.	(18) 1·2.	(19) 1·23.	(20) 14·2.
(21) 32·1.	(22) 42·1.	(23) 53·1.	(24) 212.	(25) 314·12.
(26) 16231.	(27) 1·3.	(28) 1732·1.	(29) 0·605.	(30) 0·006.

Page 104 Exercise 51

(1) 6·594	(2) 52·36.	(3) 304·8.	(4) 3948·8.	(5) 5·13.
(6) 3·45.	(7) 665·28.	(8) 37·436.	(9) 24·2204.	(10) 17·6.
(11) 40·936.	(12) 15·562.	(13) 30·58.	(14) 8·904.	(15) 80·64.
(16) 160·8.	(17) 33·44.	(18) 0·72.	(19) 15·36.	(20) 16·8.
(21) 10·24.	(22) 40·401.	(23) 2·0928.	(24) 13·081.	(25) 0·0064.
(26) 10·656.	(27) 0·000062.	(28) 0·00048.	(29) 17·3264.	(30) 0·0204.
(31) 0·001312.	(32) 602·4.	(33) 31·776.	(34) 6·2328.	(35) 0·812.
(36) 0·000156.	(37) 2412.	(38) 1688·4.	(39) 608·608.	(40) 16·12.
(41) 22·47.	(42) 54·4.	(43) 143·22.	(44) 151·5.	(45) 64·8032.
(46) 0·0101.	(47) 60·2602.	(48) 0·05661.	(49) 723·6.	(50) 0·3322.
(51) 0·0202.	(52) 6006.	(53) 12012.	(54) 69·104.	(55) 0·00008.
(56) 0·00056.	(57) 484·8.	(58) 0·0625.	(59) 0·000169.	(60) 4·5796.

Page 104 Exercise 52

(1) 0·1.	(2) 1.	(3) 0·16.	(4) 0·3.	(5) 3·3.
(6) 0·4.	(7) 16.	(8) 12000.	(9) 0·26.	(10) 60.
(11) 0·2.	(12) 1·1.	(13) 150·5.	(14) 10.	(15) 10.
(16) 100.	(17) 0·1.	(18) 30.	(19) 7·7.	(20) 200.
(21) 0·909.	(22) 1·2.	(23). 3100.	(24) 0·004.	(25) 0·666.
(26) 4·8.	(27) 6000.	(28) 1·2.	(29) 0·002.	(30) 0·09.

Page 106 Exercise 53

(1) 13.	(2) 11.	(3) 8.	(4) 9.	(5) 11.
(6) 1·1.	(7) 0·13.	(8) 0·13.	(9) 130.	(10) 1300.
(11) 1·3.	(12) 0·25.	(13) 1·3.	(14) 1·3.	(15) 13.
(16) 1·4.	(17) 0·013.	(18) 0·8.	(19) 1·1.	(20) 1·5.
(21) 5.	(22) 0·3.	(23) 0·2.	(24) 2·14.	(25) 22.
(26) 0·08.	(27) 0·2.	(28) 254.	(29) 2·1.	(30) 1·5.

Page 107 Exercise 54

(1) 0·031 25. (2) 0·062 5. (3) 0·125. (4) 0·25. (5) 0·375
(6) 0·5. (7) 0·625. (8) 0·4. (9) 0·875. (10) 0·1
(11) 0·3. (12) 0·7. (13) 3·031 25. (14) 2·1875. (15) 1·125.
(16) 2·3125. (17) 4·35. (18) 2·6. (19) 3·8. (20) 3·625.
(21) 0·333 (22) 0·666. (23) 0·833 3. (24) 0·142 857 (25) 0·428 571.
(26) 0·444 (27) 1 (28) 0·181 8 (29) 0·636 3 (30) 0·307 692.

Page 108 Exercise 55

(1) 2·1, 2·06, 2·065.
(2) 12·2, 12·15, 12·150.
(3) 3·5, 3·51, 3·511.
(4) 0·0, 0·02, 0·017.
(5) 4·1, 4·06, 4·056.
(6) 5·1, 5·13, 5·125.
(7) 2·0, 2·00, 1·999.
(8) 0·1, 0·05, 0·051.
(9) 7·1, 7·09, 7·092.
(10) 18·5, 18·53, 18·529.

Page 109 Exercise 56

(1) 1. (2) 1. (3) 1. (4) 2. (5) 2.
(6) 3. (7) 2. (8) 3. (9) 2. (10) 2.
(11) 4. (12) 1. (13) 5. (14) 3. (15) 1.
(16) 5. (17) 6. (18) 3. (19) 2. (20) 11.
(21) 2. (22) 2. (23) 5. (24) 6. (25) 1.
(26) 5. (27) 2. (28) 4. (29) 3. (30) 6.
(31) 1. (32) 2. (33) 2. (34) 7. (35) 2.
(36) 4. (37) 5. (38) 10. (39) 9. (40) 8.

Page 110 Exercise 57

(1) £3·15. (2) £2·47. (3) £4·15. (4) £1·69. (5) £0·59.
(6) £6·65. (7) £7·11. (8) £5·56. (9) £20·06. (10) £26·51.

Page 111 Exercise 58

(1) 14·07 g. (2) 0·022 g. (3) 430·002 g. (4) 1400 g. (5) 0·058 g.
(6) 62 g. (7) 2·15 g. (8) 4600 g. (9) 60·1 g. (10) 0·21 g.
(11) 24 m. (12) 350 m. (13) 1600 m. (14) 2·01 m. (15) 7·4 m.
(16) 0·51 m. (17) 0·037 m. (18) 0·908 m. (19) 5000·02 m. (20) 100·004 m.
(21) 42 1. (22) 103 1. (23) 7005 1. (24) 9·9 1. (25) 0·23 1.
(26) 0·045 1. (27) 20·07 1. (28) 800·9 1. (29) 7000·06 1. (30) 500·04 1.

Page 111 Exercise 59

(1) 1 kg. 2 hg. 6 dag. (2) 5 g. 6 mg. (3) 1 g. 6 dg. 1 mg.
(4) 1 kg. 6 hg. (5) 1 dg. 1 cg. 1 mg. (6) 3hg. 1 dag. 4 g. 2 dg.
(7) 1 dag. 3 g. 3 dg. 2 cg. 1 mg. (8) 200 kg. 3hg. (9) 7 hg. 4 g. 2 cg. 1 mg.
(10) 1 hg. 2 dag. 3 g. (11) 2 dm. 3 cm. (12) 14 km. 1 dam.
(13) 1 dm. 6 m. 2 dm. (14) 1 hm. 2 dam. 3 m.
(15) 8 m. 2 mm. (16) 5 m. 6 cm. (17) 10 km. 2 hm.
(18) 7 dm. 5 cm. 3·1 mm. (19) 2 m. 2 mm. (20) 1 dam. 1 mm.
(21) 7 cl. 2·4 ml. (22) 31. 2 dl. 6 cl. 1 ml.
(23) 7 dal. 3 1. 2 dl. 1 cl. (24) 1 dl. 4 cl. 1·4 ml.
(25) 1 1. 6 dl. 2 cl. 8 ml. (26) 1 hl. 7 dal. 2 1. 8 dl.
(27) 14 kl. 1 hl. (28) 3 hl. 2 dal. 6 dl.
(29) 1 dal. 4 cl. (30) 2 hl. 7 dal. 5 1. 1 dl.

Page 113 Exercise 60

(1)

(*a*) MATHS. 43·2 ENGLISH 42·5. HISTORY 39·1. GEOGRAPHY 47. SCIENCE 40·8.

(*b*) 4 above av. 4 above av. 6 above av. 5 above av. 4 above av.

(*c*) POSITIONS: (1) J. Jacobs 58 ; (2) P. Coleman 55·8; (3) J. Billings 51 ; (4) J. Forsyth 45 ; (5) B. Stroud 44·4; (6) M. Jones 42 ; (7) J. Scallop 37 ; (8) R. Benton 32·8; (9) P. Palmer 32 ; (10) G. May 27·2.

(2) 45. (3) Greig 25·53, Hendrick 26·66, Birkenshaw 31·13, Willis 35·06, Underwood 44·07, Pocock 44·42. (4) Smith 52 km/h., Jones 54 km/h. (5) Total 60 cm. Av. 5 cm. (6) Total 87 cm. Av. 7·25 cm. (7) 32·5 cm.

Page 117 Exercise 61

(1) 15 cm^2. (2) 9 cm^2. (3) 12 cm^2. (4) 72 cm^2. (5) 48 cm^2.
(6) 8 cm^2. (7) 4 cm2. (8) 20 cm^2. (9) 108 cm^2. (10) 56 cm^2.

Page 117 Exercise 62

(1) 24 cm^2. (2) 21 cm^2. (3) 36 cm^2. (4) 26 cm^2. (5) 120 cm^2.
(6) 144 cm^2. (7) 27 cm^2. (8) 18 cm^2. (9) 100 cm^2. (10) 48 cm^2.

Page 118 Exercise 63

(1) 24 cm^2. (2) 48 m^2. (3) 15 m^2. (4) 12 dm^2. (5) 30 m^2.
(6) 200 cm^2. (7) 0·6 cm^2. (8) 36 m^2. (9) 30 dm^2. (10) 0·24 m^2.
(11) 4050 m^2. (12) 8250 m^2, 4200 m^2. (13) 450 m^2. (14) 4500 m^2.
(15) 264 m^2.

Page 120 Exercise 64

(1) 140 cm^2. (2) 66 cm^2. (3) 40 cm^2. (4) 25 cm^2. (5) 42·75 cm^2.
(6) 80·75 cm^2. (7) 10 m^2. (8) 7 m^2. (9) 18 m^2. (10) 7·5 m^2..
(11) 22 m^2. (12) (*a*) 8·75 m^2, (*b*) 7 m^2. (13) (*a*) 42 m^2, (*b*) 34·25 m^2.
(14) 4500 cm^2. (15) 6 m^2. (16) (*a*) 18 m^2, (*b*) £7·65, (*c*) 22 m^2, (*d*) £35·64, (*e*) 16 m. (17) (*a*) 60 m^2, (*b*) 50 m^2, (c) 88 m^2, (d) 42 m^2 (*e*) 52 m.
(18) 2800 m^2. (19) (*a*) 2950 cm^2, (*b*) 2238 cm^2, (*c*) 712 cm^2.

Page 124 Exercise 65

(1) a^2. m^2. (2) c^2. m^2. (3) ac. m^2. (4) $2a^2$. cm^2. (5) $(x^2-y^2)cm^2$.
(6) $px-qy$. cm^2. (7) $pq-rs$. cm^2. (8) $mg-nf$. m^2. (9) $ad-bc$. m^2. (10) $5a^2$. m^2.

Page 126 Exercise 66

(1) 12 cm^2. (2) 8 cm^2. (3) 13·5 cm^2. (4) 1·875 cm^2. (5) 3·75 m^2.
(6) 2·7 m^2. (7) 1·8 m^2. (8) 16 cm^2. (9) 10 m^2. (10) 3 km^2.
(11) 40 cm^2. (12) 24 m^2. (13) 17·5 m^2. (14) 11 m^2. (15) 7 m^2.
(16) 9·35 m^2. (17) 18 m^2. (18) 4·75 m^2. (19) 3·98 m^2. (20) 6·9 m^2.
(21) 12 cm^2. (22) 5 m^2. (23) 3·225 m^2. (24) 4·48 m^2. (25) 8·37 m^2.
(26) 11 m^2. (27) 36 cm^2. (28) 17 m^2. (29) 6·15 m^2. (30) 24 m^2.

Page 130 Exercise 67

(1) $22{\cdot}5\ cm^2$. (2) $27\ cm^2$. (3) $20\ cm^2$ (4) $1{\cdot}42\ km^2$. (5) $120\ m^2$.
(6) $67{\cdot}5\ m^2$. (7) $28{\cdot}35\ m^2$. (8) $160\ dm^2$. (9) $200\ cm^2$. (10) $4{\cdot}05\ cm^2$.
(11) $7{\cdot}14\ m^2$. (12) $5{\cdot}405\ cm^2$. (13) $12{\cdot}25\ m^2$. (14) $11{\cdot}62\ m^2$. (15) $2\ m^2$.
(16) $7{\cdot}53\ m^2$. (17) $1750\ cm^2$. (18) $1312{\cdot}5\ cm^2$. (19) $0{\cdot}627\ km^2$. (20) $34\ km^2$.

Page 130 Exercise 68

(1) $400\ m^2$. (2) $52\ cm^2$. (3) $15{\cdot}5\ cm^2$. (4) $58{\cdot}5\ cm^2$. (5) $108\ cm^2$.
(6) $100\ m^2$. (7) $51{\cdot}9\ m^2$. (8) $105\ m^2$. (9) $47\ cm^2$. (10) $850\ m^2$.

Page 133 Exercise 69

(1) $48\ cm^2$. (2) $52\ cm^2$. (3) $5\ m^2$. (4) $14\ m^2$. (5) $121{\cdot}5\ m^2$.
(6) $37{\cdot}5\ cm^2$. (7) $13{\cdot}75\ cm^2$. (8) $207{\cdot}5\ m^2$.

Page 136 Exercise 70

(1) 88 cm. (2) 44 cm. (3) 22 cm. (4) 11 cm. (5) 5·5 cm.
(6) 2·75 cm. (7) 4·4 m. (8) 220 cm. (9) 154. (10) 176 cm.
(11) 220 m. (12) 176 m. (13) 88 m. (14) 44 m. (15) 22 m.
(16) 13·2 m. (17) 66 cm. (18) 26·4 m. (19) 528 cm. (20) 17·6 m.

Page 137 Exercise 71

(1) 24·75 cm. (2) 15 t. (3) 1760 cm. (4) 35·2 m. (5) 1760 m.
(6) $10{,}000\frac{4}{7}$ km. (7) 4·54 t. (8) 6·4 r. (9) 454·5 r. (10) 10·6 r.
(11) 80 r. (12) 25 hr. (13) $43\frac{1}{2}$ p. (14) 176 cm. (15) $14\frac{2}{3}$ cm.
(16) 352 m, 224 m. (17) Less (18) 88 chairs. (19) 157 cm.
(20) 157 plants.

Page 140 Exercise 72

(1) $3850\ m^2$. (2) $2464\ cm^2$ (3) $616\ m^2$. (4) $154\ cm^2$.
(5) $38\frac{1}{2}\ m^2$. (6) $9\frac{5}{8}\ cm^2$. (7) $346\frac{1}{2}\ cm^2$. (8) $6{\cdot}16\ m^2$.
(9) $24{\cdot}6\ m^2$. (10) $38{\cdot}5\ cm^2$. (11) $0{\cdot}2826\ cm^2$. (12) $1{\cdot}5386\ cm^2$.
(13) $6{\cdot}1544\ cm^2$. (14) $24{\cdot}6176\ cm^2$. (15) $530{\cdot}66\ cm^2$. (16) $19{\cdot}625\ cm^2$.
(17) $7{\cdot}065\ cm^2$. (18) $452{\cdot}16\ mm^2$. (19) $200{\cdot}96\ dm^2$. (20) $254{\cdot}34\ m^2$.

Page 140 Exercise 73

(1) $38{\cdot}5\ m^2$. (2) $3{\cdot}14\ cm^2$. (3) $490{\cdot}625\ cm^2$. (4) $200{\cdot}96\ cm^2$.
(5) $115{\cdot}5\ m^2$. (6) $47\ m^2$. (7) $77\ m^2$. (8) $50{\cdot}28\ cm^2$.
(9) $28{\cdot}875\ m^2$. (10) $962{\cdot}5\ m^2$.

Page 141 Exercise 74

(1) $10\frac{1}{2}\ cm^2$. (2) $2\frac{5}{8}\ m^2$. (3) $5{\cdot}3156\ cm^2$. (4) $181\frac{1}{7}\ cm^2$.
(5) $1{\cdot}884\ cm^2$. (6) $126\ m^2$. (7) $217\ cm^2$. (8) $134\frac{3}{4}\ m^2$.
(9) $43\frac{5}{16}\ cm^2$. (10) $462\ cm^2$.

Page 143 Exercise 75

(1) $r = \sqrt{\frac{A}{\pi}}$. (2) $h = \frac{2A}{b}$. (3) $l = \frac{A}{b},\ b = \frac{A}{l}$. (4) $r = \frac{C}{2\pi}$.

(5) $d = \frac{c}{\pi}$. (6) $P = 2(l + b);\ b = \frac{P - 2l}{2}$. (7) 12 cm.

(8) $a^2\ cm^2$. (9) 7 cm., 9 cm., 10 cm. (10) 5 cm.

Page 144 Exercise 76

(1) 3 m. (2) 8 dm. (3) 9 m. (4) 35 m. (5) 40 m, 120 m.
(6) 8 m. (7) 8 cm. (8) 16 cm. (9) 16 cm. (10) 6 cm.
(11) 31·8 m. (12) 47·7 m. (13) 1·25 cm. (14) 70 cm. (15) 1·4 m.
(16) 7 m. (17) 42 m, 132 m. (18)(*a*) 13·1 cm (19) 4400 m.
(20)(*a*) 10 cm. (*b*) 15·714 cm.
(18)(*b*) 20·6 cm. (*c*) Round one.
(21) 63 cm^2. (22) 10·5 dm^2. (23) 259·875 m^2. (24) $157\frac{1}{7}$ cm^2. (25) 274 cm^2.

Page 146 Exercise 77

(1) 39 m^2. (2) 70 m^2. (3) 36 cm^2. (4) 480 m^2. (5) 640 m^2.
(6) 390 m^2. (7) 576 m^2. (8) 202 m^2. (9) $78\frac{5}{8}$ cm^2. (10) $111\frac{9}{14}$.

Page 152 Exercise 78

(1) $25c + 37s + 5h$. (2) $66l + 25c + 12p$. (3) $6b + 20p + 2l$.
(4) $3s + 7w + 4f$. (5) $6f + 12d + r$. (6) $12E + 9S + 8W$.
(7) $6c + 10d + 5i + 6e$. (8) $5a + r + e + w$. (9) $5t + 2s + 2b + y$.
(10) $11r + 6l + 8s$.

Page 154 Exercise 79

(1) $4a$. (2) $3b$. (3) $7d$. (4) $8c$.
(5) $13f$. (6) $22p$. (7) $20s$. (8) $15s$.
(9) $26f$. (10) $24e$. (11) $4a + 3b$. (12) $6a + 3b + 2c$.
(13) $6g + 8d + f$. (14) $11y + 5x$. (15) $12m + 5n$. (16) $12p + 5q + 3r$.
(17) $5r + 4s + 15t$. (18) $22n + 9m$. (19) $13b + 13h$. (20) $10x + 8y + 7z$.

Page 155 Exercise 80

(1) 13. (2) −1. (3) −1. (4) −5. (5) 5.
(6) 1. (7) −5. (8) 0. (9) −1. (10) −3.

Page 156 Exercise 81

(1) $10a$. (2) $2b$. (3) $2x$. (4) $4r$. (5) $5m$.
(6) $4y$. (7) $8s$. (8) $-n$. (9) $-11t$. (10) $-3z$.
(11) $7a + 3b$. (12) $6c + d$. (13) $4f + 3g$. (14) $6m + 3n$. (15) $2j - 2h$.
(16) $4x + 2y$. (17) $h - 7k$. (18) $-8r - 8s$. (19) $-7y - 7z$. (20) b.

Page 157 Exercise 82

(1) $10z$. (2) $10x$. (3) $1000g$. (4) $100m$. (5) $60r$.
(6) $10y$. (7) $24k$. (8) $100p$. (9) $100m$. (10) $1000r$.
(11) $1000d$. (12) $1000e$. (13) $1000r$. (14) $52t$. (15) $2m$.
(16) $60z$. (17) $10y$. (18) $1000f$. (19) $7p$. (20) $100f$.

Page 157 Exercise 83

(1) $\frac{b}{100}$. (2) $\frac{n}{100}$. (3) $\frac{k}{1000}$. (4) $\frac{f}{1000}$. (5) $\frac{p}{60}$.
(6) $\frac{y}{10}$. (7) $\frac{p}{1000}$. (8) $\frac{k}{100}$. (9) $\frac{m}{7}$. (10) $\frac{n}{1000}$.
(11) $\frac{d}{1000}$. (12) $\frac{s}{60}$. (13) $\frac{n}{50}$. (14) $\frac{y}{12}$. (15) $\frac{c}{10}$.
(16) $\frac{m}{1000}$. (17) $\frac{n}{100}$. (18) $\frac{n}{100}$. (19) $\frac{t}{24}$. (20) $\frac{j}{100}$.

Page 158 Exercise 84

(1) 12 cm^2.	(2) $4a$ cm^2.	(3) ab cm^2.	(4) $2xy$ cm^2.
(5) $12ab$ cm^2.	(6) 36 mn cm^2.	(7) $(2ab + b)$ cm^2.	(8) $(2xy + 2y)$ cm^2.,

Page 159 Exercise 85

(1) mn.	(2) pq.	(3) $2y$.	(4) er.	(5) gf.
(6) bdh.	(7) acf.	(8) dnr.	(9) hkm.	(10) wyz.
(11) $6ab$.	(12) $8pr$.	(13) $3yz$.	(14) $16mn$.	(15) $6st$.
(16) $15yz$.	(17) $8prs$.	(18) $24fgh$.	(19) $12acd$.	(20) $30abc$.

Page 160 Exercise 86

(1) a^4.	(2) m^6.	(3) n^8.	(4) p^3.	(5) r^5.
(6) t^4.	(7) s^9.	(8) d^{10}.	(9) e^{10}.	(10) q^{10}.
(11) b^9.	(12) g^{10}.	(13) k^{18}.	(14) u^{13}.	(15) c^6.
(16) y^{14}.	(17) h^{15}.	(18) z^{13}.	(19) x^{15}.	(20) f^{16}.

Page 160 Exercise 87

(1) p^2.	(2) y^2.	(3) z^3.	(4) m^3.	(5) $2z^2$.
(6) $2y^2$.	(7) $2x^2$.	(8) $4a^2$.	(9) x^3.	(10) x^4.
(11) p^3.	(12) p^4.	(13) $6p^4$.	(14) $3a^4$.	(15) $4a^5$.
(16) $3a^7$.	(17) $4t^5$.	(18) $2t^4$.	(19) $6t^4$.	(20) $8g^5$.
(21) $2a^4b^2$.	(22) a^5.	(23) a^3b^3.	(24) p^2t^2.	(25) f^3n.
(26) f^3n^2.	(27) $36a^4$.	(28) $32x^5$.	(29) m^5y^5.	(30) $18a^3b^3$.
(31) g^3h^7.	(32) y^5z^7.	(33) $p^7q^4r^6$.	(34) $a^7b^5c^4$.	(35) $12a^3b^3$.
(36) $10c^7d^5$.	(37) $21m^7n^5$.	(38) $60r^3s^4t^6$.	(39) $24f^5g^3h^3$.	(40) $30b^8c^5d^7$.

Page 162 Exercise 88

(1) a^4.	(2) m^2.	(3) r^3.	(4) s.	(5) e^2.
(6) y^3.	(7) $\frac{g^3}{h}$.	(8) m^2n.	(9) $\frac{s}{t}$.	(10) $2b^2$.
(11) $\frac{2c^2}{d}$.	(12) $2fg^2$.	(13) $2h^2k$.	(14) $\frac{4z^2}{3y}$.	(15) $\frac{4s^2}{3r}$.
(16) $\frac{5m^4}{3n}$.	(17) $\frac{8x}{5}$.	(18) $\frac{3b}{4a}$	(19) $2pq$.	(20) $\frac{5v}{3w^2}$.
(21) a^2.	(22) b^2.	(23) b.	(24) abc.	(25) ac.
(26) mpc.	(27) ts^2.	(28) xy.	(29) a^3b.	(30) $\frac{1}{3f}$.
(31) $4no$.	(32) $\frac{4g}{h}$.	(33) $4q^2pt$.	(34) $5n^2m$.	(35) $\frac{6b}{a}$.
(36) $4rs^2t$.	(37) $5m^2n$.	(38) $\frac{t}{3k}$	(39) $2p^2t$.	(40) $\frac{2bc^2}{a}$.

Page 164 Exercise 89

(1) 68.	(2) 47.	(3) 73.	(4) 86.	(5) 78.
(6) 84.	(7) 64.	(8) 83.	(9) 36.	(10) 66.
(11) 43.	(12) 37.	(13) 48.	(14) 76.	(15) 34.
(16) 63.	(17) 46.	(18) 87.	(19) 38.	(20) 44.

Page 165 Exercise 90

(1) 24.	(2) 20.	(3) 14.	(4) 36.	(5) 15.
(6) 6.	(7) 4.	(8) 18.	(9) 6.	(10) 55.
(11) 10.	(12) 27.	(13) 35.	(14) 72.	(15) 18.
(16) 54.	(17) 12.	(18) 25.	(19) 21.	(20) 12.

Page 165 Exercise 91

(1) 12. (2) 14. (3) 5. (4) 3. (5) 1.
(6) 0. (7) –4. (8) 0. (9) 20. (10) –6.
(11) 43. (12) 19. (13) 4. (14) 46. (15) 28.
(16) 20. (17) 11. (18) 9. (19) –3. (20) 11.

Page 166 Exercise 92

(1) 20. (2) 50. (3) 100. (4) 490. (5) 350.
(6) $1\frac{3}{7}$. (7) $\frac{4}{7}$. (8) 40. (9) $36\frac{3}{4}$. (10) $19\frac{1}{21}$.
(11) $\frac{40}{49}$. (12) $14\frac{2}{7}$. (13) 56. (14) 1. (15) $2\frac{9}{20}$.
(16) $3\frac{3}{4}$. (17) $1\frac{71}{125}$. (18) 1. (19) $2\frac{1}{7}$. (20) 4.

Page 167 Exercise 93

(1) 3. (2) 5. (3) 6. (4) 2. (5) 4.
(6) 11. (7) 15. (8) 5. (9) 13. (10) 12.
(11) 5. (12) 8. (13) 11. (14) $4\frac{1}{2}$. (15) $2\frac{1}{2}$.
(16) $2\frac{3}{4}$. (17) $3\frac{1}{2}$. (18) $3\frac{1}{7}$. (19) $6\frac{2}{5}$. (20) $4\frac{1}{6}$.

Page 167 Exercise 94

(1) 42 kg. (2) 10 m. (3) $x = 3$. (4) $f = 16\frac{1}{2}$p. (5) $b = 9$.
(6) 1000 g. (7) $a = 8$ min. (8) 0·04 m (4 cm). (9) d = 400 n miles.
(10) $h = 4$.

Page 168 Exercise 95

(1) $r = 80$ cm. (2) $b = 975$ g. (3) $c = 3$p. (4) 26·8 l. (5) $w = 19$ cm.
(6) $m = 1·18$ m. (7) $g = 16$ l. (8) $c = 35$ m. (9) $p = 19·5$ m. (10) $x = 15$.
14 km./l.

Page 170 Exercise 96

(1) 3. (2) 4. (3) 5. (4) 9. (5) 2.
(6) 3. (7) 4. (8) 5. (9) 3. (10) 4.
(11) 7. (12) 9. (13) 7. (14) 10. (15) 12.
(16) 10. (17) 10. (18) 15. (19) 8. (20) 18.

Page 171 Exercise 97

(1) – 1. (2) 11. (3) 6. (4) 19 (5) 20.
(6) 28. (7) 40. (8) 18. (9) 9. (10) $10\frac{1}{2}$.

Page 171 Exercise 98

(1) 3. (2) 2. (3) 4. (4) 1. (5) 3.
(6) 8. (7) 5. (8) 5. (9) 8. (10) 4.
(11) 8. (12) 6. (13) 7. (14) 11. (15) 12.
(16) 4. (17) 3. (18) 9. (19) 7. (20) 4.
(21) 9. (22) 6. (23) 4. (24) 9. (25) $2\frac{1}{2}$.
(26) $1\frac{1}{3}$. (27) $4\frac{1}{2}$. (28) $2\frac{1}{3}$. (29) $6\frac{1}{2}$. (30) $3\frac{1}{5}$.

Page 172 Exercise 99

(1) 5. (2) 4. (3) 4. (4) 6. (5) 5.
(6) 3. (7) 4. (8) 5. (9) 2. (10) 7.
(11) 8. (12) 5. (13) 4. (14) 3. (15) 3.
(16) $\frac{2}{3}$. (17) 2. (18) 8. (19) 4. (20) 2.

Page 173 Exercise 100

(1) 12. (2) 10 (3) 21. (4) 6. (5) $1\frac{1}{3}$.
(6) $4\frac{2}{3}$. (7) 6. (8) $3\frac{1}{3}$. (9) $6\frac{2}{3}$. (10) $3\frac{1}{2}$.
(11) $3\frac{1}{2}$. (12) $3\frac{1}{3}$. (13) $1\frac{3}{4}$. (14) $4\frac{4}{5}$. (15) 9.
(16) 6. (17) 22. (18) 21. (19) 26. (20) 20.

Page 176 Exercise 101

(1) 4 cm. (2) 6·4 cm. (3) 4 cm. (4) 2·2 cm. (5) 2 cm.
(6) 3 cm. (7) 10 cm. (8) 5 cm. (9) 16 cm. (10) 60 cm.
(11) 4 cm. (12) 2 cm. (13) 2·5 cm. (14) 11 cm. (15) 10 cm.
(16) 20 g. (17) 26 g. (18) 8 kg. (19) 1 g. (20) $2\frac{1}{2}$ kg.

Page 181 Exercise 102

(1) $e = f - d$. (2) $p = r - q$. (3) $m = t + n$. (4) $\overset{l}{a} = d + b + c$.
(5) $w = r - s - t$. (6) $b = a - x - y$. (7) $m = \frac{r - n}{2}$. (8) $b = a + 3c$.
(9) $s = 3r - 2p + q$. (10) $j = \frac{4g + 2h - 2k}{3}$. (11) $b = \frac{a}{l}$.
(12) $d = \frac{c}{p}$. (13) $h = \frac{2t}{p}$. (14) $l = \frac{V}{bh}$. (15) $b = \frac{3p}{lh}$.
(16) $m = pr$. (17) $t = \frac{rs}{u}$. (18) $b = \frac{c}{nd}$. (19) $h = \frac{gk}{i}$.
(20) $q = \frac{str}{p}$.

Page 181 Exercise 103

(1) $r = \frac{c}{2\pi}$; $2\frac{5}{22}$. (2) $V = \frac{m}{s}$; 5. (3) $v = \frac{k}{p}$; 4.
(4) $d = \frac{wo}{m}$; 12. (5) $f = ur$; 2. (6) $t = \frac{v - u}{a}$; $8\frac{4}{5}$.
(7) $L = ae$; 25. (8) $r = \frac{D}{d}$; 10. (9) $v = c(u + t)$; 11·6.
(10) $r = 2f$; 21. (11) $t = \frac{s}{u + v}$; 2. (12) $p = \frac{a}{(1 + r)}$; 100.
(13) $x = \frac{p}{L + b}$; $4\frac{2}{3}$. (14) $p = \frac{I.100}{rt}$; 100. (15) $t = \frac{I.100}{pr}$; 2.

Page 189 Exercise 104

(1) 32 kg. (2) $119\frac{1}{4}$ cm. (3) 59 runs. (4) 5 mm.
(5) (*a*) November. (*b*) August. (*c*) No.
(6) (*a*) May and August. (*b*) January and February. (*c*) Australia or N. Zealand. (*d*) Less. (7) (*a*) £138. (*b*) £11·50. (*c*) April.

Page 191 Exercise 105

(1) (*a*) 9 a.m. (*b*) 2 p.m. (2) (*a*) 10·7 (*b*) 6.
(3) (*a*) Northern Hemisphere. (*b*) May-September.
(4) (*a*) Straight line. (*b*) varies. (5) (*a*) 136·3 km/h. (*b*) 6 laps.
(6) No. (7) (*a*) 6·5 km. approx. (*b*) 7 m. approx.
(8) (*a*) 47 m. (*b*) 60 km/h. (9) (*a*) 34·3 m/sec., 46·1 m/sec. (*b*) $5\frac{1}{2}$.

Page 196 Exercise 106

(1) 130 km. (2) (*a*) 50 km. (*b*) 20 km/h.
(3) (*a*) 11·49 a.m. (*b*) 3 km. (*c*) 20 km/h.
(4) (*a*) 40 km. (*b*) 2 p.m. (*c*) 1 hr. 50 mins. (*d*) 60 km/h.
(5) (*a*) 2 hrs. 30 min. (*b*) 4·30 p.m. (*c*) 3·40 p.m. (*d*) 27 km. (*e*) 30 min.
(6) (*a*) 1 hr. 56 mins. (*b*) 74 km. (*c*) 1 hr. (7) (*a*) 11·30 a.m. (*b*) 88 km.

Page 199 Exercise 107

(1) (*a*) 5·8 km/h. (*b*) 6·7 m.p.s. (*c*) 15 km/h. approx. (*d*) 14 m/s., 19 m/s.
(2) (*a*) 36·8°C. (*b*) 50°, 140°. (3) (*a*) $4\frac{1}{2}$ l, (*b*) £1·26, (*c*) 81 p., (*d*) 35p. less.
(4) (*a*) 9·42 cm., 22 cm., 25·12 cm., 34·54 cm. (*b*) 2 cm., 9 cm., 5 cm., 7 cm.

Page 204 Exercise 108

(11) (2, 4) (12) (2, 5) (13) (–1, 0) (14) (4, 10) (15) (1, 2)

Page 206 Exercise 109

(2) $112\frac{1}{2}$ sq. units. (3) 89 sq. units. (4) 64 sq. units. (5) 174 sq. units.
(6) 75 sq. units. (7) $112\frac{1}{2}$ sq. units (8) 126 sq. units.

Page 208 Exercise 110

(1) 7 paces S.W. (2) 10 paces S. (3) 10 paces W. (4) 5 paces W.
(5) 14 paces S.W.

Page 208 Exercise 111

(1) 14 paces N.W. (2) 5 paces N. (3) 15 paces S. (4) 5 paces W.
(5) 2 paces E.

Page 208 Exercise 112

(1) 17 paces 260°. (2) 15 paces 273°. (3) (*a*) 12 paces. (*b*) 180°.
(4) 270° 10 paces. (5) (*a*) 15 paces. (*b*) 180°.

Page 209 Exercise 113

(1) 15 paces 090°. (2) 22 paces 298°. (3) 15 paces 335°. (4) 17 paces 255°.
(5) 30 paces 270°.

Page 211 Exercise 115

(1) 30°. (2) $a = 24°$. (3) $a = 15°$. (4) $x = 20°$. (5) $x = 30°$.

Page 212 Exercise 116

(1) $a = 23°$. (2) $x = 25°$.
(3) (*a*) 90°. (*b*) 144°. (*c*) 126°. (*d*) 117°. (*e*) 180°.
(4) (*a*) $13\frac{1}{3}°$. (*b*) 30°. (*c*) 40°. (*d*) $53\frac{1}{3}°$. (*e*) 10°.
(5) (*a*) 10°. (*b*) 20°. (*c*) 30°. (*d*) 50°. (*e*) 15°.

Page 213 Exercise 117

(1) 60°. (2) 26°. (3) 110°. (4) 80°. (5) 15°.

Page 213 Exercise 118

(1) 30°. (2) 20°. (3) 20°. (4) 40°. (5) $x = 90°$, $y = 60°$.

Page 216 Exercise 123

(1) $4\frac{5}{12}$. (2) $2\frac{4}{5}$. (3) $5\frac{7}{8}$. (4) $3\frac{1}{2}$. (5) $2\frac{5}{8}$.
(6) $3\frac{1}{4}$. (7) $\frac{11}{18}$. (8) $1\frac{7}{12}$. (9) $\frac{3}{8}$. (10) $1\frac{11}{24}$.

Page 216 Exercise 124

(1) $4\frac{3}{4}$. (2) $6\frac{1}{30}$. (3) $5\frac{5}{24}$. (4) $\frac{7}{8}$. (5) $2\frac{17}{30}$.
(6) $2\frac{19}{30}$. (7) $2\frac{1}{12}$. (8) $1\frac{7}{12}$. (9) $1\frac{11}{12}$. (10) $4\frac{23}{24}$.

Page 216 Exercise 125

(1) $\frac{25}{27}$. (2) $1\frac{9}{13}$. (3) $\frac{13}{22}$. (4) 2. (5) 2.
(6) 1. (7) $3\frac{18}{19}$. (8) $5\frac{1}{3}$. (9) $\frac{8}{27}$. (10) $\frac{13}{18}$.

Page 216 Exercise 126

(1) $\frac{3}{4}$. (2) $1\frac{1}{14}$. (3) $\frac{5}{6}$. (4) $1\frac{7}{23}$. (5) $1\frac{1}{2}$.
(6) $2\frac{4}{7}$. (7) 1. (8) $\frac{10}{11}$. (9) $4\frac{1}{13}$. (10) $3\frac{4}{13}$.

Page 217 Exercise 127

(1) 2·3165. (2) 0·2244. (3) 33·712. (4) 0·0503. (5) 70·72.
(6) 3·241. (7) 1·2. (8) 3·4. (9) 3·6. (10) 0·034.

Page 217 Exercise 128

(1) 3·57. (2) 30·24. (3) 0·105468. (4) 254·254. (5) 31·3626.
(6) 3·2. (7) 13·6. (8) 0·03. (9) 2·31. (10) 6·4.

Page 217 Exercise 129

(1) 2304 m. (2) 342·6 dm. (3) 34·1 dam. (4) 2010·3 cm. (5) 1·2603 km.
(6) 2·054 m. (7) 341 dm. (8) 2·00071 hm. (9) 0·15 dam. (10) 2006·4 m.

Page 217 Exercise 130

(1) 3 km. 7 hm. 5 dam. 1m.
(2) 2 m. 4 dm. 3 cm. 7 mm.
(3) 1 dam. 4 m. 2 dm. 2 mm.
(4) 3 hm. 4 dam. 2 m.
(5) 2 dam. 1 m. 5 cm.
(6) 1 hm. 7 m. 4 dm.
(7) 3 dam. 1 dm. 4 cm.
(8) 2 dm. 1 cm. 7 mm.
(9) 3 km. 1 hm. 2 m.
(10) 2 hm. 3 dm. 4 cm.

Page 217 Exercise 131

(1) 2·1 m^2. (2) 114 m^2. (3) 20·4 cm^2. (4) $6\frac{1}{2}$ cm^2. (5) 3 m^2.

Page 218 Exercise 132

(1) $57\frac{3}{4}$ cm^2. (2) 196 cm^2. (3) $13\frac{3}{7}$ m^2. (4) 23 cm^2. (5) $39\frac{3}{8}$ cm^2.

Page 219 Exercise 133

(1) $x + y$. (2) $b - 7a$. (3) $9a - b$. (4) $4b - 2a$.
(5) $2a + 2b + 4c$. (6) $2x - 3y$. (7) $4b - 2a$. (8) $6r - s$.
(9) $s - p - 4n$. (10) $y - z$.

Page 219 Exercise 134

(1) $3p - 2n$. (2) $8p - 7q$. (3) $e + j$. (4) $5t - 5s$. (5) $3v - 3f$.
(6) $-2a - 2b$. (7) $2x - y - c$. (8) a. (9) $4b + 2c$. (10) $-x + y + z$.

Page 219 Exercise 135

(1) $6a^5b^3c^4$. (2) $8p^5q^4$. (3) $36x^4y^6$. (4) $8m^7n^3$. (5) $10r^4s^5t^3$.

Page 219 Exercise 136

(1) $6a^3b^5c^3$. (2) $6x^5y^4$. (3) $20f^4g^3h^3$. (4) $6s^2t^6u$ (5) $8c^6d^5$.

Page 219 Exercise 137

(1) $\frac{2c^2}{d}$. (2) $\frac{4f^3}{g}$. (3) $2mn$. (4) $2xy$. (5) $2p^2r$.

Page 219 Exercise 138

(1) $5mn$. (2) $\frac{5}{2cd}$. (3) $2rs^2$. (4) $2a^2$. (5) $4gh^2$.

Page 219 Exercise 139

(1) 6. (2) -1. (3) 2. (4) 2. (5) $\frac{2}{3}$.
(6) 3. (7) 5. (8) 3. (9) 8. (10) 4.

Page 219 Exercise 140

(1) 5. (2) 4. (3) 4. (4) 6. (5) 5.
(6) 3. (7) 4. (8) 5. (9) 2. (10) 7.

Page 220 Exercise 141

(1) 4 cm. (2) 0·5 cm. (3) 4 cm. (4) 9 cm. (5) 6 cm.

Test 1

(1) $\frac{7}{8}$, $\frac{5}{6}$. $\frac{3}{4}$, $\frac{2}{3}$, $\frac{1}{2}$, $\frac{5}{12}$. (2) $3\frac{7}{8}$. (3) $\frac{1}{4}$. (4) $2\frac{1}{2}$.
(5) $\frac{1}{2}$ of $2\frac{2}{3}$. (6) 108. (8) 150° or 210°. (9) $4a - 6b$. (10) 26.

Test 2

(1) $\frac{9}{10}$, $\frac{17}{20}$, $\frac{4}{5}$, $\frac{3}{4}$, $\frac{7}{10}$, $\frac{1}{2}$. (2) $7\frac{3}{4}$. (3) $\frac{3}{4}$. (4) $\frac{2}{3}$. (5) 3·87 cm.
(6) 40°. (7) $67\frac{1}{2}^\circ$. (8) 315°. (9) $5p + 2q + 5r$. (10) 60.

Test 3

(1) $4\frac{1}{5}$. (2) $10\frac{2}{3}$. (3) 6. (4) 15·236. (5) 181·5.
(6) 44. (7) 17°. (9) $6m - 6n$. (10) $a^3 b^2$.

Test 4

(1) $1\frac{7}{24}$. (2) $\frac{6}{7}$ (3) 15·737. (4) 9·48. (5) 0·75, 0·666.
(7) 35°. (9) 2. (10) a^6.

Test 5

(1) $5\frac{7}{30}$. (2) 15. (3) 66·315. (4) 80. (5) £4·90.
(6) 115°. (7) 030°, 220°, 115°, 333°. (8) $4d - c - 4e$. (9) 8.
(10) a^3b^3.

Test 6

(1) 4. (2) 0·875, 0·416. (3) Mary 11 cm. (4) £2·67.
(5) 62. (6) 30°. (9) p^4q^3. (10) $x = 2$.

Test 7

(1) $\frac{5}{6}$, $\frac{4}{5}$, $\frac{7}{10}$, $\frac{2}{3}$, $\frac{8}{15}$, $\frac{1}{2}$. (2) 2·441. (3) 48·6. (4) Bettv 2000 g.
(5) 31 kg. (6) N.30°E., S.17°W., S.38°E., S.84°W. (9) $m = 4$.
(10) $1\frac{1}{2}$ p.

Test 8

(1) $5\frac{3}{8}$. (2) Joan, 483 cm. (3) 0·104. (4) Anglia.
(5) 13·86 cm^2. (8) $3d + 6e + 6s + 6w$. (9) $x = 1$. (10) $r = p - q + s$.

Test 9

(1) 22. (2) 0·1875, 0·16. (3) 127 cm. (4) 50 cm^2.
(5) $7\frac{1}{2}$ sq. cm. (6) 65°. (7) 6·5 m^2. (8) 21·8 m.
(9) $b = \frac{c}{ad}$. (10) $x = 1, y = 4$.

Test 10

(1) 7. (2) 2·581. (3) £0.87. (4) $\left\{\begin{array}{l} 2025 \\ 962\frac{1}{2} \\ 1062\frac{1}{2}. \end{array}\right.$ dm^2 (5) 132 cm.
(6) $22\frac{1}{2}^\circ$. (8) $-2r - 5t$. (9) $x = 4$. (10) $d = \frac{c}{p}$.